IUTAM Symposium on Impact Biomechanics: From Fundamental Insights to Applications

SOLID MECHANICS AND ITS APPLICATIONS
Volume 124

Series Editor: G.M.L. GLADWELL
Department of Civil Engineering
University of Waterloo
Waterloo, Ontario, Canada N2L 3GI

Aims and Scope of the Series

The fundamental questions arising in mechanics are: *Why?, How?,* and *How much?* The aim of this series is to provide lucid accounts written by authoritative researchers giving vision and insight in answering these questions on the subject of mechanics as it relates to solids.

The scope of the series covers the entire spectrum of solid mechanics. Thus it includes the foundation of mechanics; variational formulations; computational mechanics; statics, kinematics and dynamics of rigid and elastic bodies: vibrations of solids and structures; dynamical systems and chaos; the theories of elasticity, plasticity and viscoelasticity; composite materials; rods, beams, shells and membranes; structural control and stability; soils, rocks and geomechanics; fracture; tribology; experimental mechanics; biomechanics and machine design.

The median level of presentation is the first year graduate student. Some texts are monographs defining the current state of the field; others are accessible to final year undergraduates; but essentially the emphasis is on readability and clarity.

For a list of related mechanics titles, see final pages.

IUTAM Symposium on

Impact Biomechanics: From Fundamental Insights to Applications

Edited by

M. D. GILCHRIST

University College
Dublin, Ireland

Springer

A C.I.P. Catalogue record for this book is available from the Library of Congress.

ISBN 978-90-481-6968-9 (PB)
ISBN 1-4020-3796-1 (e-book)
ISBN 978-1-4020-3796-2 (e-book)

Published by Springer,
P.O. Box 17, 3300 AA Dordrecht, The Netherlands.

www.springeronline.com

Printed on acid-free paper

In memoriam

Professor K. Bertram Broberg

1925 – 2005

TABLE OF CONTENTS

SESSION 2

SESSION 3

PREFACE

Substantial fundamental work has been undertaken in the different aspects of impact biomechanics over the past three decades. Much of this has been motivated and undertaken by the automotive industry in their efforts to improve transport safety. More recently, however, it has become apparent that the multidisciplinary synergies which are realised by interactions between engineers, scientists and clinical practitioners will ultimately lead to a greater understanding of the complex interacting phenomena within the human body after it has sustained an impact. In turn, this greater depth of knowledge will provide more fundamental insights into the analysis, diagnosis, treatment and prevention of impact injuries across a broader spectrum of accident environments.

The scientific focus of this IUTAM symposium is to address those topics that are centrally important to the biomechanics of impact. These can be grouped into those that are concerned with the different causes of accidents (e.g., transport, occupational and sports injuries), the mechanics involved in accident analysis (e.g., accident investigation, computational modelling techniques), the different types of resulting traumatic injuries (including musculoskeletal, organ, spinal and head injuries), methods of assessing the extent of injury (e.g., injury assessment, injury criteria, constitutive laws for human tissue), and providing protection during an impact (e.g., injury prevention, energy absorption materials, and safety devices). For example, fatalities and serious injuries resulting from sports accidents can be reduced in future years if real-world accidents are analysed using accident reconstruction techniques such as multibody dynamics simulations or instrumented dummy simulations to quantify the magnitude and duration of forces distributed over the human body. This force-velocity-acceleration-time history can be compared against clinical injury severity indices, established from forensic pathology investigations for different organs and biological tissue, whilst alternative future designs of different protective clothing and safety equipment can be compared to characterise their relative efficacy in absorbing impact energy.

This symposium could not have been organised without the excellent and dedicated efforts of the members of the International Scientific Committee and the Local Organising Committee as well as the scientific referees who carefully reviewed the manuscripts contained within this volume. It is a sincere pleasure for me to acknowledge their endeavours and on their behalf to express gratitude to the Principal Sponsors who have generously provided financial support for the symposium. I would particularly like to record my appreciation to the following individuals for their enthusiastic assistance in producing this volume:

- Jolanda Karada-van Kuijk, Karada Publishing Services
- Debra Heeney, University College Dublin
- Nathalie Jacobs, Springer
- Anneke Pot, Springer
- Graham Gladwell, University of Waterloo

INTERNATIONAL SCIENTIFIC COMMITTEE

Faris Bandak (U.S.A.)
Max Cameron (Australia)
Dick Van Campen (The Netherlands), **IUTAM Representative**
Jeff Crandall (U.S.A.)
Michael D. Gilchrist (Ireland), **Chairman**
Per Lövsund (Sweden)
Gerald Nurick (South Africa)
Koshiro Ono (Japan)
Rémy Willinger (France)

LOCAL ORGANISING COMMITTEE

K. Bertram Broberg (University College Dublin)
David FitzPatrick (University College Dublin), **Secretary**
Michael D. Gilchrist (University College Dublin), **Chairman**
Clive Lee (Royal College of Surgeons in Ireland)
Padraic O'Donoghue (NUI-Galway)
Patrick Prendergast (Trinity College Dublin)
Ciaran Simms (Trinity College Dublin)
Denis Wood (Denis Wood Associates)

PRINCIPAL SPONSORS

International Union for Theoretical and Applied Mechanics, IUTAM
Science Foundation Ireland
University College Dublin
Enterprise Ireland
National Standards Authority of Ireland
Institution of Mechanical Engineers
Springer (formerly Kluwer Academic Publishers)
The Turf Club

The objective of the symposium is to provide a single forum at which the latest knowledge can be presented, current gaps in understanding identified and future research challenges suggested. In particular, it is intended that researchers from the traditionally distinct backgrounds of engineers, scientists and clinicians will bring their respective expertise to bear in addressing this multidisciplinary research topic.

Michael Gilchrist
University College Dublin, 2005

SESSION 1

THE CONTRIBUTION OF ACCIDENT INVESTIGATION RESEARCH TO BIOMECHANICS

Murray Mackay
Mull House, Bride, Isle of Man IM7 3EA, U.K.
Tel: 44 1624 880 762 Fax: 44 1624 880 107
mmackay@netcomuk.co.uk

Abstract. The evolution of accident investigation procedures is traced from early railway accidents, through the work by De Haven on light aircraft accidents to the first in depth studies of road crashes. Individual multi-disciplinary teams addressed many biomechanical issues in the 1960s and 70s, leading to larger cooperative projects such as NASS and CCIS. Examples are given of the linking of field accident research with experimental biomechanics, and the importance of population variations is emphasised. The optimisation of vehicle crashworthiness requires good accident investigation of the mechanisms of injury as well as detailed crash data. The limitations of current test tools for road user protection is discussed.

Key words: accident investigation, crashworthiness, optimisation, standards.

1. EARLY BEGINNINGS

Although accidental injury has been occurring since the origins of man, the application of the scientific method to these events is surprisingly recent. Intuitively mediaeval suits of armour illustrate good biomechanical principles; protect the vital parts of the human frame, the head, neck and chest, and distribute the applied loads over as large an area as possible, deflect sharp pointed objects and remain on your horse or in your vehicle to

Michael D. Gilchrist (ed.), IUTAM Proceedings on Impact Biomechanics: From Fundamental Insights to Applications, 3–16.

avoid being ejected to strike the ground. These general principles clearly apply to road vehicles today, but the application and evaluation of these principles has been absent until the last century.

The growth of the railways in the 19th century led to the first systematic accident investigations of major railway disasters, but the whole focus of those studies was causation rather than injury mitigation. Nevertheless the early reports of railway accidents are useful reading today as they illustrate the evolving recognition of the complex interaction of engineering design, the perceived environment and the human response. In particular the current policy in many countries is now aimed to produce the "forgiving highway", a philosophy which acknowledges the limitations of road users and our propensity to make mistakes and incorrect decisions. This approach places the onus more on engineers to design and manages fail safe road networks and vehicles; it can be traced to the concepts developed by railway engineers two hundred years ago.

An example of an unrealistic expectation of the human response comes from an investigation of a major accident in Reading in 1840. It was caused by a train driver who failed to comply with the following regulation:

"A Signal Ball will be seen at the entrance to Reading Station when the Line is Right for the train to go in. If the Ball is not visible the train must not pass it."[1].

Early accident investigations therefore focused almost exclusively on causation, applied to the growing number of accidents in the evolving modes of road traffic and aircraft. It was in aviation that the analyses of injury producing mechanisms were first conducted. This was pioneered by Hugh De Haven in the United States. As a young cadet in the Canadian Air Force in 1917 he was involved in a midair collision at about 500 feet. He survived albeit with serious leg fractures and a ruptured liver and pancreas, the latter he attributed to the hostile design of the buckle of his seat belt. But he also realised that his survival was due to the cockpit maintaining its integrity, and the seat belt preventing his ejection and limiting contacts with the cockpit interior.

He subsequently conducted a study of survivors of falls from great heights, and produced some of the first data points for the tolerance limits for whole body deceleration. He did this by measuring and estimating the stopping distances of people landing flat on soft earth, or being cushioned by snow and other materials [2].

De Haven subsequently set up the Crash Injury Research Project in 1942, funded by the military, but in the department of public health in the

Cornell University Medical College, to examine the mechanisms of injuries in light aircraft accidents. That programme evolved after the war into the first major study of injuries and their causes in car crashes. That early research developed data bases from specific samples of crashes examined in detail by the police, using special recording forms designed by the researchers, and it provided the first field evaluations of anti-burst door latches, improved laminated glass for windscreens, seat belts, instrument panel design and steering columns. More importantly, by producing representative samples of crashes, that research gave quantitative descriptions of the frequencies of various collision characteristics. By prioritising the most frequent types of crashes and injury mechanisms, it allowed rational crashworthy design priorities to be developed and introduced [3].

Other pioneers in this early period included Sir Hugh Cairns who demonstrated substantial benefits of compulsory helmet use by army dispatch riders during World War II, and pointed out that most blows to the head were to the front and the side rather than on the crown, reflected some 30 years later in the evolution of jet style and full face helmets [4]. An elegant study by Sheldon, a physician in Wolverhampton, "On the Natural History of Falls in Old Age" showed how the routine observations of a practicing clinician could lead to new insights into the aetiology and mechanisms of injury, particularly of long bone fractures in the elderly [5].

2. THE RISE OF DATA BASES

The 1960s and 70s saw a large growth in accident investigation studies around the world, spurred largely by the first generation of United States Federal Motor Vehicle Safety Standards, promulgated around 1968, which introduced the concept of crashworthy design as a requirement. The introduction of those standards was largely due to William Haddon, the first director of the National Highway Safety Bureau, the forerunner of N.H.T.S.A. [6]. Suddenly it became important for car manufacturers to actually evaluate the injury reducing potential of their designs, and other governments around the world to introduce like standards and prioritise the countermeasures which reduce traffic deaths and injuries.

This impetus generated many small scale in-depth accident investigation groups around the world. Their rationale was that police and insurance data, about the only sources of information on road accidents at that time, provided an inadequate source of information, particularly in terms of crash performance and injury outcomes. It was necessary for the researchers themselves to investigate crashes. Secondly the subject was clearly interdisciplinary, mainly requiring both engineering and medical inputs. This movement resulted in a significant increase in our knowledge of crashes, but the very nature of such research is small scale and limited to specific geographical regions. Good in-depth crash investigation is labour intensive and therefore each case examined is relatively expensive. On the other hand, to produce findings which have some general validity requires relatively large numbers of cases, careful experimental design and sampling plans, and often matching control data. Too often there were inadequate numbers to explore the issues of interest satisfactorily, and biases in case selection limited the general validity of the conclusions drawn. Nevertheless many new findings, directly illustrating the response of real people to actual collision forces were made, and the effectiveness and limitations of all aspects of crashworthy design were examined. Thus ejection risks [7], injuries from toughened and laminated windscreens [8], energy absorbing steering column design [9] seat belt effectiveness [10], pedestrian injury mechanisms [11] and many other issues were described and evaluated.

However, caught between the need for a good quality, comprehensive investigation of each case, often requiring 2-3 man-days per case, and the aim of having an acceptable number of cases to justify statistically valid conclusions, the limitations of small ad hoc teams became clear, and a consolidation took place. In the United States this gave rise to the National Crash Severity Study and subsequently the National Accident Sampling System and Fatal Accident Reporting System programmes. FARS is a census of all fatal traffic crashes in the US, whilst NASS is a stratified sample of some 5000 crashes per year, selected using a complex weighted sampling plan, to represent crashes across the whole country. Similar but smaller scale programmes developed in the U.K. with the Cooperative Crash Injury Study, currently examining some 1500 cases per year, and the GIDAS study in Germany.

The NASS programme in particular has made a major contribution to our knowledge of crash characteristics and injury outcomes. It has been running since 1978 and is freely accessible to all. As a result many useful research papers have been produced, comparisons with other data bases made, and biomechanical analyses conducted on specific injuries which would otherwise have been impossible. Proponents of proprietary research should note that such openness and generosity is necessary if maximum use is to be made of relatively expensive data collection systems.

3. CURRENT CRASH INVESTIGATION TECHNIQUES AS APPLIED TO BIO-MECHANICAL ISSUES

A serious injury crash investigation usually begins with an examination of the vehicles involved. Measurements of the extent and nature of exterior damage is the starting point for analytical work to establish the severity of the impact to the vehicle. Change in velocity is the current variable preferred for assessing the severity of a collision, and there are various programmes such as EDCrash and SMAC which contain data bases of make and model specific vehicle crush characteristics. These facilitate the calculations to arrive at a given change in velocity for the case under investigation. However, the change in velocity at the centre of gravity of the car is only a very approximate surrogate for the specific crash forces applied to a given occupant. If high rates of rotation are involved as a result of the collision, then each sitting position can have its own, different change in velocity. If there is extensive intrusion, as in side impacts for example, then an occupant's contact velocity with such structures as an intruding door will be greater than the change in velocity of the centre of gravity of the car in which he is sitting. In a frontal collision with a restrained occupant, if for example there are intrusions of the toe board and instrument panel resulting in knee contacts, then the feet, the knees and the torso are all experiencing different localised changes in velocity. Careful surrogate work and comparisons with experimental collisions which may match the relevant characteristics of a specific crash can give insights into the severities of these localised contacts.

Event data recorders (EDRs) have the potential of supplying more detail on crash severity, including the whole time/deceleration history of a collision, but at present overall change in velocity, mean acceleration and estimates of peak deceleration (at the vehicle's C.G.) are the current normal measures for collision severity.

It helps greatly if at the time of a vehicle examination the number of occupants, their sitting positions and something of their injuries are known. Small dents and scuffs on interior trim, crash load marks on seat belt webbing and hardware, fabric and lipstick transfer marks onto airbags, and deformation of seat backs and bases, all help in establishing the likely contacts which have caused an injury, and also in reconstructing the kinematics of an occupant during the crash phase.

The analysis then considers the nature of the injuries and involves clinicians, particularly radiologists and surgeons of all specialties. By matching, for example, fracture patterns in lower limb injuries to intrusion, pedal contacts and knee contacts, detailed descriptions of the mechanisms of the injuries can be developed [12].

Pedestrian and motorcycle accidents present different issues. A detailed investigation of these events can only be done using an on the spot approach, where the scene is visited within a few minutes of the accident occurring. Apart from witness assessments, skid marks are about the only technical evidence that can be used for assessing speed, although the throw distance of a pedestrian can give some approximations [11]. Sorting out the source of these road users' injuries between vehicle and ground contacts is often difficult, and needs detailed inspections at the scene, coupled with good descriptions of the injuries, particularly of superficial, external abrasions and bruises. This requires detailed liaison with the clinicians, and can often be frustrating in that such minor injuries, of no clinical significance, go unreported in many normal hospital records. Some elegant work on brain injury has been done by reproducing the dents made by pedestrians' heads on car bonnets, using Hybrid 3 dummy heads, and comparing the nature and severity of the actual injuries with the dummy response [13].

For assessing the severity of injury the Abbreviated Injury Scale is the universally accepted parameter. However, although it is a good descriptor of the nature of tissue damage by body region, and useful in assessing the threat to life, it does not address long term disabilities. The 2004 version of the AIS will however, include a disability measure, the Functional Capacity Index [14].

Each individual crash investigation is only a single point in an accident research data base. The usefulness of such data depends on careful experimental design, sampling appropriately to allow enough cases of specific interest to be collected, checking that the sampling plan does indeed provide a representative cross section of whatever categories of severity are to be considered, and the development of weighting factors when biases are introduced into the procedures. All too often too much is expected from a small number of cases, often selected in an unknown or varying manner from a poorly defined geographical area. With such difficulties it is often impossible to draw conclusions which have general validity. However, the longitudinal programmes such as NASS, CCIS and GIDAS are increasingly useful as they grow in size.

4. WHAT HAS BEEN ACHIEVED AND SOME CURRENT ISSUES

Large scale national accident data sources, usually based on police reports, provide an overall numerical picture of traffic injuries, but only at a superficial level. The two basic parameters of accident investigation, the severity of the event and the outcome in terms of consequent injuries are essentially absent, so that only broad demographic characteristics and frequencies can be analysed using such data. International comparisons show wide differences in definitions even for a relatively obvious parameter such as a traffic death. Some countries include suicides and deaths from natural causes, others do not. The working definition of a death across EU countries varies from one occurring within 24 hours of a collision, to up to 30 days (the standard UN/WHO definition). Correction factors are applied to arrive at a 30 day equivalent, introducing some uncertainties. Hence even for fatalities within the EU there is a range of uncertainty of some ±4% [15]. For the survivors, the definitions of the slight and serious categories vary enormously between countries. For example the ratio of serious to fatal cases across the EU15 countries varies by a factor of 6, which can only be explained by differences in definitions of "serious" injury. Hence such data are inadequate for examining biomechanical issues in the real world.

Thus the more detailed accident investigation data bases are the main source of our epidemiological knowledge. By providing such knowledge, priorities for countermeasures can be set, regulatory tests can be justified, and the benefits and limitations of improved crashworthiness of vehicles can be evaluated. At a more detailed level, specific mechanisms of injury are described and their frequency and consequences assessed. Just as an example, detailed crash investigation studies of US Indy car crashes, where the cars are fitted with comprehensive crash recorders, have suggested that the current thoracic criteria may be overly conservative, although the population of crashes in that sample is not typical in that it is restricted to very fit young males [16].

Many current studies link field accident data with experimental biomechanics. Studies of the brain's greater susceptibility to diffuse injury when struck laterally rather than frontally have been confirmed in detailed accident investigation studies. Head and neck injuries to small females from airbags gave an impetus to the development of the current out of position test requirements. Detailed accident investigation is a way of examining the biofidelity of our current dummies and modelling techniques. By comparing the kinematics of real people in real collisions with that of dummies in comparable collisions, the limitations of our current test tools can be examined. These are just a few examples of how detailed field accident investigations contribute to biomechanical knowledge.

4.1 Population Variations

Of particular interest currently is the area of population variations. Historically vehicle crash performance has been specified by requiring that certain parameters measured in a given dummy in a given crash test do not exceed certain values. This has encouraged a pass/fail process in that if those values are met, then a given car is deemed to be "safe".

The reality is that the each injury criterion is merely one data point of a risk curve, see Figure 1. This set of curves, developed by Prasad and Mertz [17] shows that an HIC of 1000 is equivalent to an 18% probability of a severe (AIS 4) head injury, a 55% probability of a serious (AIS 3) injury and a 90% probability of a moderate (AIS 2) head injury, to the average adult. How those probabilities shift across the spectrum of the population is largely unknown.

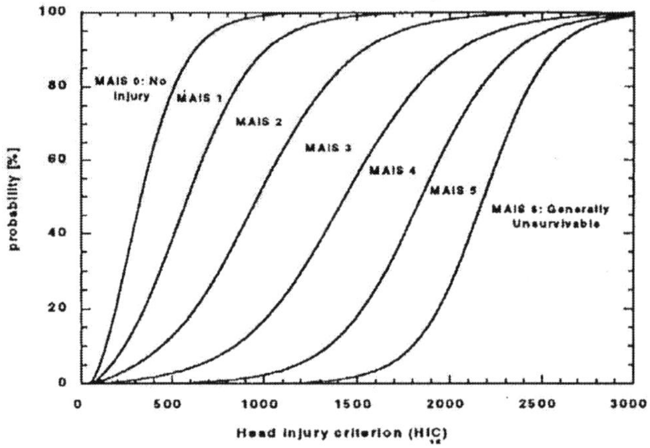

Figure 1. Probability of Head Injuries for HIC values [17].

Experimental biomechanics has documented the variations across the population for bone strength in particular, and its relationship to bone demineralisation and age effects [18, 19]. As a generalisation, fracture tolerance varies by a factor of between 3 and 5 for the strongest and weakest in the active population exposed on the roads.

For injuries unrelated to fractures, population variation is less well researched. In the absence of actual biomechanical tolerance data on children and small females for example, scaling techniques have been used to produce injury criteria tolerance levels for small females and children [20]. Geometrical and material properties (length, mass, modulus, etc.) are used to scale down to the various smaller sizes from the 50%ile adult male. This technique has resulted in the various head, neck, chest and femur tolerance levels specified in the most recent FMVSS 208 requirements governing out of position situations, with the various sized dummies. Such techniques involving mathematical procedures and "engineering judgment" have allowed progress to be made in specifying the requirements of advanced restraint systems. However, their relationship to the real populations at risk is still to be tested. Accident investigation studies over time will provide some evaluation of the validity of such techniques.

For the brain, abdominal organs, the neck, and soft tissue muscle and skin trauma there is little documentation of actual population variations. Females are shown to be more susceptible to soft tissue neck injury. Skin in older people is somewhat more resistant to threshold lacerations, perhaps a unique biomechanical case of the advantages of growing old.

The injury outcome from a traffic crash is the result of combinations of many factors, both intrinsic and extrinsic to the casualty, and permanent or temporary in nature. For a car occupant these can be categorised within the three parts of a crash; the pre-crash, crash and post-crash phases, as follows:

- Pre-crash Variables: Sex, Age, Height, Weight, BMI, Pre-existing medical conditions, Biomechanical tolerance, Muscle tone, Stomach contents, Bladder volume, Alcohol, Drugs, Clothing, Seat position in the vehicle, Sitting posture, Belt position, Pre-impact braking.
- Crash Variables: Impact direction, Velocity change, Peak vehicle deceleration, Pulse duration, Peak belt load, Posture at peak belt load, Airbag interactions, Point in the cycle of the heart, Loads and Durations of localised contacts, Rear loading, Interaction with other occupants.
- Post-Crash Variables: Severity of injuries, Combinations of injuries, Response time of EMS, Quality of diagnosis of injuries, Quality of treatment, Resulting disabilities.

Given the number of such variables it is perhaps surprising that relatively crude accident data, which post facto can only address a small number of these variables, can detect trends at all. Age and sex for example are both sensitive to crash severity. See the following figures [21].

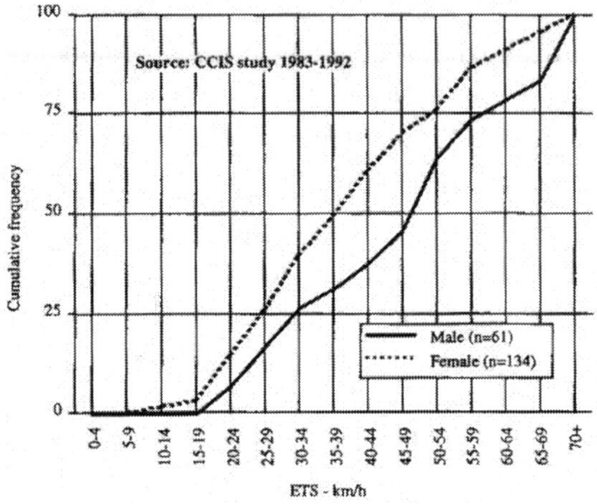

Figure 2. The cumulative frequency curves for restrained front seat occupants in frontal collisions receiving AIS 2+ injuries, by change in velocity, for males and females.

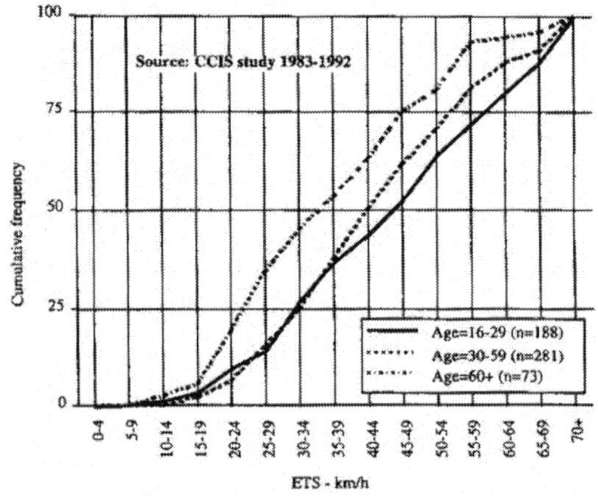

Figure 3. The cumulative frequency curves for restrained front seat occupants receiving AIS 2+ injuries in frontal collisions by change in velocity, for three age groups.

These data illustrate that older age groups, particularly those over 60 years of age, and females, are in collisions some 10 km/hr less in terms of change in velocity than younger age groups. About the same difference exists between females and male across the whole age range. Expressed differently, males and the young, require over 30% more kinetic energy than females and those over 60 years of age, to produce the same injury outcomes in say a 40 km/hr collision.

However, there are probably subtle combinations of some of the other factors listed above which lead to especially high or low levels of risk [22]. Small female drivers currently have to sit close to the steering wheel, and thus have been found in field studies to be at higher risk of airbag related injuries. Tall, thin males have been found to have a greater number of belt related chest injuries than smaller males, probably because the path of the shoulder belt is lower across the rib cage and is thus loading the lower ribs, which fracture more easily than those higher in the rib cage where the rib attachments to the sternum and spine are more substantial and less cartilaginous [23].

4.2 Crash Severity Distributions and Regulatory Requirements

Vehicle design currently is optimised around either the US Federal Standards or the various NCAP specifications in Europe, Australia, the United States and Japan. For example the EuroNCAP frontal crash test at 64 km/hr into an offset deformable barrier sets the gold standard for car design in Europe. Car manufacturers clearly design and tune their cars to do well in that test. However, accident data show that most collisions which cause AIS 3+ injuries to restrained occupants in frontal crashes, are at crash severities much lower that the EuroNCAP condition. Figure 4 shows that the mean value for the change in velocity is around 38 km/hr [24].

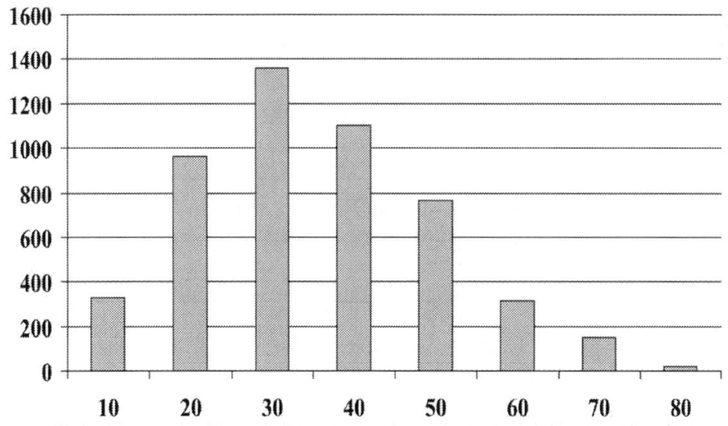

Figure 4. Distribution of restrained front seat occupants in frontal collisions who receive AIS 3+ injuries by change in velocity in km/hr.

These data suggest two things. First the crash severity of the test represents an extreme condition compared to the real world of serious injury collisions. This may well be producing vehicles which are not optimal for the mainstream serious injury crashes, being unnecessarily stiff for the numerous cases in the groups between 20 and 40 km/hr. Secondly and more importantly, it suggests that because the majority of AIS 3+ casualties are occurring below the conditions specified in the EuroNCAP test, in vehicles which very predominantly meet the test requirements, the injury tolerance levels in the test are inappropriate and should be lowered. Perhaps an HIC < 750, chest accelerations < 45 g, and femur loads < 750 kg would be appropriate. To address this issue properly requires a more detailed analysis of the in-depth accident data available, but also the development of injury probability curves for the age bands of the adult crash involved population.

The situation is different for the EuroNCAP side impact test requirement. Accident studies show that the typical side impact Delta V is substantially in excess of the current regulatory and NCAP requirements. For mainstream cars the change in velocity in the test is around 25 km/hr., whereas the mean Delta V for AIS 3+ injuries is some 45 km/hr [25]. Hence meeting those test requirements does not address the majority of the serious injuries in real world side impacts.

5. CONCLUSIONS

From this short review it is clear that accident investigation is an integral part of the subject of biomechanics. The test tools of experimental biomechanics, dummies with their current instrumentation, multi-body and FEM computer models, are all still relatively poor representations of the human frame. In particular, modelling of soft tissues is not adequate. Hence there must always be validation against the real world of crashes.

Beyond these specifics, well structured in-depth accident investigation provides early warning of emerging biomechanical problems, insights into specific mechanisms of injuries, puts dimensions to population variations, and addresses the validity and effectiveness of regulations and vehicle design changes.

Future research should strengthen the links between crash investigation, clinical studies and experimental biomechanics. Each will enhance the other.

REFERENCES

1. Rolt, L.T.C., *Red for Danger,* The Bodley Head Ltd., London, 1955.
2. De Haven, H., Mechanical Analysis of Survival in Falls from Heights of Fifty to One Hundred and Fifty Feet, *War Medicine*, **Vol. 2**, 1942, pp. 586–596.
3. Schwimmer, S. and Wolf, R.A., Preliminary Ranking of Injury Causes in Automobile Accidents, *Proceedings of Fifth Stapp Conference*, Minneapolis, Minnesota, USA, September 14 – 16, 1961, pp. 6–31.
4. Cairns, H., Crash Helmets, *British Medical Journal,* No. 4213, 1946, pp. 465–471.
5. Sheldon, S.H., On the Natural History of Falls in Old Age, *British Medical Journal,* 1960, December 10, pp. 1685–1690.
6. Haddon, W., Energy Damage and the Ten Countermeasures Strategies, *Proceedings of Ircobi Conference,* Amsterdam, Netherlands, June 26–27, 1973, pp. 489–505.
7. Huelke, D.F. and Gikas, P.W., Ejection – The Leading Cause of Death in Automobile Accidents, *Proceedings of the Tenth Stapp Conference,* Holloman, New Mexico, USA, November 8–9, 1966, pp. 260–294.
8. Mackay, M., Siegel, A.W. and Hight, P.V., Tempered versus HPR Laminated Windscreens, *Proceedings of Fourteenth Stapp Conference,* Ann Arbor, Michigan, USA, November 17–18, 1970, pp. 369–387.
9. Gloyns, P.F. and Mackay, M., Impact Performance of Some Designs of Steering Assembly in Real Accidents and under Test Conditions, *Proceedings of Eighteenth Stapp Conference,* Ann Arbor, Michigan, USA, December 4–5, 1974, pp. 1–28.
10. Thomas, C., Faverjon, G., Henry, C., Tarriere, C., Got, C. and Patel, A., Comparative Study of 1624 Belted and 3242 Non-Belted Occupants, *Proceedings of 24th A.A.A.M. Conference,* Rochester, New York, USA, October 7–8, 1980, pp. 422–436.
11. Ashton, S. and Mackay, M., Pedestrian Injuries and Death, *The Pathology of Violent Injury,* Mason, J.K., Edward Arnold, London, 1978, pp. 38–55.
12. Crandall, J.R., Martin, P.G., Kuhlmann, T.P., Pilkey, W.D., Burgess, A. and Quinn, T., The Influence of Vehicle Structural Intrusion on Lower Extremity Response and Injury in Frontal Crashes, *Proceedings of 39th A.A.A.M. Conference,* Chicago, Illinois, USA, October 16–18, 1995, pp. 269–286.
13. McLean, A.J., Brain Injury without Head Impact?, J. Neurotrauma, Vol. 12, 1995, pp. 621–625.
14. Gennarelli, T. and Wodzin, E. Introducing the Abbreviated Injury Scale 2004, *Proceedings of the Fifth Measuring the Burden of Injury Conference,* Baden bei Wien, Austria, June 3–4, 2004, pp. 16–19.
15. Mackay, M., National Differences in European Mass Accident Data Bases. *Proceedings of the Ircobi Conference 2003,* Lisbon, Portugal, 2003, pp. 6–15.
16. Melvin, J., Baron, K.J., Little, W.C. and Gideon, T. Biomechanical Analysis of Indy Race Car Crashes, *Proceedings of the 42nd Stapp Conference,* Tempe, Arizona, USA, November 2–4, 1998, pp. 247–266.
17. Prasad, P. and Mertz, H.J., The Position of the United States Delegation to the ISO Working Group 6 on the Use of HIC in the Automotive Environment. *S.A.E. Government Industry Meeting,* Washington DC, USA, S.A.E. 851246, 1985.
18. Pintar, F.A., Yoganandan, N., Hines, M., Maltese, M., McFadden, J., Saul, R, Eppinger, R., Khaewpong, N. and Kleinberger, M., Chestband Analysis of Human Tolerance in Side Impact. *Proceedings of 41st Stapp Conference,* Buena Vista, Florida, USA, November 13–15, 1997, pp. 63–74.
19. Kent, R., Patrie, J., Poteau, F., Matsuoka, F. and Mullen, C., *Proceedings of the Enhanced Safety Vehicle Conference,* Tokyo, Japan, June, 2003.

20. Kleinberger, M., Sun, E., Eppinger, R., Kuppa, S. and Saul, R., Development of Improved Injury Criteria for the Assessment of Advanced Restraint Systems, N.H.T.S.A., Washington DC, USA, September, 1998.
21. Mackay, M., Hassan, A. and Hill, J., Adaptive Restraints – Their Characteristics and Benefits, *Proceedings of Automotive Environmental Impact and Safety Conference,* Instn. Mech. Eng., London, England, November 4–6, 1997, pp. 37–52.
22. Ydenius, A., Influence of Crash Pulse Duration on Injury Risk in Frontal Impacts Based on Real Life Crashes, *Proceedings of Ircobi Conference 2002,* Munich, Germany, September 18–20, 2002, pp. 155–166.
23. Hill, J., Mackay, M. and Morris, A., Chest and Abdominal Injuries Caused by Seat Belt Loading, *Accident Analysis and Prevention,* Vol. 26, No. 1, 1994, pp. 11–26.
24. Mackay, M. and Hassan, A., Age and Gender Effects on Injury Outcome for Restrained Occupants in Frontal Impacts, *Proceedings of 44th A.A.A.M. Conference,* Chicago, Illinois, USA, October 2–4, 2000, pp. 75–92.
25. Mackay, M., The Characteristics of Lateral Collisions and the Evolution of Test Requirements, *Conference of Legislative Proposals for Occupant Protection,* Instn. Mech. Eng., London, England, April 28, 1989, pp. 1–16.

MODELING THE EFFECTS OF BLAST ON THE HUMAN THORAX USING HIGH STRAIN RATE VISCOELASTIC PROPERTIES OF HUMAN TISSUE

Emily E. Ward, Michael Kleinberger, Andrew M. Lennon and Jack C. Roberts
Johns Hopkins University Applied Physics Laboratory, 11100 Johns Hopkins Road, Laurel, Maryland, USA; phone: 443-778-4614, fax: 443-778-6916, emily.ward@jhuapl.edu

Abstract. An anatomically detailed finite element model (FEM) of the male human torso has been generated using geometry obtained from a model of the human anatomy developed for the computer graphics industry. The model represents a 5th percentile male human based on anthropometric data from the US Army. The geometry was used as a starting point to create finite element models of all the anatomic components, including the skeletal structure (vertebral column, ribs, cartilage, and sternum), stomach, lungs, liver, heart (including the aorta), muscles, and skin. The model also includes personal body armor representing a Level II vest. LS-DYNA was used to analyze the 245,000 element model simulating air blast. Material properties for the various soft tissues were obtained from high strain rate experiments on human organ tissue samples at rates ranging from 200-3000 s^{-1} using a modified Kolsky (split-Hopkinson) bar. This model has been used to evaluate pressures and deformation in specific areas of the thorax in response to blast loading, and to compare the results to existing injury criteria.

Key words: thorax, blast loading, Finite Element Method (FEM), human tissue data.

Michael D. Gilchrist (ed.), IUTAM Proceedings on Impact Biomechanics: From Fundamental Insights to Applications, 17–24.

1. INTRODUCTION

Human injury resulting from air blast occurs mainly by three mechanisms. These mechanisms include blast overpressure, displacement of the body, and projectiles caused by the explosion imposed on the body. Direct overpressure effects on the human body include eardrum rupture, contusions of the gastrointestinal tract, and lung hemorrhage due to alveoli tearing or shearing, the latter two frequently leading to mortality [**1**]. Although these are the more common air blast injuries, little is known about the effects of the pressure wave as it travels through the body during a blast event. The criteria most often used for predicting injury in humans are the Bowen curves which are based on experimental tests on animals [**2**]. Therefore, computational models of the human torso may provide new insight into humans subjected to air blast.

During the early years of blast studies, shock tube tests were developed and used to study different types of blast wave forms and their effects on animals [**3**, **4**]. There have been further studies supported by the Department of Defense developing analytical equations to determine criteria for personnel exposed to air blasts. In 1986, the criteria were reevaluated to define new terms and conditions. The updated criteria included data collected and reported from animal testing that included goats, sheep, dogs, and monkeys as well as additional human exposure data collected from accidental detonation. The data from these tests were compiled to predict and estimate human injury [**5**]. In the mid 1990s, a test device was developed that provided peak chest wall accelerations that can be compared to that of a porcine thorax. This device was used to extrapolate benefits of protective gear for humans [**6**]. There are also computer programs that incorporate the analytical models to calculate blast parameters for given blast environments. One such program is the Blast Effects Computer (BEC) program [**7**]. This program concentrates on the characteristics of the blast wave while it calculates and predicts the probability of damage to structures including glass breakage and human injury such as ear drum rupture and blast lung.

The purpose of this research is to develop a finite element model of the human torso using anatomically correct geometry and human tissue properties generated from high strain rate testing. This finite element model (FEM) will determine pressures in the critical organs affected by open air blast overpressure. The results will be compared to injury criteria curves and tables for human injury from past studies.

2. METHODS

2.1 Geometry and Mesh

A complete model of the human male was obtained from Digimation [**8**]. The model represents a 5th percentile male human based on anthropometric data from the US Army [**9**]. Only the thorax was used in this study. It consists of the complete skeletal structure along with the skin, heart, lungs, stomach and liver. Non-uniform Rational B-Spline (NURBS) surfaces, an industry standard for surface manipulation, of these components were created from the imported point clouds using Geomagic [**10**]. From these surfaces, individual models were made for each major component in the human torso: skin/muscles, ribs and vertebral column, lungs, heart, liver, and stomach. These entities were then exported from Geomagic into the finite element code IDEAS [**11**]. The surface model from Geomagic was then used as a template to construct cross-sections and create new simplified geometry in IDEAS. The cross-sections generated from the imported file and the lofting function resulted in smoother surfaces for the ribs and removal of some of the complex detail of the vertebral column.

After all the component sets were successfully imported into IDEAS, they were assembled together. All the surfaces were then meshed with triangular linear shell elements. These shell elements were then used to generate the solid tetrahedral elements for each individual organ. The skin was modeled with shell elements while all the other organs and components were modeled with solid tetrahedral elements. The body armor was modeled with eight noded solid elements based on the thickness of a Level II bullet proof vest with a thickness that ranges from 9-10 mm thick [**12**]. The components of the torso model are exposed in Figure 1(b). The skeletal

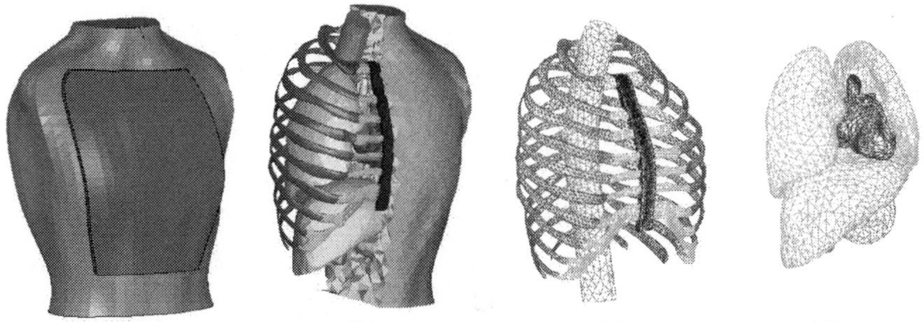

(a) (b) (c) (d)

Figure 1. Human Thorax Finite Element Model (a) Skin with level II protective armor (b) Assembled components with cutout (c) Skeletal system mesh(c) internal organs mesh.

system was meshed as an assembly to allow the interface of the sternum, cartilage, ribs and vertebral column to share element faces. The ribs and vertebral column were further divided into ribs, cartilage, sternum, and vertebral column, as shown in Figure 1(c). The linear triangular shell elements were used as boundaries to generate solid tetrahedral elements in the negative space between the components to represent the mediastinum, but includes elements for the diaphragm and all the remaining space not represented by the internal organs or skeleton. All of the shell elements except the skin were then deleted. A graphical representation of these elements representing the mediastinum membrane and diaphragm can be seen in the cutout view in Figure 1(b).

2.2 Material Characterization

Material properties for the various soft tissues were obtained from high strain rate experiments on human organ tissue from post-mortem human subjects (PMHS) at rates ranging from 200-3000 s^{-1} using a modified Kolsky (split-Hopkinson) bar. The soft tissues analyzed were obtained from post mortem human subjects. Heart, liver, lung, and stomach tissue was tested in shear using a double-lap shear specimen configuration, and for bulk stiffness using a radial confinement cell to generate volumetric strains. The results for each organ were then averaged to generate a linear bulk modulus and one or two shear stress versus shear strain curves (depending on the spread of the data as a function of strain rate). An example of the experimental results from can be seen for the heart in Figure 2. The shear stress versus shear strain results from the modified Kolsky bar tests were reduced to one or two

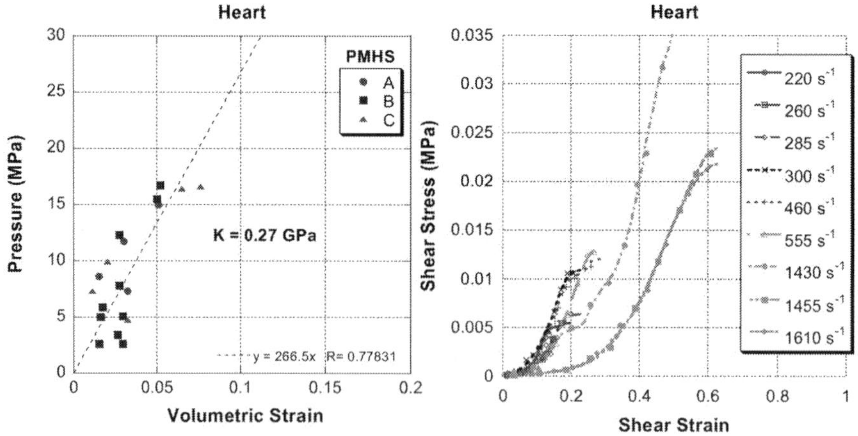

Figure 2. Experimental Data (a) Results from multiple PMHS labeled A, B and C for Bulk Modulus data (b) Experimental shear response from heart tissue.

distinct load deflection curves for input into the strain rate sensitive material model #181 in LS-DYNA [**13**]. The mediastinum and lungs were characterized using the basic viscoelastic material model #6 with data collected from the high strain rate experimental tests. The material inputs for the mediastinum were assumed to be similar to that of the liver except that the density was taken from the literature [**14**]. Material model #6 requires a short term shear modulus (G_0) and a long term shear modulus (G_∞), as well as a decay constant (β) and time (t) to describe the shear relaxation as seen in (1).

$$G(t)=G_\infty + (G_0 - G_\infty)\, e^{-\beta t} \tag{1}$$

The shear modulus and bulk modulus from the modified Kolsky bar tests were used as input for organ properties. To apply the experimental data for the lungs and the mediastinum the short term shear modulus was calculated from the final slope of the shear stress versus shear strain data collected.

Table 1. Number and type of elements and material properties for thorax components.

Component	No. of Elem.		Material Properties						
	Shell	Solid	Mat Model No.	ρ [kg/mm^3] [**14—15**]	E [GPa] [**14—15**]	ν [**15**]	K [GPa]	Strain Rates [s^{-1}]	[s^{-1}]
Ribs		15048	1	1.08 E -6	11.5	.2			
Sternum		4919	1	1.25 E -6	11.5	.25			
Cartilage		6238	1	1.07 E -6	.0245	.4			
Vertebral Column		8651	1	1.07 E -6	.0245	.4			
Heart		6841	181	1.0 E -6			.27	350	1500
Lungs #		11262	6	.6 E -6			.13	500	2100
Stomach		2314	181	1.05 E -6			.4	625	1500
Liver		3377	181	1.06 E -6			.41		
Mediastinum*		182781	6	2.07 E -6			.27		
Skin	3948		1	1.08 E -6	.0055	.3			
Armor		486	1	7.3 E-6	8.6	.36			
TOTAL	3948	241917							

Inputs based on experimental lung data (G_o= 81 kPa , G_∞= 80 kPa , β= .1)
* Inputs based on experimental heart data (G_o= 11 kPa , G_∞= 10 kPa , β= .1)

Because the blast event is of short duration, the long term shear modulus was set nearly equal to the short term shear modulus to satisfy the code for the values to be different while reducing the influence of the decay constant. This allowed the material to behave more elastically in the linear region while using the experimentally obtained shear results. The densities for the soft tissue and other components came from the literature [**14, 15**]. The material parameters for the body armor were approximated as a linear elastic material. The material properties are provided in Table 1.

3. RESULTS AND DISCUSSION

The simulation is based on an equivalent blast load of 10 kg of TNT detonated 7.5 meters away. This level was chosen to be near the threshold of lung damage, while still providing the human with above 99% probability of survival from the Bowen curves [**2**]. The BEC also predicts a zero percent probability of fatality due to lung damage but it calculates a 39% and a 90% probability of eardrum rupture based on two different theory calculations from Mercx and Eisenberg, respectively [**7**]. The FEM simulation does not include a modeled entity of air as a medium to transmit and reflect the blast wave, however, the theory behind the blast loading function in LS-DYNA is based on empirical air-blast models.

BEC predicted that 10 kg of TNT discharged at 7.5 m would arrive at the target with an incident pressure of 85.1 kPa. LS-DYNA calculated the blast wave to hit the target with an incident pressure of 82.7 kPa. Pressure and displacement sequence results can be seen in the heart in Figure 3. The path of the pressure is disrupted as it travels through organs with different material properties as in Figure 3(c).

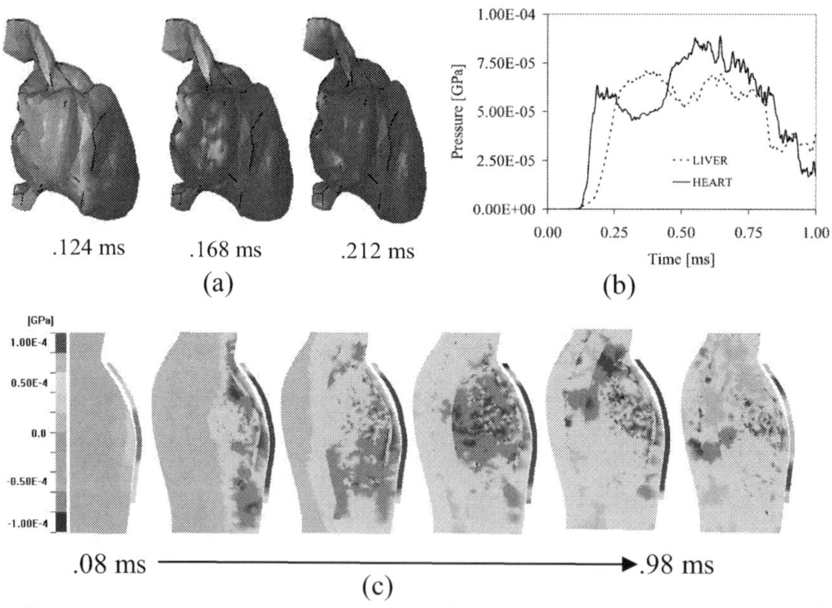

Figure 3. Pressure wave (a) Time sequence of heart as the pressure propagates from right to left (b) Pressure time history of a central element in heart and liver (c) Time sequence of thorax as the pressure wave propagates from right to left.

4. CONCLUSIONS

The thorax model appears to behave in a manner consistent with the environment to which it has been exposed. The pressure propagates through the body changing speed and direction as it travels though different defined materials. The results of the simulation give insight into the pressure profile seen in the thorax. However, many improvements need to be made before it can accurately predict internal injury in a blast simulation.

The geometry in this simulation is simplified for the benefit of a shorter runtime, but the detail should be enhanced in order to obtain more realistic results. A 50th percentile male model has been constructed with the addition of a bronchial tree in the lungs, addition of large and small intestines and enhanced detail in all other parts of the model. A distinct diaphragm has also been included along with separate musculature. Expansion of the digestive system to include the small and large intestine may be useful since gas filled organs are at greater risk of injury during exposure to a blast wave. The introduction of air to the surrounding environment, as well as inside the lungs, will allow for modeling of the blast wave as it travels through the thorax and through gas containing organs. The addition of fluid to various organs such as the heart and digestive system could be critical to a blast simulation. These improvements can be applied by placing the model in an air envelope and the various organs with empty cavities in fluid envelopes. However, this additional detail will greatly increase the runtime of the simulation because of the fluid/structure interaction that is imposed on the model.

ACKNOWLEDGEMENTS

The authors would like to express their appreciation to the Office of Naval Research (ONR) under contract no. N00024-03-D-6606 for funding this work.

REFERENCES

1. Elsayed, N.M., Toxicology of Blast Overpressure, *Toxicology*, **vol. 121**, 1997, 1—15.
2. Bowen, I.G., Fletcher, E.R. and Richmond, D.R., Estimate of Man's Tolerance to the Direct Effects of Air Blast, *Report DASA-2113*, Defense Atomic Support Agency, Washington, D. C. Oct. 1968.
3. Richmond, D.R., Clare, V.R., Goldizen, V.C., Pratt, D.E., Sanchez, R.T. and White, C.S., Biological Effects of Overpressure: II. A Shock tube Utilized to Produce Sharp-

rising Overpressures pf 400 Milliseconds Duration and Its Employment in Biomedical Experiments, *Aerospace Medicine*, **vol. 32**, 1961, 997—1008.

4. Richmond, D.R., Goldizen, V.C., Clare, V.R., Pratt, D.E., Sherping, F., Sanchez, R.T., Fischer, C.C. and White, C.S., The Biologic Response to Overpressure: III. Mortality in Small Animals Exposed in a Shock Tuber to Sharp-rising Overpressures of 3 to 4 Msec Duration, *Aerospace Medicine*, **vol. 33**, 1962, 1—27.
5. Richmond, D.R., Yelverton, J.T. and Fletcher, E.R., New Air Blast Criteria for Man, Presented at the *Twenty-Second DoD Explosive Safety Seminar*, Anaheim Marriott Hotel, Anaheim CA, 26-28 August 1986.
6. Cooper, G.J., Pearce, A.J., Sedman, I.S. and Oakley, C.W., Experimental Evaluation of a Rig to Simulate the Response of the Thorax to Blast Loading, *The Journal of Trauma*, **vol. 40 no. 3**, 1996, S38—S41.
7. Swisdak, M.M. and Ward, J.M., DDESB Blast Effects Computer Computer Version 5.0, *DDESB TP 17*, Department of Defense Explosive Safety Board, Alexandria, VA, May 2003.
8. Digimation © 1996-2004 Digimation, Inc. 100 James Drive, Suite 150, St. Rose, Louisiana, USA 70087
9. Gordeon, C.C., U.S. Army Anthropometric Survey Database: Downsizing, Demographic Change, and Validity of the 1988 Data in 1996, *Science and Technology Directorate*, TR NATICK/TR-97/003, October 1996.
10. Geomagic © 1996-2001, *Raindrop Geomagic*, Inc, P. O. Box 12219, Research Triangle Park, NC 27709.
11. Ideas NX10 © 1979-2002. *Unigraphics Solutions, Inc.*, 10824 Hope Street, Cypress, CA 90630.
12. Ballistic Resistance of Personal Body Armor: NIJ Standard 0101.04, *NCJ 183651*, National Law Enforcement and Corrections Technology Center, P.O. Box 1160, Rockville, MD 20849, June 2001.
13. LS-DYNA © 1992-2003, Version 970, Rev 5434, *Livermore Software Technology Corp,* Livermore, California, CA 94551.
14. Duck, F.A., Physical Properties of Tissue, Academic Press, Harcourt Brace Jovanovich, Publishers, London, 1990.
15. Wang, H.C., Development of a Side Impact Finite Element Human Thoracic Model, PhD Dissertation, Wayne State University, Detroit, MI, 1995.

PRELIMINARY ANALYSIS OF BLUNT BALLISTIC IMPACTS TO THE ABDOMEN

Cynthia Bir and Jerome Eck
Wayne State University, 818 W. Hancock, Detroit, MI 48201, USA;
Phone: 313-577-3830, Fax: 313-577-8333, E-mail: cbir@wayne.edu

Abstract. Law enforcement agencies, as well as military personnel, are relying on non-lethal technology to diffuse potentially life-threatening situations without the use of lethal force. However, a standardized method for testing the health effects of these non-lethal devices has not been established. In recent studies, the abdominal region has been cited as the area most impacted during the deployment of kinetic energy rounds [1]. In the current study, the first phase of biomechanical testing involved the determination of biomechanical corridors. A total of six post-mortem human subjects were tested. An impact condition of a 45 g mass at 60 m/s resulted in 4,741 N of force generated with 22 mm of deflection. The second phase involved the determination of a valid injury criterion based on a pre-established tolerance of injury in a porcine model. Injuries observed were classified according to AIS. Logistic regression analysis of the injury data, in combination with an empirically derived criterion, facilitated this process. The predictive ability of several key injury criteria was evaluated.

Key words: abdominal injury, blunt ballistic impacts, less-lethal kinetic energy rounds.

1. INTRODUCTION

The use of kinetic energy rounds by law enforcement agencies has increased over the past several years. These rounds afford the officers with an alternative to lethal force but have not been well defined in terms of risk of injury. Kinetic energy rounds or less-lethal impact munitions are designed to incapacitate individuals without causing serious or fatal injuries. These munitions include 12 gauge, 37 mm and 40 mm rounds. Materials of the

Michael D. Gilchrist (ed.), IUTAM Proceedings on Impact Biomechanics: From Fundamental Insights to Applications, 25–32.

submunition vary from hard rubber to soft foam; however one of the more popular rounds in the United States is the "bean-bag" round.

The National Institute of Justice (NIJ) recently published a report based on data collected from 373 incidents in which kinetic energy rounds were deployed [1]. The study delineated the region of impact and noted that the impact to the abdomen resulted in the higher percentage of impacts (34%). The thorax was the second highest at 19%. Although 94% of the impacts to the abdomen were related to some level of injury, no injuries were fatal.

Abdominal injuries associated with the deployment of kinetic energy rounds have been previously reported [2]. In a recent study, Suyama et al. [2] reported on injury patterns observed in 25 patients treated after a period of civil unrest in an urban emergency department. The abdominal region was cited as having the highest significant morbidity with 75% of all cases requiring admission. Injuries to the abdominal region included: liver laceration, abdominal wall contusions, and rectus sheath hematoma. Previous research has also indicated that there is a high risk of AIS 3 type injuries, which included bowel perforations and splenic lacerations, associated with impacts to the abdomen [3]. This was based on a review of 155 cases resulting in 172 injuries that occurred during the period between July 7 and 15, 1996 in Northern Ireland.

Mahajna et al. [4] reviewed 152 medical records of individuals treated during the Israeli-Arab conflict. Injuries were documented according to body region, mechanism of injury, severity of injury as well as other outcome parameters. A total of twelve impacts were documented to the abdominal region. Of these injuries, ten were blunt and two penetrating. Contusions of the abdominal wall, laceration of the spleen, perforations of the bowel were all reported injuries to the abdomen [4].

Although a biomechanical analysis has been completed for the thoracic region [5], each region of the body has a unique response to specific impacts. Therefore, the results are not transferable to other anatomical locations or from varying impact conditions. Therefore, a biomechanical analysis of blunt ballistic impacts to the abdomen is essential.

1.1 Biomechanical Assessment

The first step taken in a biomechanical assessment is to determine how the body responds to a given impact. The role of Post-Mortem Human Subjects (PMHS) is critical when establishing biomechanical response corridors. Force-time curves and deflection-time curves taken from each specimen are used to establish the respective corridors. These corridors serve two purposes. First, they describe how the body responds (deforms) to a given impact (force). The second purpose is to provide data for the development of biomechanical, mathematical and computer surrogates.

Due to the variability between specimens normalization of the PMHS data is often required. Eppinger et al. [6] described a methodology based on equal stress-equal velocity. The developed techniques assume a constant mass density and constant modulus of elasticity between subjects. A scaling factor, λ, is determined by assuming geometric similarity. For most biomechanical testing, data are scaled to the 50_{th} percentile male.

After data are normalized, all individual response curves are plotted on the same graph to determine the average response in the form of corridors. Two methodologies have been previously employed. Bir et al. [5] developed corridors by maintaining the general effect of the impact by including the majority of the individual responses of all specimens within the corridor and segregating outliers. Cavanaugh et al. [7] determined the mean and standard deviation values for each time point. Corridors were developed by adding and subtracting the standard deviation from the mean.

1.2 Injury Criteria

The next critical step in the biomechanical assessment of an impact is the determination of limits for specified injury data or the development of injury criteria. These injury criteria are related to biomechanical parameters that are physically measured during the impact event. Viano and King [8] have described injury criteria as the numerical relationship of a measurable engineering parameter; i.e. force, to a given risk of injury; i.e. rib fracture. Although risk to the bony structures is possible with the cadaveric model, other physiologic injuries are not possible. Therefore, a biological model is often used for this purpose.

The validation of injury criteria involves determining both a biomechanical parameter and related level of injury. The evaluation of varying parameters is explored using logistic regression to determine the best fit. Logistic regression determines the relationship between the probability of an event occurring (injury) and the value of an independent variable (engineering parameter). The desired injury level must be established prior to evaluation. A dichotomous state of occurrence versus nonoccurrence is established based on this level for the physiological parameters or the dependent variables. The independent variables are the continuous values of the injury criteria being studied. The probability of injury p(x) is related to a response parameter x based on a statistical fit to the sigmoidal function $p(x)=[1+\exp(-\alpha-\beta x)]^{-1}$, where α and β are parameters fit to the responses from the injury data.

Research has shown that the sus scrofa (swine) most closely resembles the human anatomy in terms of girth and organ distribution [9]. Therefore, it is the most appropriate surrogate for the assessment of blunt ballistic impacts to the abdomen. To perform a complete biomechanical assessment of blunt ballistic impacts to the abdomen, both cadaveric and swine models were employed in the current study. Impact conditions were chosen to represent those seen in the field with varying locations of impact selected due to the inhomogeneity of the abdominal region.

2. METHODOLOGY

2.1 Biomechanical Corridor Development

To develop biomechanical corridors, six post-mortem human subjects (PMHS) were tested. Specimens were obtained through the Wayne State University Body Bequest Program and the University of Michigan – Anatomical Donations Program. All specimens were treated according to ethical standards previously established **[10]**.

Prior to testing, each specimen was examined using both radiographs and previous medical history to identify any existing anomalies. Impacts were accomplished with a 45 gram rigid projectile at 60 m/s. The diameter of the impactor was 37 mm which allowed for the attachment of a 7270A-20K (Endevco) accelerometer. Impact force was calculated based on acceleration measurements and the known mass of the impactor. High-speed video was obtained at 20,000 fps.

Each PMHS was suspended in a harness system located 14.5 inches from the launcher and was allowed to swing freely upon impact. The specimen was positioned so that the impact would occur at the epigastric region. Since the impacts being studied are concentrated in the upper abdomen, the anterior-posterior diameter was used as the basic scaling factor.

2.2 Injury Criteria Evaluation

A total of 17 sus scorfa (swine) were used for testing. Specimens weighing approximately 100 pounds were chosen in order to most closely match the thoracic dimension of the 50th percentile male human. Testing was conducted in cooperation with the Surgical Research Services (SRS) and Division of Laboratory Animal Resources (DLAR) at Wayne State University. SRS provided full veterinary support throughout the procedure as well as technician support. Full anesthesia and analgesia was used throughout the experiments with no resuscitation. The full animal protocol was approved by the Animal Care Committee.

After a plane of surgical anesthesia was achieved, a pressure transducer was placed in the right atrium to monitor central venous pressure. An additional catheter was placed at the intersection of the carotid artery and the aorta to monitor mean arterial pressure. An ECG was connected to the specimen for continuous monitoring. The specimen remained in a spinehorizontal position in an effort to prevent traumatic apnea.

A vertical air cannon was positioned under the specimen to deliver impacts to the liver and bowel regions. Multiple impacts were conducted on each specimen in a randomized order using a 50 gram projectile at varying velocities. An impactor similar to that used in the PMHS testing with a diameter of 37 mm was used.

A comprehensive necropsy was performed to assess the severity of injuries due to each impact. Injuries were categorized according to the AIS guidelines. In an effort to have both injurious and non-injurious events, the input energy was increased or decreased accordingly. Logistic regression analysis was performed using the determined injury data from necropsy reports and the instrumentation and video data from the specimen. Explored injury criteria include: Viscous Criterion (VC) **[11]**, Abdominal Injury Criterion (AIC) **[9]**, Blunt Criterion (BC) **[12]** and Energy Dissipated.

The VC represents the maximum value of the time-dependent product of velocity and compression with units of m/s. The AIC is the product of impact velocity and maximum normalized compression with units of m/s. The injury criterion of BC represents the natural logarithm of the impact energy divided by the product of the specimen mass to the 1/3 power, wall thickness of the specimen, and area of impact. The final criterion, Energy Dissipated, is calculated by subtracting the kinetic energy at the end of the impact from the kinetic energy at the beginning of impact.

3. RESULTS

3.1 Biomechanical Corridors

Six specimens were used to develop the force-deflection corridors for impacts to the epigastric region (Figure 1). The average peak force was 4741 + 553 N which occurred within .25 ms. The average peak deflection of 22 mm occurred over the first millisecond of the impact event.

Injuries that occurred due to the liver impacts varied from no injury to multiple liver lacerations. The results for the injury level associated with each injury criterion evaluated can be found in Table 1. A total of four injury parameters are presented along with the associated AIS ranking.

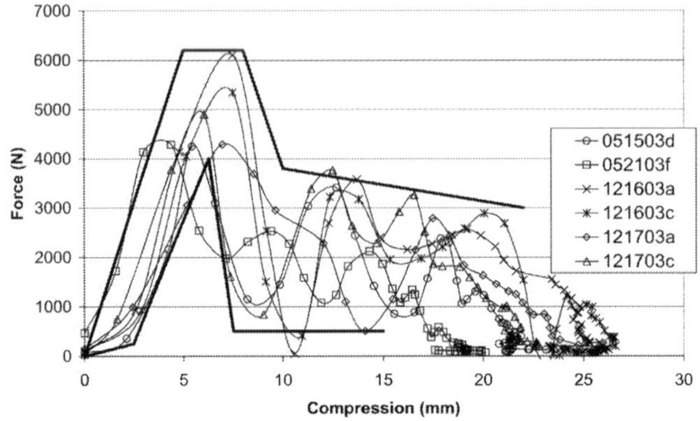

Figure 1: Force-deflection corridors for epigastric region.

Table 1..Injury criterion results for liver impacts with associated AIS injury ranking.

Test Number	Velocity (m/s)	VC max (m/s)	AIC (m/s)	BC	Energy Dis max (Nm)	AIS Injury
P6055A	60.82	3.94	10.44	0.96	70.97	2
06065A	61.56	3.86	11.16	1.08	71.39	2
P6064C	62.47	4.59	12.95	1.05	80.51	2
P6068C	62.70	2.62	9.81	1.03	56.08	3
P6069A	60.32	2.62	8.91	0.86	56.19	2
P6067B	58.31	2.60	8.65	0.79	51.20	2
P6066B	54.56	2.95	5.66	0.59	33.43	2
P6073A	55.94	2.09	7.23	0.66	46.05	0
P6072C	57.20	2.70	9.15	0.56	50.53	0
P6071C	54.67	2.65	8.28	0.86	45.45	2
P6074A	55.75	1.89	6.64	0.54	41.31	2
P6075B	51.15	1.92	5.66	0.38	38.86	2
P6077A	50.14	2.19	6.88	0.77	41.23	2
P6097C	45.77	2.24	6.81	0.44	36.59	0
P6096C	45.24	1.41	4.73	0.35	25.18	0

The results for the occurrence versus nonoccurrence of AIS 2 predicted by each criterion can be found in Table 2. Results indicated that BC ($\chi^2 = 5.754$ p = 0.016) had the best predictive ability of all criteria examined. Based on this analysis, a BC of .51 will result in a 50% chance of sustaining a liver injury of AIS 2-3.

Table 2. Results of logistic regression analysis of independent variable AIS > 2.

Injury Criterion	α	β	Chi-Square	P-Value
Vcmax	-3.95	2.05	3.57	0.059
AICmax	-2.25	0.42	1.87	0.171
BC	-4.00	7.83	5.75	0.016
Energy Dissipated max	-3.22	0.09	3.10	0.078

Injuries that occurred due to the impacts to the bowel varied from no injury to small hemorrhages in the subserosal layers. The results for the injury level associated with each injury criterion evaluated can be found in Table 3. A total of four injury parameters are presented along with the associated AIS ranking.

Table 3. Injury criterion results for bowel impacts with associated AIS injury ranking.

Test Number	Velocity (m/s)	VC max (m/s)	AIC (m/s)	BC	Energy Dis max (Nm)	AIS Injury
P6065C	64.12	4.36	15.54	1.07	69.18	0
P6064B	64.61	4.05	13.46	0.62	66.39	0
P6068A	60.82	3.16	12.04	1.17	53.88	0
P6069B	60.39	3.35	12.04	0.87	57.92	0
P6067A	64.86	3.17	11.95	1.06	59.05	0
P6066C	65.69	5.14	15.68	0.90	58.14	0
P6073C	73.37	5.68	20.14	1.40	89.50	0
P6072B	74.23	4.73	18.71	1.39	84.08	2
P6071A	75.44	5.17	18.56	1.53	84.45	2
P6074B	76.35	5.31	21.12	1.60	94.44	2
P6075C	76.46	5.01	18.57	1.60	90.63	2
P6077C	76.58	5.33	22.09	1.50	100.02	2
P6097A	67.62	4.02	17.82	1.04	69.38	0
P6096B	68.62	5.47	17.40	1.32	74.98	2

The results for the occurrence versus nonoccurrence of AIS 2 predicted by each criterion can be found in Table 4. Results indicated that BC (χ^2 = 13.63 p = 0.000) had the best predictive ability of all criteria examined. Based on this analysis, a BC of 1.32 will result in a 50% chance of sustaining a bowel injury of AIS 2-3.

Table 4. Results of logistic regression analysis of independent variable AIS > 2.

Injury Criterion	α	β	Chi-Square	P-Value
Vcmax	-11.23	2.31	6.48	0.011
AICmax	-13.79	0.77	8.48	0.004
BC	-21.74	16.41	13.64	0.000
Energy Dissipated max	-13.94	0.18	10.12	0.001

4. CONCLUSIONS

The current research represents a preliminary biomechanical assessment of blunt ballistic impacts to the abdomen. Further research is currently being explored in regards to proper filtering techniques to further validate the results. Both biomechanical corridors and injury criteria were developed.

Basic theoretical principles and laboratory techniques were applied. These principles and techniques have been employed in the field of occupant protection and crashworthiness for over sixty years and have led to significant changes in automotive design. The same knowledge has been applied in the area of sports biomechanics and has led to rule changes, protective gear development and an overall reduction in injuries. Applying these principles and techniques to blunt ballistic impacts will allow for critical decisions to be made in terms of risk of injury with deployment of a given round.

ACKNOWLEDGEMENTS

The authors would like to acknowledge Robert Hammond, Don Sherman, Marianne Wilhelm and Elizabeth Dawe for their assistance with the testing. This research was funded by the National Institute of Justice.

REFERENCES

1. Hubbs, K. and D. Klinger, Impact Munitions Data Base of Use and Effects, 2001.
2. Suyama, J., et al., Injury patterns related to use of less-lethal weapons during a period of civil unrest, *J. Emergency Med.,* **25**, 2003, 219-227.
3. Steele, J.A., et al., Plastic Bullet Injuries in Northern Ireland: Experiences during a week of Civil Disturbance, *J. Trauma,* **46**, 1999, 711-714.
4. Mahajna, A., et al., Blunt and penetrating injuries caused by rubber bullets during the Israeli-Arab conflict in October, 2000: A retrospective study, *Lancet,* **359**, 2002, 1795-1800.
5. Bir, C., D. Viano, and A. King, Development of biomechanical response corridors of the thorax to blunt ballistic impacts, *J. Biomech.*, **37**, 2004, 73-79.
6. Eppinger, R., J. Marcus, and R. Morgan, Development of dummy and injury index for NHTSA's thoracic side impact protection research program, in: *SAE GovernmentIndustry Meeting* , 1984, pp. 983-1011.
7. Cavanaugh, J., et al., Lower abdominal tolerance and response, in: *30th Stapp Car Crash Conference*, 1986, pp. 41-63.
8. Viano, D.C. and A.I. King, Biomechanics of Chest and Abdomen Impact, in: *The Biomedical Engineering Handbook*, Bronzino (eds.), CRC Press, Boca Raton, 1995, pp. 24(1-12).
9. Rouhana, S.W., Biomechanics of abdominal trauma, in: *Accidental Injury: Biomechanics and Prevention*, A.M. Nahum and J.W. Melvin (eds.), Springer-Verlag, New York, 1993, pp. 398-428.
10. King, A.I., et al., Humanitarian benefits of cadaver research on injury prevention, *J Trauma,* **38**, 1995, 564-569.
11. Viano, D.C. and I.V. Lau, A viscous tolerance criterion for soft tissue injury assessment, *J Biomech.*, **21**, 1988, 387-399.
12. Bir, C. and D.C. Viano, Design and injury assessment criteria for blunt ballistic impacts, *J Trauma*, **57**, 2004, 1218-1224.

DEVELOPMENT OF AN ASSESSMENT METHODOLOGY FOR LOWER LEG INJURIES RESULTING FROM ANTI-VEHICULAR BLAST LANDMINES

J. Manseau[1] and M. Keown[2]

[1]*Defence R&D Canada – Valcartier, 2459 Pie-XI Blvd North, Val-Bélair, QC, Canada, G3J 1X5, phone: (418) 844-4000 ext. 4470, fax: (418) 844-4502, email: josee.manseau@drdc-rddc.gc.ca;* [2]*Biokinetics and Associates Ltd., 2470 Don Reid Drive, Ottawa, ON, Canada, K1H 1E1, phone: (613) 736-0384, fax: (613) 736-0990, email: keown@biokinetics.com.*

Abstract: Data were generated to assist development of an injury assessment methodology for risk evaluation of lower leg injuries caused by anti-vehicular blast landmines. An instrumented Hybrid III lower leg and frangible DRDC Complex Lower Leg (CLL) were each subjected to dynamic axial impacts at various severity levels. Biofidelity of the CLL was evaluated and the effect of military boot on injury severity was studied.

Key words: lower leg, Complex Lower Leg (CLL), Hybrid III, injury criteria, anti-vehicular blast landmine, military boot, light-armored vehicle.

1. INTRODUCTION

Anti-vehicular (AV) blast landmines are a constant threat during troop transportation. When a light-armoured vehicle is subjected to a strike from an AV blast landmine, passengers may survive the impact but the vulnerability of their lower legs predispose them to a high risk of debilitating injuries. During a mine strike, the vehicle structure dynamically deforms and may result in significant axial loading to the lower leg. To test and improve protection systems to mitigate this effect, a suitable test surrogate and an appropriate injury assessment methodology are required. This work was

Michael D. Gilchrist (ed.), IUTAM Proceedings on Impact Biomechanics: From Fundamental Insights to Applications, 33–40.

realized with the use of the frangible DRDC Complex Lower Leg (CLL) and a 50th percentile male Hybrid III lower leg. The first objective of the study was to verify the biofidelity of the CLL under dynamic blunt axial impacts. The second objective was to collect data to develop a transfer function between injury severity and tibia force response as measured by a Hybrid III leg. Finally, the third objective of this study was to evaluate the effect of military footwear on severity of foot/ankle injuries.

2. THE COMPLEX LOWER LEG

The DRDC Complex Lower Leg (CLL) was developed by DRDC Valcartier to evaluate lower leg injuries sustained by anti-personnel mine and is now distributed by Biokinetics and Associates Ltd. [1, 2]. The leg, shown in Figure 1, is comprised of polymeric bones (that represent tibia/fibula, talus, and calcaneus), a nylon tendon, silicone rubber cartilage pads, a silicone rubber heel pad, ballistic gelatin (representing the flesh), and a latex skin. Since the objective of the CLL was to model the injury path up through the heel into the tibia, the forefoot is not considered and the leg has two calcaneus.

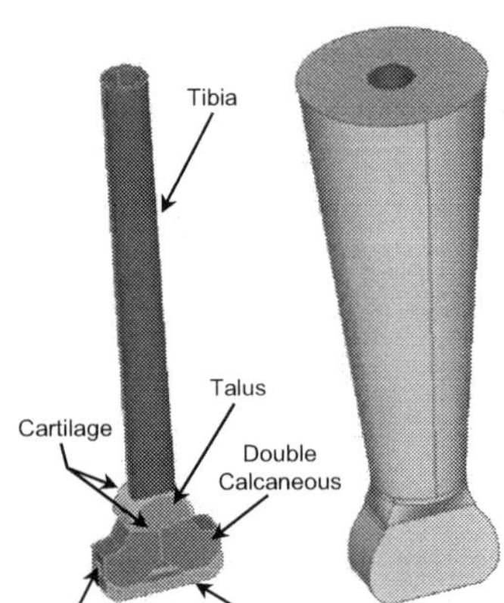

Figure 1. The Complex Lower Leg (CLL).

3. EXPERIMENTAL SET-UP

An air cannon and sled/rail system was developed to strike the base of the foot of either a CLL or Hybrid III instrumented lower leg. To evaluate CLL biofidelity, the test configurations of Owen [3] (non-injurious testing) and

Funk [4] (injurious testing) were approximated to compare injury response between CLL and Post Mortem Human Surrogates (PMHS). For Owen-style testing, a semi-cylindrical, sliding impact head was installed on the front of the sled. For Funk-style testing, a rectangular steel plate was used as the impact surface. As with the original tests by Funk, a foam layer was secured on the plate to prevent direct impact between foot and impactor. As shown in Figure 2, the Hybrid III and CLL tibias were installed on a Hybrid III knee that was fixed to a table attached to the floor.

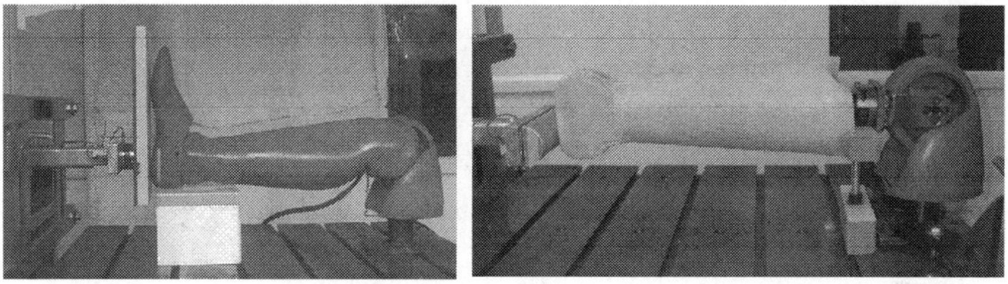

Figure 2. Funk set-up with Hybrid III (left) and Owen set-up with CLL (right).

4. TESTING CONDITIONS

The input conditions were defined using the Hybrid III output response. To reach Hybrid III tibia axial force response comparable to the peak value and rise time of the Owen testing, the impactor, of 1.5 kg, struck the heel at a velocity of 5.6 m/s. Funk input conditions were determined by comparing PMHS tibia force corridor (provided by the author) and Hybrid III tibia force response. An 18.5 kg impactor struck the plantar surface of the foot at 5 m/s. The peak force value measured on the Hybrid III tibia was approximately two times greater than the maximum peak value measured on PMHS. This loading configuration was believed to be optimal to generate CLL injuries, knowing that Hybrid III tibia force response is higher than of the response of a PMHS tibia [1]. Finally, the 'mine' loading conditions were obtained by removing the foam from the impact plate of the Funk-style set-up. This simple modification reduced the rise time and loading duration, and increased the peak force values, which resulted in a Hybrid III tibia response comparable to that of full-scale mine testing. The impactor velocity was the same as in the Funk-style testing (i.e. 5 m/s). Testing was completed using each of the four configurations (Owen, Funk, 'mine' without boot and 'mine' with a boot) using both the Hybrid III lower leg and the CLL, and each test was conducted three times for repeatability.

5. RESULTS

As shown in the literature [3, 5], human and standard Hybrid III tibias do not have the same mechanical properties. For this reason, the correlation between Hybrid III tibia force response and CLL injury severity is of great interest for the development of an injury assessment method based on Hybrid III tibia response. Table 1 shows a summary of results obtained during Hybrid III and CLL testing. Average values for Hybrid III tibia peak axial force and impulse, and CLL injuries are presented.

Table 1. Summary of results.

Test	HIII tibia force (kN)	HIII tibia impulse (N·s)	Tibia fx*	Talus and/or calcaneus fx	Soft tissue injuries
Owen-style	2.9	11	No	No	No
Funk-style	13.9	127	No	Yes	Yes
'Mine' w/o boot	17.4	120	No	Yes	Yes
'Mine' w/ boot	13.3	129	No	No	Yes

*fx = fracture

As expected, the CLL subjected to Owen-style testing did not sustain any injury while those subjected to Funk-style testing did sustain fractures. Funk-style testing resulted in minor calcaneus fracture and soft tissue injuries. Soft tissue injuries include heel pad lacerations and cartilage damage. 'Mine' testing without a boot showed severe calcaneus and/or talus fractures as well as soft tissue injuries. Tests with a military boot only produced soft tissue injuries without any fracture. None of the tests in the current study produced tibia fractures. Figure 3 shows one of the three CLL after being submitted to Owen-style and Funk-style testing, and Figure 4 shows CLL after 'mine' testing without and with military boot.

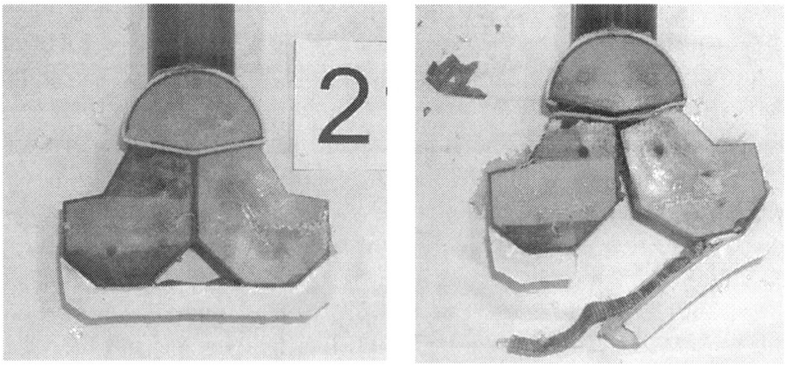

Figure 3. CLL resulting injuries for Owen-style (left) and Funk-style (right) testing.

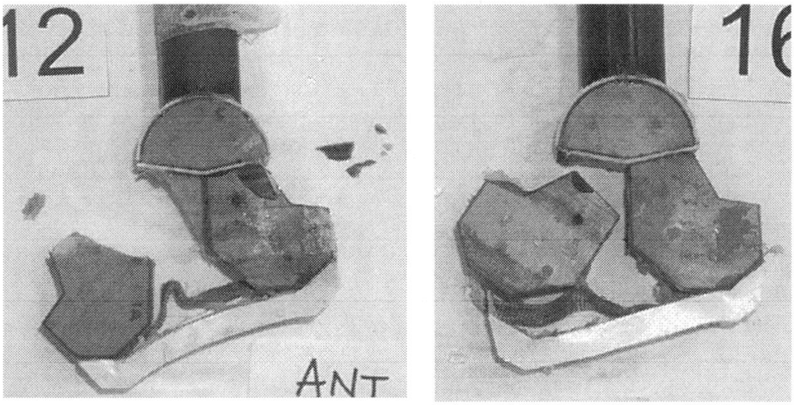

Figure 4. CLL resulting injuries for 'mine' testing without boot (left) and 'mine' testing with boot (right).

Figure 5 shows the Hybrid III tibia axial forces recorded during each of the four types of testing. These results quantify, in terms of peak values, the effect of the military boot on the loading transferred to the leg.

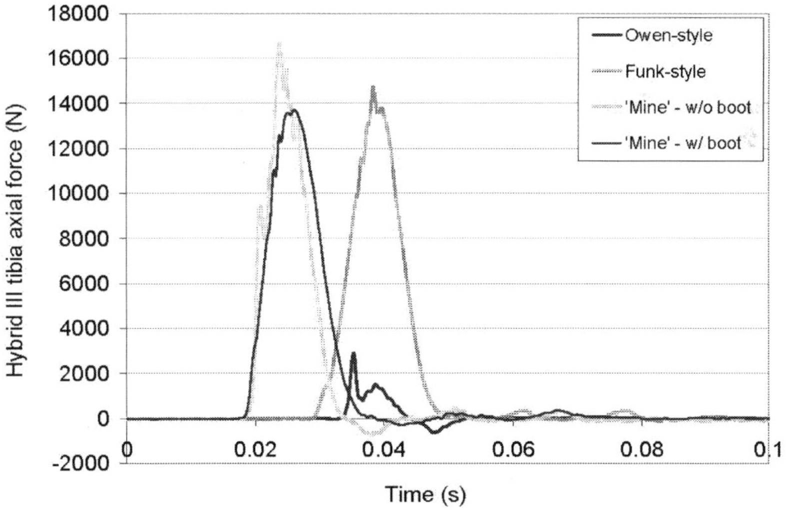

Figure 5. Hybrid III tibia axial force

6. DISCUSSION

The first objective of this study was to verify the biofidelity of the CLL under blunt axial impact. To reach this goal, the approach was to reproduce Owen and Funk test conditions to generate non-injurious and injurious loading, respectively, and to compare PMHS and CLL response. As with the PMHS in the Owen study, the CLL did not sustain any injury in this study. The CLL, similar to the PMHS, sustained foot/ankle fractures under Funk loading conditions. Based on these results, the CLL gave satisfying results in terms of injury response.

The second objective of this study was to collect data in order to develop a transfer function between Hybrid III tibia force and injury severity predicted by the CLL. Each style of testing (Owen-style, Funk-style, 'mine' without boot, 'mine' with boot) was repeated with both a Hybrid III lower leg and CLL in order to generate such data. Peak tibia force and impulse values were considered as parameters that may correlate with CLL injury severity. Based on results presented in Table 1, the Hybrid III peak force seems proportional to CLL injury severity but, when comparing 'mine' testing results with and without a boot, peak impulse does not seem to follow this same trend.

Finally, the third objective of this study was to evaluate the effect of military footwear on injury severity. The injuries sustained by the CLL for 'mine' testing showed that the military boot has a significant effect on the CLL injury severity. In that case, wearing a boot protected against foot/ankle fracture and thus, reduced significantly the risk of suffering disabling injuries.

7. CONCLUSIONS

This study was done to generate preliminary results for an R&D project on the development of an injury assessment method for lower leg injuries resulting from anti-vehicular blast landmines. The biofidelity of a new frangible lower leg surrogate was verified under axial impact loading. Injurious and non-injurious tests were performed to verify injury response of the Complex Lower Leg. The CLL showed satisfying results in terms of injury response when compared to PMHS results presented in literature [3, 4]. The limitation of the CLL is that it represents a young male lower leg, which has a relatively high tolerance to fracture with respect to the older PMHS typically available. Its utilization for civil application, where the age of the studied population includes a broader range, may be limited.

However, the CLL is believed to be a good tool to evaluate lower leg injuries caused by axial impact when PMHS testing is not available.

After biofidelity verification testing, the loading severity was increased to produce an impact similar to that seen during anti-vehicular blast mine detonation. Tests with and without military boot showed that footwear has a significant influence on lower leg protection against disabling injuries. The effect of boot on injury severity demonstrates the importance of including a boot in procedures for full-scale vehicle mine testing. Loading produced with the current set-up was close to an AV blast mine load in terms of peak value and rise time, but it was not optimal. In future work, the air cannon system will undergo a redesign to generate loading that better represents AV blast landmine loading measured by dummies in a light-armoured vehicle. To reach this goal, focus will be on improving the loading duration and rise time.

Finally, data were collected in an effort to develop a transfer function between Hybrid III tibia response and injury severity. Results show that peak axial force correlates well with injury severity. In future work, other parameters, such as loading duration, rise time and impulse, will be considered to determine which, if any, have a significant influence on lower leg injury tolerance. To study these parameters, a tibia load cell normally used on a Hybrid III will be installed on the proximal end of the CLL. Tibia force measurement on the CLL will again be compared to PMHS force corridors to increase confidence of its biofidelity under axial impact loading.

ACKNOWLEDGEMENTS

Marike van der Horst and Piet-Jan Leerdam from TNO Defence, Security and Safety (The Netherlands), and Frank Dosquet from WTD 91 (Germany), for their scientific and technical advice on this project.

REFERENCES

1. Williams, K., Bourget, D., Cronin, D., Bergeron, D. and Salisbury, C., *Simplified Biofidelic Lower Leg Surrogate*, U.S. Provisional Patent 60/406.949, August 30, 2002.
2. Biokinetics and Associates Ltd., *The Complex Lower Leg (CLL)*, July 2003.
3. Owen, C., Lowne, R. and McMaster, J., Requirements for the Evaluation of the Risk of Injury to the Ankle in Car Impact Tests, *17th International Technical Conference on the Enhanced Safety of Vehicles,* Amsterdam, The Netherlands, National Highway Traffic Safety Administration, Washington, D.C., 2001.
4. Funk, J.R., Crandall, J.R., Tourret, L.J., MacMahon, C.B., Bass, C.R. and Patrie, J.T., The Axial Injury Tolerance of the Human Foot/Ankle Complex and the Effect of Achilles Tension, *Journal of Biomechanical Engineering*, vol. 124, 2002, 750 - 757.

5. Kuppa, S.M., Klopp, G.S., Crandall, J.R., Hall, G., Yoganandan, N., Pintar, F.A., Eppinger, R.H., Sun, E., Khaewpong, N. and Kleinberger, M., Axial Impact Characteristics of Dummy and Cadaver Lower Limbs, *16th International Technical Conference on the Enhanced Safety of Vehicles*, Windsor, Ontario, Canada, National Highway Traffic Safety Administration, Washington, D.C., 1998.

OCCUPANT LOWER LEG INJURY ASSESSMENT IN LANDMINE DETONATIONS UNDER A VEHICLE

M.J. van der Horst[1], C.K. Simms[2] , R. van Maasdam[1] and P.J.C. Leerdam[1]
[1] TNO Defence, Security and Safety, PO Box 45, 2280 AA, Rijswijk, The Netherlands
vanderhorst@pml.tno.nl, phone +31 15 284 3329
[2] Centre for Bioengineering, Dept of Mechanical Engineering, Trinity College, Dublin, Ireland

Abstract. Occupant safety has a high priority in peacekeeping military operations. During an anti tank (AT) mine detonation under a vehicle, the global and local impulse load threaten occupant safety. The proximity of the occupant's feet to the vehicle floor (through direct contact, or via pedals/foot rests) means that lower leg injuries occur frequently in AT mine strikes. Analysis of these injuries has proceeded with surrogate legs originally developed for automotive impacts. However, none of these has been validated for the loading conditions present in a mine strike. Recently, a combined experimental and numerical approach was used to obtain greater confidence in the injury assessment method with lower leg surrogates used in AT mine strike tests. Blast tests were performed using the standard Hybrid III dummy to study the differences between the standard Denton leg and the Thor lower leg model (Thor-Lx). Computational models of the test setups were created using a commercial multi-body code (Madymo) to investigate whether virtual testing could be used for future parameter studies in vehicle design. The current study focused on the validation of the Hybrid III Denton leg. The simulations showed good correspondence to the experimental data and are therefore suitable for use in injury assessment.

Key words: lower leg, experiment, simulations, vehicle mine protection.

1. INTRODUCTION

Vehicle mine protection is an emerging field of international research and new occupant models and injury criteria are under development. At present,

Michael D. Gilchrist (ed.), IUTAM Proceedings on Impact Biomechanics: From Fundamental Insights to Applications, 41–49.

the standard Hybrid III crash test dummy is the most frequently used human surrogate in assessing occupant loading in a vehicle subject to an anti-tank (AT) mine strike. Previous research has shown that the lower legs are very vulnerable to an AT mine strike under a vehicle and many studies have focussed on the lower leg response [3, 4, 6, 7]. However, the models used were developed for automotive research and have not been validated for mine strike loading conditions. This poses a potential research weakness as AT mine strikes generally result in shorter duration and higher loading magnitudes than automotive collisions.

Since human data for validation for AT mine loading conditions are not available, experimental studies using different types of lower leg models together with numerical studies are used to improve confidence in the injury assessment for vehicle mine protection studies.

In 2003 the Test Rig for Occupant Safety System (TROSS) developed by IABG (Lichtenau, Germany) was used to study the lower leg response using the Denton leg and the Thor lower leg (Thor-Lx). In addition, simulations of these tests were performed using the commercial crash simulation code Madymo. The main goal of this paper is to present the results of the validation of the Madymo model of the Hybrid III Denton leg using the experimental data from the TROSS tests. Also a parameter study on the initial lower leg position and its influence on the leg loads will be presented. Finally, these results will be used to discuss the issue of injury assessment.

2. METHOD

Data for validation of the Madymo Hybrid III Denton leg model for axial impact caused by an AT mine detonation under a vehicle are taken from TROSS tests performed in Germany [2].

2.1 Experimental Data

The Test Rig for Occupant Safety Systems (TROSS) developed by IABG is used for the leg model tests. Using this test rig, it is possible to provide repeatable loads caused by detonation of explosive charges beneath a membrane baseplate. There is a known relation between the charge used and real AT mine ordnance, resulting into comparable loads to the human body. The test rig is closed by a box which is decoupled from the membrane baseplate and thus free of shock. In order to prevent transmission of high frequency shock loading to the legs, a stiff footplate is mounted on the baseplate. The footplate is sufficiently stiff to resist local deformations.

The dummy's feet were positioned in the middle of this plate (see Figure 1). To prevent loading of the upper torso, the seat is fixed to the shock free box. The dummy was strapped to a military vehicle seat using a 4-point belt system. The following channels were measured: foot accelerations, tibia accelerations, lower tibia loads, femur loads, lumbar spine loads and pelvis accelerations. The data were filtered using standard low-pass filtering according to SAE J211/1. Tests were performed for three loading conditions of increasing severity of explosive loads (db1, db2, db3) and with and without shoes (db, dns). The results show reproducible Hybrid III dummy responses [2].

TROSS set-up

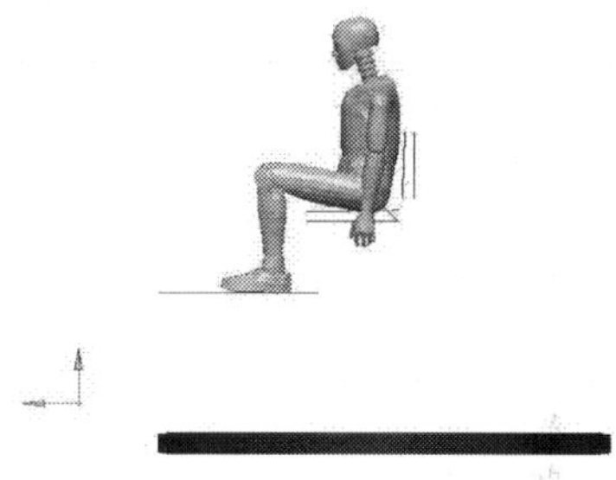

Simulation set-up

Figure 1. Test set-up as well as simulation set-up.

2.2 Simulations

The Madymo Hybrid III 50th percentile male dummy model with Denton leg was placed into a simplified model of the TROSS (see Figure 1). The seat was non-deforming and rigidly attached to the reference space while the rigidly modelled footplate can move vertically. The dummy model was positioned according to photographs of the test set-up. The simulations were organised such that prior to the start of the footplate displacement (main simulation) the dummy models were allowed to settle into the seat under the influence of gravity to find an equilibrium position. Also equilibrium in the contact between feet and footplate was reached. The arms were not repositioned after the presimulation, since they have no effect on the lower leg response. The main simulations were performed by prescribing the experimentally measured vertical footplate displacements as well as the acceleration field due to gravity.

TROSS test db2 (booted dummy loading condition 2) was used as the baseline setup for validation as this test had approximately the mean severity level. A foot–footplate contact function was derived in the absence of experimental data, providing a good fit for loading case db2 (Figure 3). In the simulations the effects of initial positioning, friction, integration timestep, elastic contact stiffness and contact damping were evaluated. The final input parameters used for db2 were then applied to simulations db1 and db3. The principal simulation output parameters for comparison with the experimental data were the foot and tibia vertical accelerations because loading of the dummy occurs via the footplate to the feet and then to the tibia and femur. Therefore, accurate modeling of the foot kinematics followed by tibia kinematics is the best approach for the validation process. In addition, the compressive load in the tibia is also important as this has direct implications for injury. The advantage of choosing accelerations rather than displacements or velocities for validation is that the acceleration pulse is more sensitive to kinetic changes than position or velocity.

For the parameter study, four initial seating positions were simulated, in which only the lower leg position was varied (see Figure 2). In all four positions the same vertical displacement of the footplate was simulated.

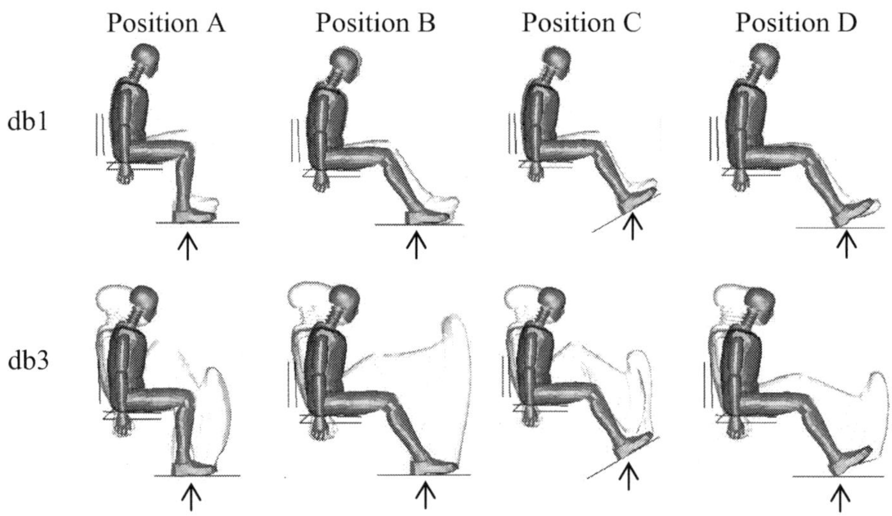

Figure 2. Tracing view using Altair Hyperworks: crash dummy motion during pure vertical footplate displacement for four initial positions at two different loading conditions.

3. RESULTS & DISCUSSION

The global occupant motions are presented in Figure 2. Position A was used for the model validation. The other positions were used for the parameter study. The model predicted all dummy responses but only a few results are shown here (see Figures 3-6). A negative tibia z force refers to axial compression, while the tibia x force refers to shear in the lower leg.

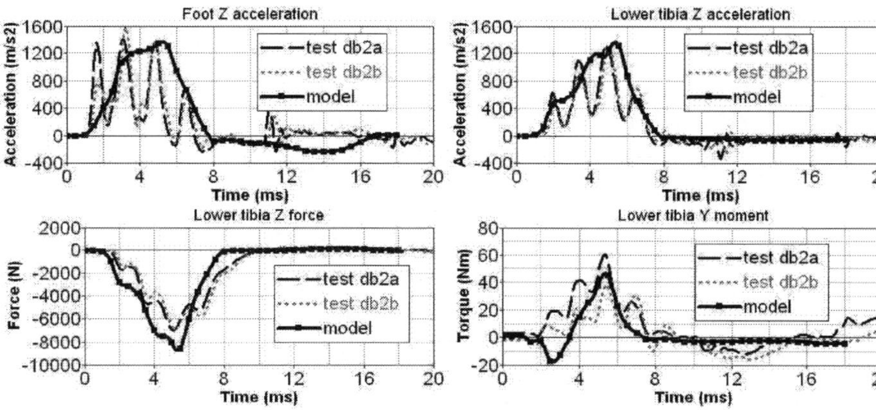

Figure 3. db2 results for position A.

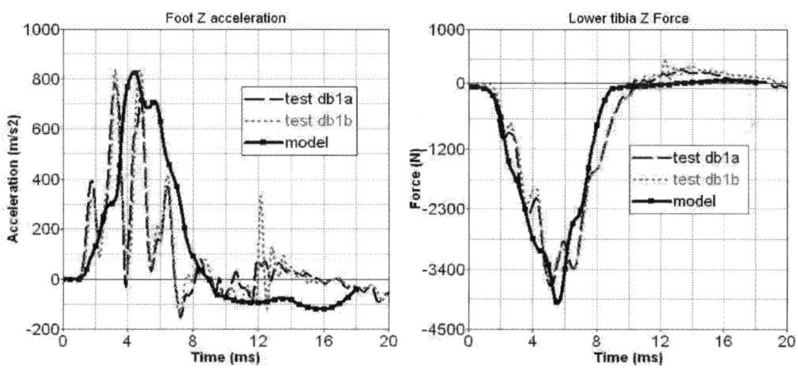

Figure 4. db1 results for position A.

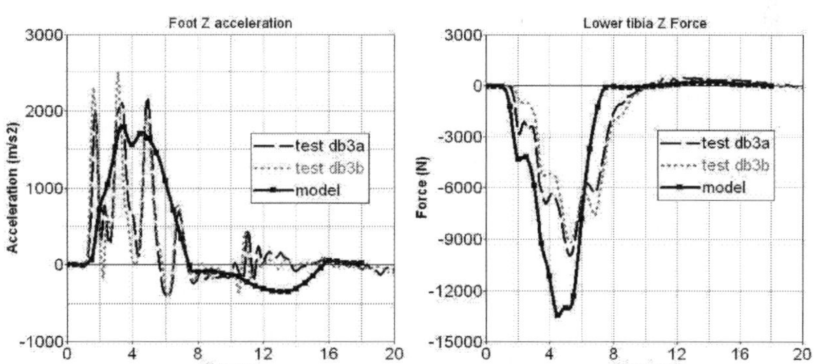

Figure 5. db3 results for position A.

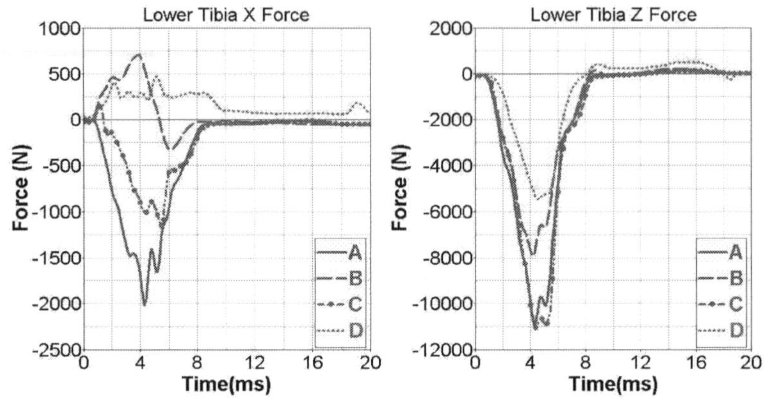

Figure 6. db3 results (Tibia Fx and Fz) for position A,B,C and D.

The Madymo Hybrid III booted dummy simulations for TROSS were successful. Varying the contact friction was not significant, and low damping at the foot-footplate contact yielded the best results. Foot and tibia Z direction accelerations (Figure 3) and femur and tibia joint reaction Z forces (not shown) match well for condition db2. The femur and tibia joint moments (not shown) are poor – but this has been noted previously in automotive research and as moments are not used for injury assessment in vehicle mine protection [2], it was decided to accept this deviation.

Load conditions db1 and db3 simulated using the same model configuration as db2 (Figures 4 and 5) yielded similar trends: for higher loads (db3) the body motions and loads are increased. The differences in lower leg motion are much clearer for db3 compared to the lower input simulations (db1). For db3, the tibia forces are too high compared to the experimental results because of the increased effects of damping at higher loading.

The current work was fit on foot and tibia accelerations, because these are more sensitive to kinetic changes than tibia velocity or motion. For future work it is suggested to use the tibia loads for tuning, because the load measurements are more reliable than the measured acceleration data. Based on the current study and the expected improvements when the tuning is focussing on the tibia loads it can be concluded that the Madymo Hybrid III Denton model is valid for this application.

Different initial lower leg positions were simulated to study the influence of positioning a dummy on injury assessment. The response differences between positions A,B,C and D are clearer for the db3 simulations than for db1. The same trend is seen for the axial tibia forces/accelerations for the lower and upper tibia as well as the upper tibia index: the highest values are seen for position A & C, while in all cases D shows the lowest values.

However, this may not mean a lower injury risk because the injury mechanism differs in these orientations.

4. GENERAL DISCUSSION

Research on cadaver lower leg tolerance to fracture under dynamic axial impacts performed by Yoganandan et al. [10], Seipel et al. [8], Kuppa et al. [5] and Funk et al. [1], are very useful. In general, the loading mechanism acting on the lower leg during an AT mine strike and in frontal car crashes is comparable [2]. In the latter, axial loading is applied via the pedal or the toepan. Within a NATO working group [2] it has been decided to use the axial tibia load injury model developed by Yoganandan [10], which recognizes that age is an important factor in lower leg fracture tolerance. Because the specimens tested are usually old, the injury risk models developed without considering this factor [5,8] give severe tolerance values. Funk's risk model, including age, gender and weight, was also satisfactory, but the Yoganandan model was used because of the larger sample size (52 specimens compared to 30) and the large age range of the specimens (27-85 compared to 41-74) used in its development. However, both correlate well at age = 45 years, giving confidence in the tolerance value of 5.4kN for 10% risk of AIS2+ injuries on the lower leg [2]. This means that db2 and db3 result in critical cases. The Yoganandan model uses the tibia force for a pure axial impact on the whole surface of the foot and may therefore not be valid for positions B & D (Figure 2), because the loading mechanism differs due to the initial leg orientation. Although the tibia forces for position B and D are lower the injury risk might be higher due to this different loading mechanism. In all cases, axial tibia compression force Fz was dominant.

Although injury tolerance is time-dependant due to visco-elastic effects [9], existing models do not consider loading duration. The available risk data for lower legs are from staged car crashes, which are usually longer events than mine strikes. Since Hybrid III tibia loading durations in AT mine tests are very short (sometimes < 10 ms), the proposed tolerance value may be too severe. WTD '91, Germany has started to develop a lower leg injury criterion based on tibia loading duration [2], but this has not yet been validated.

There is no human data for vehicle mine detonations. Previous automotive crash studies have shown that the Hybrid III leg is too stiff causing loads that might lead to a conservative injury assessment for these mine strikes. Therefore, more biofidelic models such as the Thor-Lx or the Madymo human body model are needed. The human body model is directly related to the human body, this is in contrast with the dummy models which are numerical models of mechanical crash dummy models. As part of the current research, the Madymo human body model was implemented in the same simulation environment of the TROSS test set-up. However, the model foot could not be maintained flat on the footplate, and this significantly altered the loads transmitted to the tibia. Separately, the Thor-Lx has been tested successfully and the relation between the Hybrid III Denton leg and Thor-Lx was studied experimentally, showing similar relationships seen in automotive studies [2]. Preliminary simulation results are also directing to the same relationship.

A major drawback of this work is the lack of post mortem human subject (PMHS) data for AT mine loading conditions, therefore international discussions started on proposals for PMHS testing. However, so far current experimental and numerical models will continue to be used for injury assessment for AT mine loading conditions.

5. CONCLUSIONS

The Hybrid III dummy was developed for automotive applications but can be used successfully to measure occupant loading during a mine strike in cases where the dummy is shielded from the direct blast but is loaded through vehicle deformation. The TNO study in co-operation with Trinity College Dublin indicated that simulations of the physical test setup using the commercial code Madymo showed very good correspondence to the experimental data and this approach is therefore suitable for future use in injury assessment. A parameter study showed considerable influence of occupant positioning on lower leg loading during a mine strike. Further validation in this field urgently requires PMHS test data.

ACKNOWLEDGEMENTS

Josée Manseau from Defence R&D Canada - Valcartier for research on lower leg injury assessement in the field of vehicle mine protection.
Frank Dosquet from WTD 91, Germany and Marcel Müller from IABG-Lichtenau, Germany for the use of the experimental data.
HFM, NATO/RTO Human Factors and Medicine Panel 090 Task Group 25 on test methodology for protection of vehicle occupants against anti-vehicular landmine effects.

REFERENCES

1. Funk, J.R., Crandall, J.R., Tourret, L.J., MacMahon, C.B., Bass, C.R., Patrie, J.T., Khaewpong, N. and Eppinger, R.H. (2002) The Axial Injury Tolerance of the Human Foot/Ankle Complex and the Effect of Achilles Tension, Journal of Biomechanical Engineering, 124: 750-757.
2. HFM (2002-2004) NATO/RTO Human Factors and Medicine Panel 090 Task Group 25 on test methodology for protection of vehicle occupants against anti-vehicular landmine effects, minutes and material from meetings.
3. Horst, M.J. van der and Leerdam, P.J. (2002) Experimental and Numerical Analysis of Occupant Safety in Blast Mine Loading under Vehicles, in Proceedings of International IRCOBI Conference on the Biomechanics of Impact. Munich, Germany.
4. Horst, M.J. van der (2002) Numerical Analysis of Occupant Safety in Vehicle Mine Protection, First European Survivability Workshop, Germany.
5. Kuppa, S. et al. (2001) Lower Extremity Injuries and Associated Injury Criteria, 17th International Technical Conference on the Enhanced Safety of Vehicles (ESV)", National Highway Traffic Safety Administration USA (NHTSA), Amsterdam, the Netherlands.
6. Leerdam, P.J.C. (2002) Research Experiences on Vehicle Mine Protection, First European Survivability Workshop, Germany.
7. Radonic,V. et al. (2004) Injuries from Antitank Mines in Southern Croatia, Military Medicine, 169, 4: 320-324.
8. Seipel et al. (2001) Biomechanics of Calcaneal Fractures, Clinical Orthopaedics and Related Research, 338: 218-224.
9. Yoganandan, N. et al. (1989) Dynamic Response of Human Cervical Ligaments, Spine, 14: 1102-1110.
10. Yoganandan, N., Pintar, F.A., Boyton M., Begeman, P., Prasad, P., Kuppa, S.M., Morgan, R.M. and Eppinger, R.H., (1996) Dynamic Axial Tolerance of the Human Foot-Ankle Complex, 962426, Society of Automotive Engineers, Warrendale, PA, USA.

THE SIMULATED RESPONSE OF CORTICAL AND CANCELLOUS BONE TO VARIOUS RATES OF LOADING – A PRELIMINARY NUMERICAL STUDY

S.A. Oerder[1,2] and G.N. Nurick[2]

[1]*School of Mechanical, Industrial and Aeronautical Engineering, University of the Witwatersrand, Johannesburg, Private Bag 3, PO WITS, 2050, South Africa;*
e-mail: saoerder@mech.wits.ac.za

[2]*Blast Impact and Survivability Research Unit (BISRU), Department of Mechanical Engineering, University of Cape Town, PB Rondebosch, 7701, Cape Town, South Africa;*
e-mail: gnurick@ebe.uct.ac.za

Abstract. The simulated responses of cortical and cancellous bone subjected to two different loading conditions are presented. These are impact type and blast type pressure loads. Three different models are simulated under the two loading conditions; they are a hollow cortical bone, a solid cancellous bone and a combination of cortical and cancellous bone. The strength model is based on Von Mises Yield [1]. The results show a distinct difference in the human response between an impact and a blast load.

Key words: cortical bone, cancellous bone, material properties, blast, impact.

1. INTRODUCTION

Hitherto in the field of human response to impact mechanics, the emphasis has been focused on applications such as automobile and train accidents. Several conferences have been held pertaining to these applications, such as ESV and ICrash. More recently, the world has highlighted the need to identify the effects of human response to more violent effects such as blast

Michael D. Gilchrist (ed.), IUTAM Proceedings on Impact Biomechanics: From Fundamental Insights to Applications, 51–58.

loading. The loading – pressure time – of these two impacts is distinctly different. Hence an understanding of the response of human skeletal components to these different load conditions is desired. In this respect, this preliminary numerical study investigates the effects to different rates of loading on cortical and cancellous bone.

2. MODELLING

Three cases of the human tibia approximated as a cylinder of length 600 mm are investigated. In the first case the cylinder is hollow with an external diameter of 40 mm and an internal diameter of 30 mm and is composed of cortical bone. In the second case the cylinder is solid with a diameter of 30 mm and is composed of cancellous bone. In the third case the tibia is modeled as a cylinder with an inner core of diameter 30 mm consisting of cancellous bone and an outer shell of diameter 40 mm consisting of cortical bone. This combined model is utilized in investigating the influence of cortical bone on the strength of the cancellous bone core. These three cases are shown schematically in Figure 1.

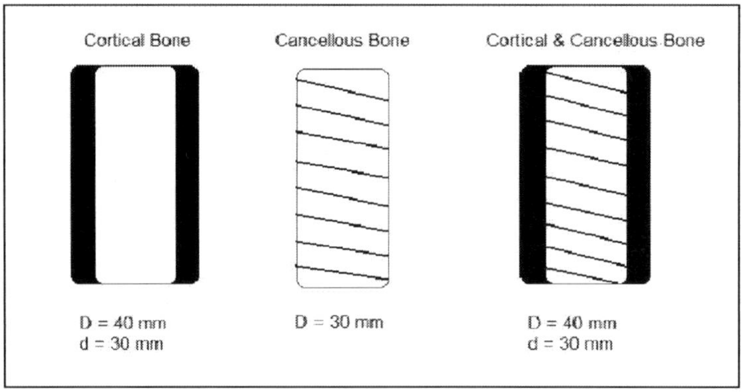

Figure 1. Human Tibia Model.

For each case two loading conditions are implemented with identical impulses. The first is a longer time span impact type pressure of 0.16 MPa ranging from 0.5 – 4.5 ms with the peak occurring at 2.5 ms [2]. The second is a short duration blast type pressure described as reaching a peak pressure of 160 MPa at 0.002 ms and lasting a total of 0.004 ms [3], as shown in Figure 2. For this study the objective is to determine the human response to two distinctly different loading conditions. Hence, for the purpose herein the loading conditions chosen for the longer timespan impact are reported by reference [2], and those for the blast type impact chosen from typical laboratory type blast experiments [3]. It is important to note the distinct difference in pressure time histories between the two loading conditions.

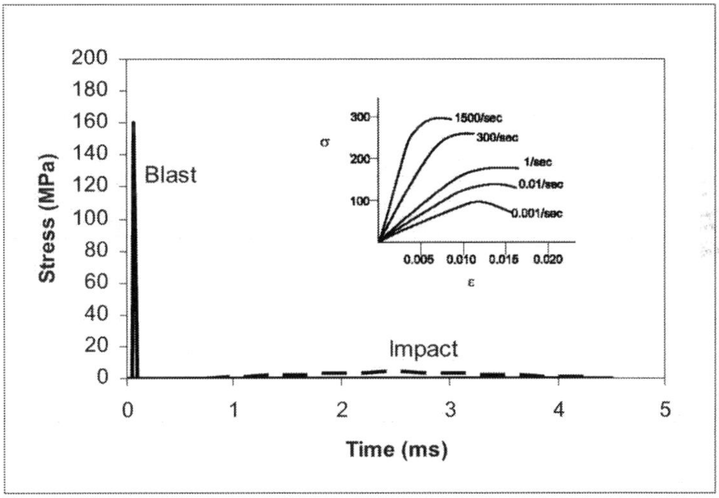

Figure 2. Two standard loading conditions.

The impact condition is modeled as a triangular stress wave directly applied to the top surface of the cylinder, indicated as position 1 in Figure 3. The blast condition is modeled as a piecewise step function also applied to position 1 in Figure 3. In both conditions the base of the cylinder, position 2 in Figure 3, is modeled as a transmit layer which enables the stress wave to propagate through the cylinder. Boundary conditions are applied to both loading conditions for the purpose of simulating the bone being constrained on a drop test rig used for impacts or a ballistic pendulum used in blast loading. In each case the structure is meshed with a rectangular mesh.

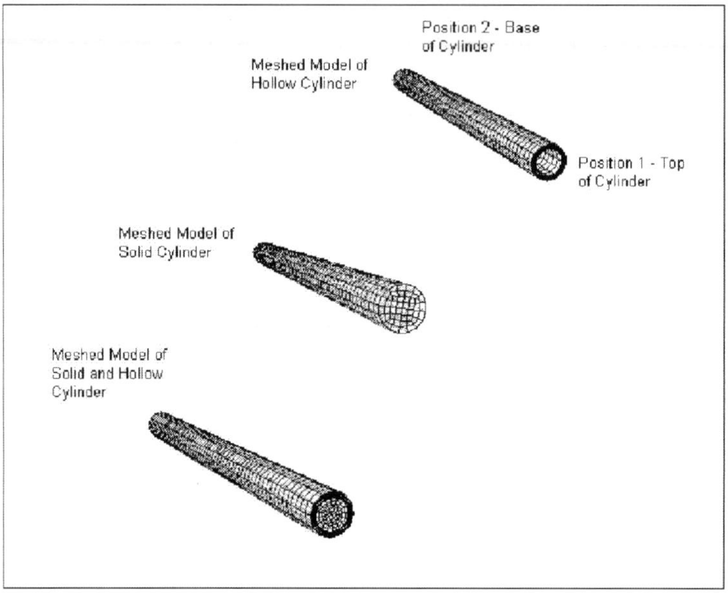

Figure 3. Meshed cylinder for all three cases.

In all three cases the bone material is modeled as a linearly elastic material with different yield conditions using Autodyn [4]. For each loading condition different material specifications are used – for the impact conditions quasi-static yield criteria is implemented [3] and for the blast pressure it is assumed the yield changes due to strain rate dependency. For the latter case the material properties for cortical bone are obtained from [5] and for cancellous bone from [6]. Material properties are given in Table 1.

Table 1. Steel and Cortical and Cancellous Bone Material Properties.

	Density [kg/m^3]	Young's Modulus [GPa]	Bulk Modulus (Calculated) [MPa]	Shear Modulus (Calculated) [MPa]	Poisson Ratio	Yield Stress [MPa] – static conditions	Yield Stress [MPa] – strain rate [~1200/s]
Cortical Bone	2120	12.2	7260	5000	0.22	170	300
Cancellous Bone	990	1.3	774	533	0.22	4	17
Steel 4340 [7]	7830	209	159000	81800	0.28	792	1000

The loading conditions and types of bone are represented in Table 2. For cases A, B and C the number of elements in the model are 14400, 14400 and 28800 respectively. For cases D, E and F the number of elements in the model are 28800, 28800 and 57600 respectively. The mesh is refined by grading it from position 1, at the top of the cylinder, to position 2. This results in a better approximation of the stress wave applied to the cylinder. Autodyn is an explicit analysis tool used in the modeling of non-linear dynamics. The program uses a finite volume formulation with exact volume integration and linear elements for solution.

Table 2.. Matrix of Loading Regime.

	Impact	Blast
Cortical Bone	A	D
Cancellous Bone	B	E
Cortical & Cancellous Bone	C	F

3. RESULTS AND DISCUSSION

Figure 4 shows the stress wave through the cylinder for loading regime A, this is an impact load applied to cortical bone. The figure shows that under the influence of an impact type load the stress wave experienced at both positions, 1 and 2, has the same peak value of – 0.16 MPa. The peak occurs at time 2.5 ms at the applied position and at 2.75 ms for the measured position.

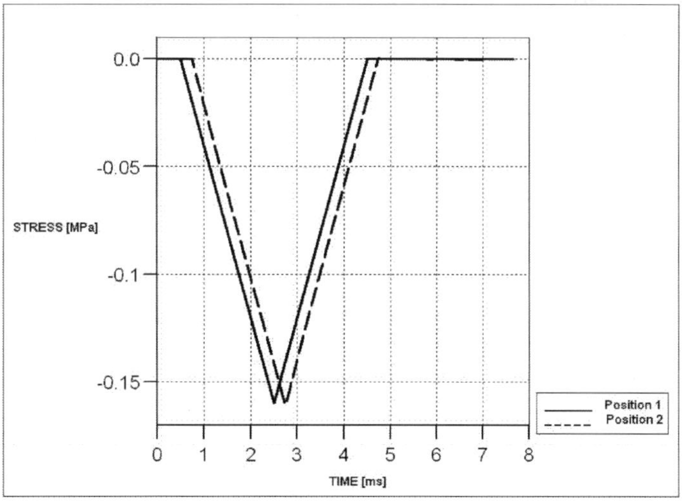

Figure 4. Stress wave through cylinder for loading regime A.

The time shift in the peak value is influenced by the speed of sound through the material. To validate the results obtained the speed of sound through bone is calculated using equation 1 [8].

$$v = \sqrt{E/\rho} \tag{1}$$

where v = speed of sound measured in [m/s], E = Modulus of Elasticity measured in [Pa] and ρ = density of material measured in [kg/m^3]. The speed of sound through cortical bone is calculated as 2399 m/s, which in turn is used to determine the time for the stress wave to reach position 2, (0.25 ms).

In the event of loading regime D, Figure 5 shows the stress wave in the cortical bone as a result of a blast wave as it travels through the cylinder. The applied stress at position 1 is – 160 MPa while the measured stress at position 2 is – 26 MPa. This difference in peak values is similar to that described in [9] for impact type loadings and this is ascribed to the material properties [9].

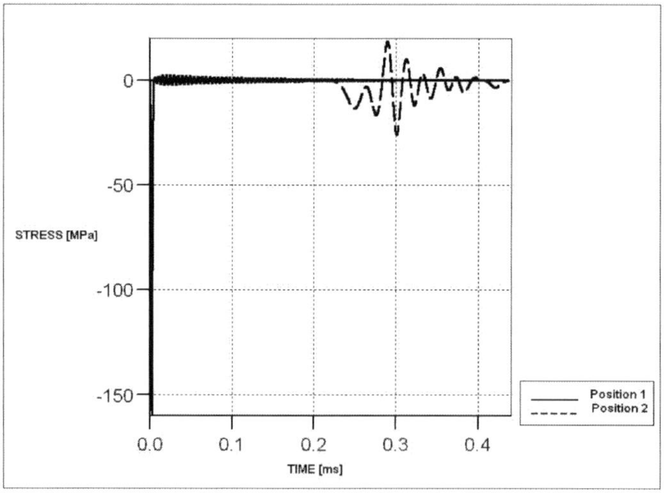

Figure 5. Stress wave through cylinder for loading regime D.

Factors, which influence the propagation of the shockwave through the cylinder, are the bone material properties, the speed of sound through bone and the length of the cylinder. Autodyn [4] automatically calculates the speed of sound through the simulated material, which is then utilized in determining the impedance at position 2 on the cylinder [10].

In this study six loading regimes are investigated, as shown in Table 3. The last two entries in Table 3 are values obtained for the simulated impact and blast loading on a conventional Steel 4340.

Table 3.. Summary of Results.

Loading Regime	Applied Stress [MPa]	Time of Applied Stress [ms]	Measured Stress [MPa]	Time of Measured Stress [ms]
A – Impact	- 0.16	2.5	- 0.16	2.75
D – Blast	- 160	0.002	- 26	0.3
B – Impact	- 0.16	2.5	- 0.16	3
E – Blast	- 158	0.002	- 7	0.55
C – Impact – Cortical	- 0.16	2.5	- 0.16	3
C – Impact – Cancellous	- 0.16	2.5	- 0.16	2.75
F – Blast – Cortical	- 160	0.002	- 26	0.3
F – Blast – Cancellous	- 158	0.002	- 7	0.5
Steel – Impact	- 0.16	2.5	- 0.16	2.7
Steel – Blast	- 160	0.002	- 25	0.1

The results show that in the case of the loading regimes A, B and C the applied and measured stresses are the same. This is also the case for the steel under the influence of the impact load. Whilst in the case of the blast load, loading regime D, the applied and measured stresses differ - the measured stress is 16% of the applied stress. In the case of loading regime E the measured stress is 4.5% of the applied stress and in the case of loading regime F the measured stress is 16% of the applied stress for the cortical bone and the measured stress is 4.5% of the applied stress for the cancellous bone.

These results show no influence on the strength of the cancellous bone core in the combined model. The values of the applied and measured stresses are identical to the values for the individual models of cortical and cancellous bone for both the impact and blast loading conditions. However, the time at which the applied stress occurs differs between the combined model for the cancellous bone (model F) and the pure cancellous bone (model E) under the influence of the blast loading.

4. CONCLUDING REMARKS

In this preliminary numerical study it has been shown that human bone responds differently for an impact load and a blast load. The bone responds according to the time-stress input applied to the cylinder, with the impact load attaining a lower stress over a longer time period and a blast attaining a higher stress for a shorter time period. The study includes some basic simulations using steel 4340 which suggest a similar response for impact and blast loadings as obtained using bone. The impact load on steel resembles that for cortical and cancellous bone with the exception of the time of the measured stress, which has been ascribed previously to the speed of sound through the material. In the case of the blast load the steel responds similarily to that of the cortical and cancellous bone and the measured stress is 16% of the applied stress. The measured stress for the impact loading is the same as that of the applied stress for all three bone materials for an impact loading whilst the measured stress is only a percentage of the applied stress for all three bone materials for a blast loading. These results are being utilized for experiments currently in the planning stage.

REFERENCES

1. Tutorial: Stress and Strain in Solids, http://biomed.wustl.edu/courses/bme_140/Readings/Stress%20and%20Strain%20in%20Solids.pdf [last accessed 11th January 2005].
2. Al-Bsharat, A., Zhou, C., Yang, K.H., Khalil, T. and King, A.I., Intracranial Pressure in the Human Head due to Frontal Impact based on a Finite Element Model, http://asme.pinetec.com/bio1999/data/pdfs/a0147874.pdf [last accessed 13th December 2004].
3. Langdon, G.S., Chung Kim Yuen, S. and Nurick, G.N., Experimental and numerical studies on the response of quadrangular stiffened plates. Part II: localized blast loading, *International Journal of Impact Engineering*, vol. 31 issue 1, 2005, 85-111.
4. AUTODYN, Century Dynamics Limited, England.
5. Chapter 2: Bone, http://dahweb.engr.ucdavis.edu/dahweb/126site/chp2.pdf [last accessed 3rd January 2005].
6. Shim, V.P.W., Yang, L.M. and Liu, J.F., Characterization of the Dynamic Mechanical Properties of Cancellous Bone from the Human Cervical Spine, presented at the 5th International Symposium on Impact Engineering, Cambridge, 2004.
7. Xue, Q., Benson, D., Meyers, M.A., Nesterenko, V.F. and Olevsky, E.A., Constitutive response of welded HSLA 100 steel. *Materials Science and Engineering*, A354, 2003, 166-179.
8. Verleger, H., *Fisika*, Universitas, Pretoria, 1960.
9. Pelker, R.R. and Saha, S., Stress wave propagation in bone, *Journal of Biomechanics*, vol. 16 no. 7, 1983, 481-489.
10. Private Communication – D. White, Century Dynamics Limited, January 2005.

SESSION 2

MULTIBODY DYNAMICS APPROACHES FOR BIOMECHANICAL MODELING IN HUMAN IMPACT APPLICATIONS

Jorge Ambrósio and Miguel Silva

Instituto de Engenharia Mecânica, Instituto Superior Técnico, Av Rovisco Pais 1, 1049-001 Lisboa, Portugal, {jorge,miguel.silva@dem.ist.utl.pt}

Abstract. The construction of multibody biomechanical models for impact is discussed here with the emphasis on the formulation aspects. First the relations between the human or the dummy anthropometric data and the rigid bodies in the model are presented. The motion restrictions between the different anatomical segments of model can be defined as kinematic joints, suitable to represent mechanical joints of dummies or a simplified kinematics of human joints, or as contact/sliding pairs, which are used to describe realistic human like anatomical joints. Another particular aspect of biomechanical models is the representation of the range of motion of the anatomical joints. This is achieved either by setting proper contact pairs between the adjacent anatomical segments or by setting resisting muscle forces or resisting moments that develop when the relative orientation between the segments reach critical values. Another fundamental aspect of the models is the ability to represent the contact geometries and the contact forces with realism. In fact, the outcome of all injury indexes predictions is strongly dependent on the quality of the representation of the contact. Contact models suitable to be used in biomechanical models, to represent the contact between anatomical segments of the biomechanical model or between these and external objects, are presented and discussed emphasizing the requirements to develop more advanced biomechanical models. The current biomechanical models either do not include muscle actions or, at the most, include a reflexive muscle contraction. It is suggested here that for the case of standing passengers it is important to include in the biomechanical models muscle models that allow for the representation of the muscle voluntary contractions and joint stiffening. It is also suggested that the evaluation of the models leading to the identification of such actions can be done by using techniques similar to those used in the evaluation of muscle force sharing in different human motions.

Key words: contact, out-of-position occupants, muscle models, voluntary muscle contraction.

Michael D. Gilchrist (ed.), IUTAM Proceedings on Impact Biomechanics: From Fundamental Insights to Applications, 61–80.

1. INTRODUCTION

The safety of occupants and their potential survival in crash events of transportation systems requires the use of various models for the structural crashworthiness, restraint systems vehicle interior or occupant biomechanics. Current design methodologies entail the use of different computer simulations of increasing complexity ranging from simplified lumped mass models [**1**], multibody models [**2**] to complex geometric and material nonlinear finite element based representations of occupant [**3**] and vehicle structures [**4**]. Some well-known simulation programs are now available: PAM CRASH [**4**], WHAMS-3D [**5**] and DYNA 3D [**6**] for structural impact and CAL3D [7] and MADYMO [**8**] for occupant dynamics. These programs are able to simulate with relative detail frontal, rear and side impact scenarios.

Most of the biomechanical models for the study of impact biomechanics rely on multibody dynamics methodologies. The outcomes the numerical models simulations are generally in close agreement with the output of the anthropomorphic test devices, or dummies, tests. This is mainly due to the fact that all the mechanical components and joints of the dummies find in the multibody formulations exact descriptions of their physics, with particular emphasis on the kinematic joints and on the numerical models of the materials and contact. The MADYMO family of dummy models for front, side, rear-end or pedestrian impact, and their variations in terms of age and gender representation, illustrates the detail with which it is possible to represent these devices using a multibody formulation [**9**]. The human body numerical surrogates have a much larger variability for their outcomes when compared with cadavers or human volunteers experimental tests. Such differences reflect, among others, the difficulties in the identification of the biomaterial dynamic response, the approximate representation of the contact models for the body anatomical segments due to geometry and compliance, deficient evaluation of the muscle actions and the complexities of experimental testing on the human body [**10**]. Therefore, improvements on multibody models of the human body requires investigations on suitable biological material models, muscle contraction models for force and activation control and the characterization of their variability with age and gender. The improvement of dummy numerical models and the design of new improved anthropomorphic testing devices require the identification of mechanical features that can represent with biofidelity the response of the human body in well defined ranges of action. Therefore, the understanding of the formulations required to achieve more advanced models is of fundamental importance.

The dynamic analysis of the vehicle occupants require that initial conditions for the biomechanical models are supplied in order to carry out the simulations [**11**]. In road vehicles it is rather predictable what the passenger positions are, and consequently the different safety systems, such as restraints and air-bags, are tuned to control the occupant kinematics and to minimize their injury risk. However, for the normal occupants in trains or buses there are no preferred positions or postures.

The most common biomechanical models are tuned for the type of kinematics experienced by the normal road vehicle occupants. The limitations in the application of current dummies and human body models to railway and bus passengers reside in the fact that the impact kinematics of these passengers do not have preferred directions, can happen with multiple impacts and can have a large variety of initial conditions. The crash event can be preceded by warning signs, take a longer time than for road vehicles and consist on multiple impacts. For these reasons special attention is paid to the need for these models to include muscle actions. In order to identify the typical postures of railway passengers and the reflexive and controlled muscle actions and joints stiffening several numerical and experimental procedures, based on gait analysis methodologies, are also proposed here [**12, 13**].

2. MULTIBODY DYNAMICS

A multibody system is a collection of rigid bodies joined together by kinematic joints and force elements as depicted in Figure 1. For the ith body in the system, $\mathbf{q}_i$ denotes a vector of coordinates [**14**]. A vector of velocities for a rigid body i is defined as $\mathbf{v}_i$. The vector of accelerations for the body, denoted by $\dot{\mathbf{v}}_i$, is the time derivative of $\mathbf{v}_i$. For a multibody system containing nb bodies, the vectors of coordinates, velocities, and accelerations are $\mathbf{q}$, $\mathbf{v}$ and $\dot{\mathbf{v}}$ that contain the elements of $\mathbf{q}_i$, $\mathbf{v}_i$ and $\dot{\mathbf{v}}_i$, for i=1, ..., nb.

The kinematic joints between rigid bodies are described by mr independent constraints:

$$\boldsymbol{\Phi}(\mathbf{q}) = \mathbf{0} \tag{1}$$

The time derivatives of the constraints yield the velocity and acceleration equations.

$$\dot{\boldsymbol{\Phi}} = \mathbf{D}\mathbf{v} = \mathbf{0} \tag{2}$$

$$\ddot{\boldsymbol{\Phi}} = \dot{\mathbf{D}}\mathbf{v} + \mathbf{D}\dot{\mathbf{v}} = \mathbf{0} \tag{3}$$

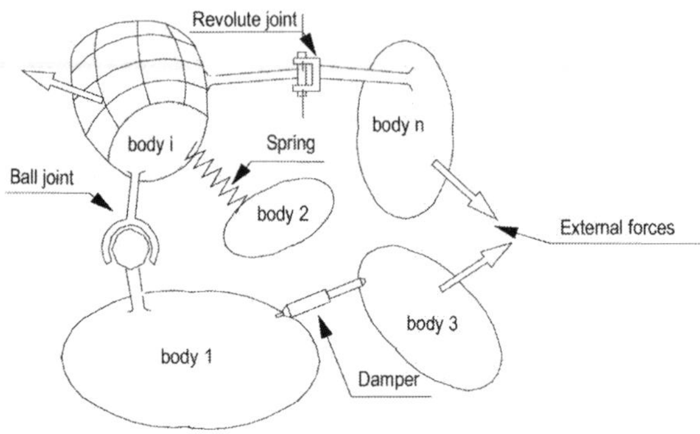

Figure 1. Schematic representation of a multibody system.

where **D** is the Jacobian matrix of the constraints. Equations (1-3) describe all kinematic restrictions between the components of the multibody system.

The equations of motion for the multibody system are written [**14**]

$$\mathbf{M}\dot{\mathbf{v}} - \mathbf{D}^T\boldsymbol{\lambda} = \mathbf{g} \tag{4}$$

where **M** is the inertia matrix, $\boldsymbol{\lambda}$ is a vector of Lagrange multipliers, and $\mathbf{g}=\mathbf{g}(\mathbf{q},\mathbf{v})$ contains the gyroscopic terms and the forces and moments that act on the bodies. Equation (4) must be solved together with equation (3) to obtain the system accelerations and the Lagrange multipliers, associated with the joint reaction forces.

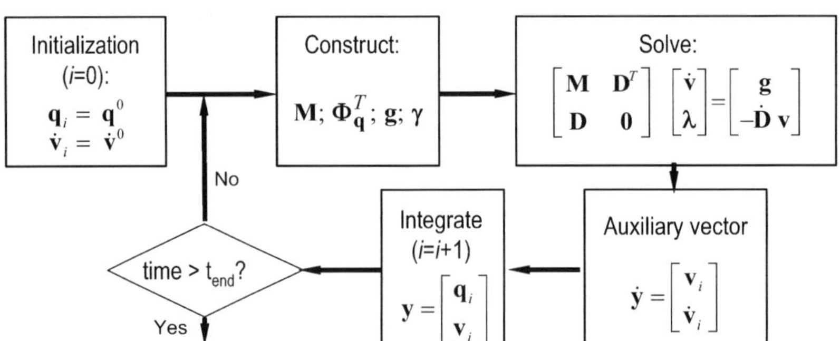

Figure 2. Solution of the forward dynamic analysis of a multibody system.

The forward dynamic analysis of a multibody system requires the initial conditions of the system, i.e. the position vector $\mathbf{q}^0$ and the velocity vector $\mathbf{v}^0$. Equations (3) and (4) are assembled and solved for the unknown accelerations and Lagrange multipliers, which are in turn integrated in time together with the velocities. This leads to the positions and velocities of the new time step. The process, shown in Figure 2, proceeds until the system response is obtained for the analysis period.

At this point it should be noted that for a biomechanical model the anatomical joints are represented by constraint equations, such as equation (1), when they are represented by mechanical joints. However, the same joints may be represented by contact elements, and consequently included in equation (4) through the force vector **g**. For anatomically correct joints the only possible representation is by using the contact pairs with appropriate geometry and compliance. The muscle actions are generally represented by internal forces in the biomechanical model, and therefore included in vector **g**. However, for voluntary muscle contraction studies these actions can be modeled through kinematic constraints for which the associated Lagrange multipliers represent the muscle forces.

3. BIOMECHANICAL MODEL FOR A VEHICLE OCCUPANTS

The multibody methodology outlined is applied to the representation of a biomechanical model of the human body or dummy suitable for impact simulations, based on the occupant model of SOMLA [**15**]. The model is general and accepts data for any individual. The information required to assemble the equations of motion of the model include the mass and inertia of the biomechanical segments, their lengths, location of the body-fixed coordinate frames and the geometry of the potential contact surfaces, as pictured in Figure 3. The data held within a database can be expanded for different individuals.

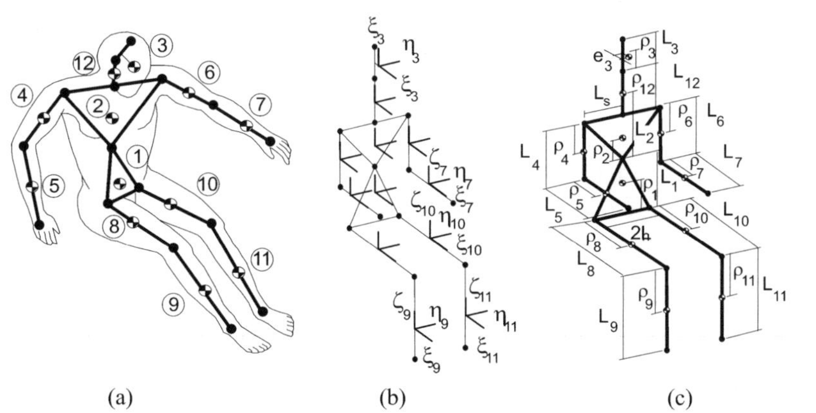

Figure 3. Three-dimensional biomechanical model for impact: (a) actual model; (b) local referential locations; (c) dimensions of the biomechanical segments.

3.1 Anatomical Segments of the Model

In impact simulations the relative kinematics of the head-neck and torso are important to the correct evaluation of injury criteria. Consequently, the head and neck are modeled as separate bodies and the torso is divided in two bodies. The hands and feet do not play a significant role in this type of problem and consequently are not modeled. The model is described using 12 rigid bodies connected by kinematic joints. Table 1 presents the location and description of the kinematic joints. For these models all anatomical joints are represented by kinematic joints and not by contact pairs.

Table 1. Kinematic joint description for biomechanical model.

Joint	Type	Description
1	spherical	Back, (12th thoracic and 1st lumbar vertebrae).
2	spherical	Torso-Neck (7th cervical and 1st thoracic vertebrae).
3-5	spherical	Shoulder.
4-6	revolute	Elbow.
7-9	spherical	Hip.
8-10	revolute	Knee.
11	revolute	Head-Neck, (at occipital condyles).

The principal dimensions of the model are represented in Figure 3(c). In most cases, the effective link-lengths between two kinematic joints are used instead of standard anthropometric dimensions based on external measurements. The set of data for the models is described in reference [**15**].

3.2 Joint Resisting Moments

In the biomechanical model, no active muscle force is considered but the muscle passive behavior is represented. Applying a set of penalty torques when adjacent segments of the biomechanical model reach the limit of their relative range of motion prevents physically unacceptable positions of the body segments. A viscous torsion damper and a non-linear torsion spring, located in each kinematic joint, describe the joint torques. Take the elbow of the model, represented in Figure 4 for instance. The total damping torque for the relative rotation of the lower and upper arm is

$$\mathbf{m}_{di} = -j_i\,\dot{\boldsymbol{\beta}}_i \tag{5}$$

where the torsion damper has a small constant coefficient j_i and $\dot{\boldsymbol{\beta}}_i$ is the relative angular velocity vector between the two bodies interconnected by joint i.

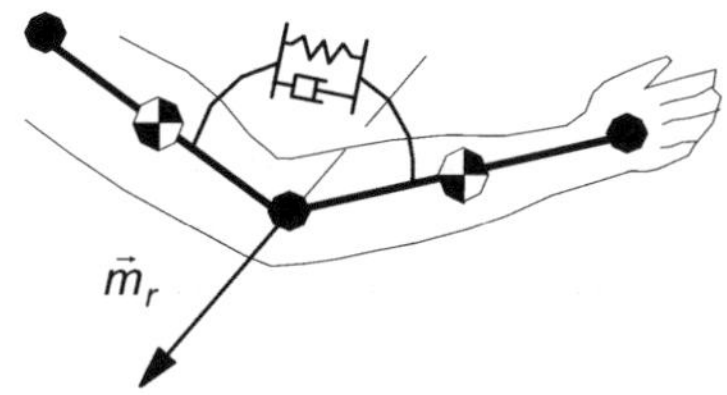

Figure 4. Joint resistance torque modeled with a non-linear torsion spring and damper.

A constant torque m_{ri} that acts resisting the motion of the joint is applied in the whole range of motion in the dummy model [**15**]. For the human joint this torque has an initial value, which drops to zero after a small angular displacement. The torque has a direction opposite to that of the relative angular velocity vector between the two bodies interconnected in the joint

$$\mathbf{m}_{r_i} = -m_{r_i}\dot{\boldsymbol{\beta}}_i \left\| \dot{\boldsymbol{\beta}}_i \right\|^{-1} \tag{6}$$

A penalty resisting torque m_{pi}, also applied at the joint, is null during the normal joint rotation but it increases rapidly, from zero to a maximum value, when the two bodies interconnected by that joint reach physically unacceptable positions. The curve for the penalty resisting moment is represented qualitatively in Figure 5.

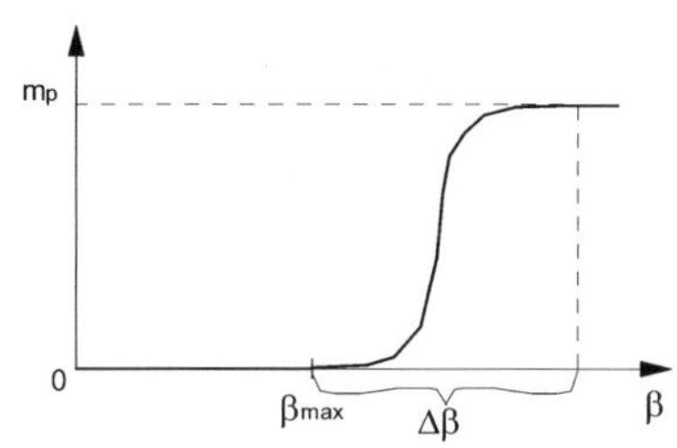

Figure 5. Penalty moment for the elbow.

The shoulder joint modeled by a spherical joint exemplifies a biomechanical joint with three degrees-of-freedom. The calculation of the penalty torque requires the construction of the cone of feasible motion. This cone has its tip in the center of a sphere with a unit radius. While the upper arm moves inside the cone its motion does not imply displacements of the upper or lower torsos. The implementation of the joint resisting torques is similar to that explained for the elbow joint and it is not detailed any further here. For more information on the procedures used the interested reader is directed to reference [**15**].

3.3 Contact Model

A set of contact surfaces is defined for the calculation of the external forces exerted on the model when the bodies contact other objects or different body segments. These surfaces are ellipsoids and cylinders with the form depicted by Figure 6. When contact between components of the biomechanical model is detected a contact force is applied to such components in the point of contact. Friction forces are also applied to the contact surfaces using Coulomb friction. Note that the characterization of the surfaces in contact is important for general applications of the biomechanical model.

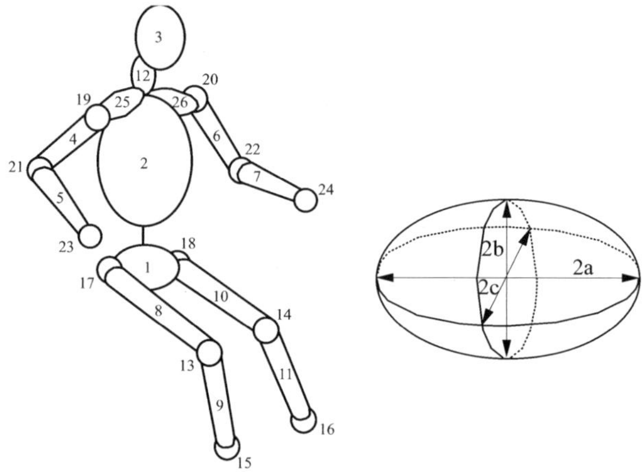

Figure 6. Representation of contact surfaces.

Based on a Hertzian description of the contact forces between two solids [**16**], Lankarani and Nikravesh [**17**] propose a continuous force contact model that accounts for energy dissipation during impact. Upon identifying the contact points between two surfaces, their relative penetration δ and penetration velocity $\dot{\delta}$ the contact force is written as

$$\mathbf{f}_{s,i} = \left(K\delta^n + D\dot{\delta} \right) \mathbf{u} \tag{7}$$

where K is the equivalent stiffness, D is a damping coefficient and $\mathbf{u}$ is a unit vector normal to the impacting surfaces. The damping coefficient is given by

$$D = \frac{3K\left(1-e^2\right)}{4\dot{\delta}^{(-)}}\delta^n \tag{8}$$

This coefficient is a function of the impact velocity $\dot{\delta}^{(-)}$, stiffness of the contacting surfaces and restitution coefficient *e*. For a fully elastic contact *e*=1 while for a fully plastic contact *e*=0. The generalized stiffness coefficient *K* depends on the geometry material properties of the surfaces in contact. For the contact between a sphere and a flat surface the stiffness is [**17**]

$$K = \frac{0.424\sqrt{r}}{\left(\frac{1-\nu_i^2}{\pi E_i} + \frac{1-\nu_j^2}{\pi E_j}\right)} \tag{9}$$

where ν_l and E_l are the Poisson's ratio and the Young's modulus associated with each surface and *r* is the radius of the impacting sphere.

The hysteresis dissipation is introduced in equation (7) by $D\dot{\delta}$. The nonlinear contact force is obtained by substituting equation (8) into equation (7), leading to

$$\mathbf{f}_{s,i} = K \quad \delta^n \left[1 + \frac{3\left(1-e^2\right)}{4} \frac{\dot{\delta}}{\dot{\delta}^{(-)}}\right] \mathbf{u} \tag{10}$$

This equation is valid for impact conditions in which the contacting velocities are lower than the propagation speed of elastic waves. Furthermore, the model for the contact force considers the material and geometric properties of the surfaces, contributes to a stable integration and accounts for some level of energy dissipation.

3.4 Advanced Features for Biomechanical Models

The biomechanical models that are used in occupant passive safety are computational representations of the anthropometric testing devices (ATD), i.e., dummies. Because of the limitations of these measuring devices to represent the whole range of kinematics of the human body for each type of impact there are several types of devices developed. Though the more realistic models of the human body can be developed in a computer environment it is not yet possible to have reliable and repeatable ATDs of 'real humans'. In this section the methods and modeling strategies required to develop computational surrogates of the human body, rather than of the ATDs, are discussed. In the process it is suggested that such human models can be used to develop more advanced and versatile ATDs that can be used in different situations and measure a wider range of responses.

3.4.1 Geometry of contact pairs and contact surfaces

The investigation of more biofidelic biomechanical models requires that some effort is put on the identification of the aspects that differentiate the posture of general vehicle occupants from that of the guided kinematics such as those of the occupants of road vehicles. In particular, a more detailed model for the spine and for the thorax, as for instance that implied by Figure 7(a), can be of major importance. Models of the different anatomical elements of the spine, including the vertebrae, discs and ligaments, as presented in Figure 7(b), are required if the correct representation of the posture is of importance. The detailed representation of the skeletal-muscle system, such as that implied by Figure 7(c) is fundamental if the biomechanics of impact for low and medium velocities is to be accurately evaluated. Furthermore, muscles and ligaments control the limits of motion on the 'human like' joints of the biomechanical models.

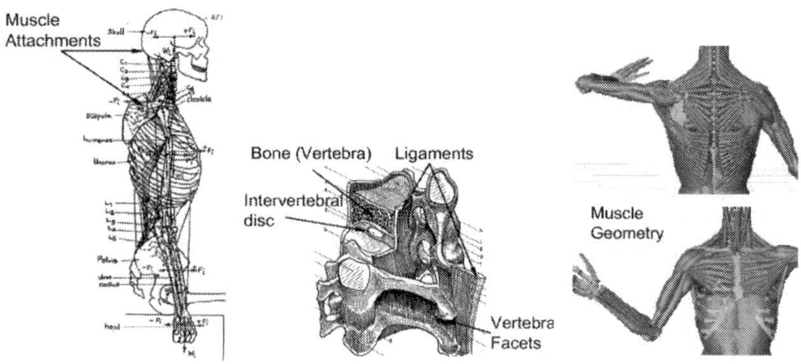

a) b) c)

Figure 7. Advanced features for occupant biomechanical models: (a) elements of the trunk [**13**]; (b) detail of the spine [**18**]; (c) Detailed representation of the muscles [**19**].

Most of the life threatening injuries of the vehicle occupant involve the spine, the torso or the head. Therefore, is no surprise that very detailed models of the human neck [**20**] and of the head [**21**] have been proposed. For less serious injuries, in other anatomical segments, it is required that the human joints, such as ankle, knee, hip, shoulder, elbow are also modelled with realism. Detailed models for the knee, ankle and shoulder have been proposed by a wide number of researchers [**22-26**]. Many of the features of these detailed models are useful for impact biomechanics studies.

Detailed models considered for anatomical joints or for the spine consider that there is contact between the joint pairs. The contact models require that: the surface geometry is described with detail in order to allow for the detection of points of contact between such surfaces; the local properties of the materials of the contacting surfaces are identified, in particular, their stiffness and damping; proper normal force and friction force models are set for the anatomical joints surfaces. For this type of realistic human joints the continuum modelling of the local deformation near the contact area is important and may require the use of contact models different from the one suggested in the previous section. Care must be exercised in order to maintain the computational efficiency of the models used. Some of the contact models with good potential for application in advanced biomechanical models are described by Stronge [**27**].

The human body has a well established anatomy independent of age, gender or size. However, geometries and material properties of the anatomical joints do not scale linearly with age or gender. It is advantageous that the features of all anatomical joints are defined in terms of parametric models that can use a database where the required information for different individuals is stored. By defining the rules of scaling of the different components of the human body in terms of age, gender or size it is possible to use the biomechanical models of the human body in a wider range of situations. Currently, there are ongoing efforts, as for instance the European Projects HUMOS II and APROSYS, to identify detailed data and models for the human body and authors, as Yamada [**28**], who provide good references on the variation of the biological material properties with gender and age.

3.4.2 Muscle activity

For realistic models of the human body the use of suitable representation of the skeletal muscles is required. Anatomical joints modelled as contact pairs only maintain their stability if the muscle and ligaments that cross them are modelled. The stiffening of the human joints, due to muscle contraction, and its consequences on the biomechanical dynamic response is another aspect that requires proper muscle models for increased biofidelity.

The problem of identifying the muscle forces requires that a detailed description of the most important muscles and muscle groups is done. The anatomical data existing today ensures that the important features required for the development of such models, including the physiological cross section area (PCSA), location of the insertion points, maximum force, geometry and physiology is available, at least for the normal adult. The data for the reflexive muscle forces, developed due to the high extension rate experienced by the muscle during impact, can be obtained also. The major problem is to identify the muscle forces developed during the voluntary contractions, as those observed for the standing passenger. The combined use of detailed muscle models, inverse dynamics analysis of biomechanical models and suitable optimization procedures provides solutions for this type of problem, already applied in many human activities such as gait analysis and athletics.

It is suggested here that the same techniques are used here for the identification of the voluntary muscle actions of the standing passenger and for the analysis of the joint stiffening during the crash events. As the crash event is generally very short, an experimental program supported by proper biomechanical models can be devised to obtain the most relevant pre-crash muscle data and body postures, without involving the actual crash event.

One of the particular features of railway and buses vehicles passengers is the mobility that they experience while travelling. Much of this mobility is possible because of the wide volumes for the motion of the occupants, the lack of any enforced use of restraining devices, such as seat belts, the different areas inside the vehicle and the furniture of its interiors that invite a more 'comfortable' posture and a better access to the entrance and exit. The simulations of particular postures and the different seating and standing positions of the railway vehicle occupants are not as biofidelic as for the occupants of road vehicles because: the kinematics of the occupant are not guided; the front and side facing occupants lead to body to body impact that is not the type of impact for which the biomechanical models have been developed; the lag of time between the warning signs that develop before the collision and the impact that follows allow for the passengers to take defensive measures; the muscle activity in the standing passengers modifies the post-impact kinematics. Furthermore, the types of impacts that these occupants are subjected to cannot be simply defined as frontal or side impact. Therefore the use of the current ATD models in these modes of transportation present technical difficulties that can be overcome with a new generation of dummies. Many of the advanced features suggested in the previous section have a high potential for the design of such devices.

4. INITIAL POSITIONS OF VEHICLE OCCUPANTS

In any practical situation, and for the development of intelligent restraint systems in particular, it is required that the anatomical characteristics of the occupants and their positions inside the vehicle can be monitored. In order to make reliable analysis tools available it is necessary that 'real life' positions of the vehicle occupants can be quantified and used in the multibody analysis program. The process of recording the human body actual motion and to extract the position of its anatomical segments for every frame is designated by spatial reconstruction.

4.1 Photogrammetric Tools

The most common techniques used for the spatial motion reconstruction are found in gait and sports motion analysis [**29**, **30**]. The selection of a technique depends upon several factors such as the purpose of the analysis, the type of motion, the available time to obtain the results and the costs involved. Photogrammetry is the most frequently used technique, being the process of acquiring and digitizing images easily automated if markers, located at the anatomical joints and extremities, are applied [**30**].

Contrary to the gait or sports applications of the photogrammetric techniques, where the motion of the subject is inside a large open volume, applications to vehicle occupants are characterized by a static position of the subjects and by the closed volumes where these are seated. A laboratory apparatus of cameras suitable for the reconstruction of the occupant spatial position is schematically represented in Figure 8.

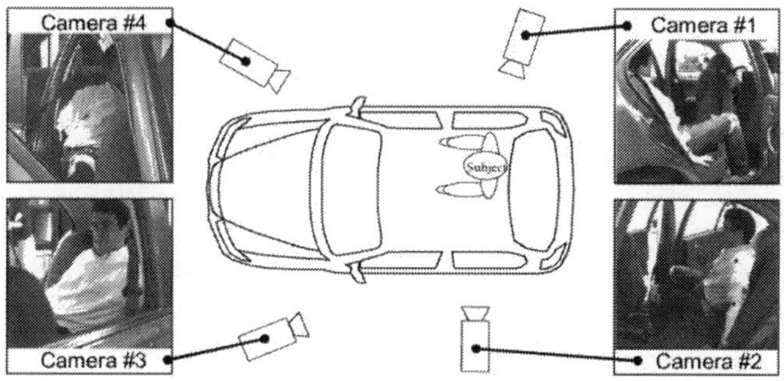

Figure 8. Vehicle and video cameras for the recording of the out-of-position occupants.

The images collected by a single camera are collections of two-dimensional information, resulting from the projection of a three-dimensional space into a two-dimensional one. Mathematically, the inverse of the transformation does not exist. Consequently, it is not possible to reconstruct the three-dimensional coordinates of a point in space from its two-dimensional projection in a single frame. Aziz and Karara [**31**], proposed a solution for the reconstruction process called Direct Linear Transformation that is used in virtually all spatial reconstruction techniques.

The biomechanical model used in this work requires the reconstruction of the spatial position of 23 anatomical points for each frame of the analysis period. This set of points is depicted in Figure 9, where the underlying kinematic structure of the model is also outlined. The spatial position and orientation of the anatomical segments of the biomechanical model are obtained from the spatial positions of these reconstructed points. Before the reconstructed model can be used it is necessary to ensure the kinematic consistency of the data with the biomechanical model used and that the data is filtered in order to eliminate the noise from the digitalization process. For more information on the reconstruction procedures the interested reader is referred to [**32**].

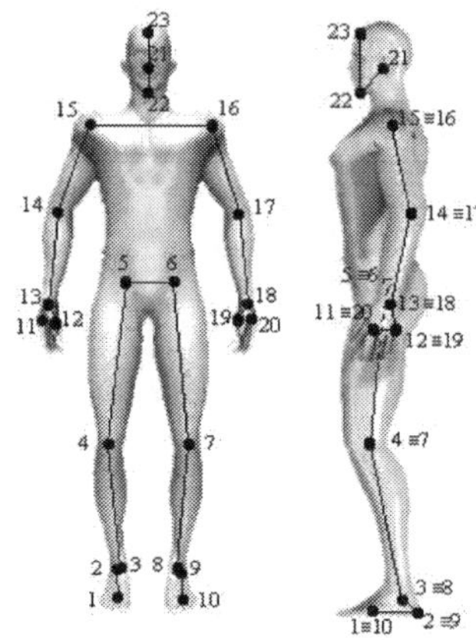

Figure 9. Set of 20 digitized points and kinematic structure.

In order to perform the dynamic analysis, the initial velocities of the anatomical points are also required. The velocities are obtained by solving equation (2). This procedure assures that the velocities obtained belong to the null space of the Jacobian matrix, making them consistent with the kinematic structure of the biomechanical model.

4.2 Out-of-Position Occupants

To demonstrate the methodology presented here, instead of a complete vehicle, a vehicle seat and steering column and wheel are use to seat the occupant, as presented in Figure 10. Notice that using this setup prevents many meaningful positions for the occupant from being obtained.

Figure 10. Vehicle seat used in the experimental setup to find the occupant initial conditions.

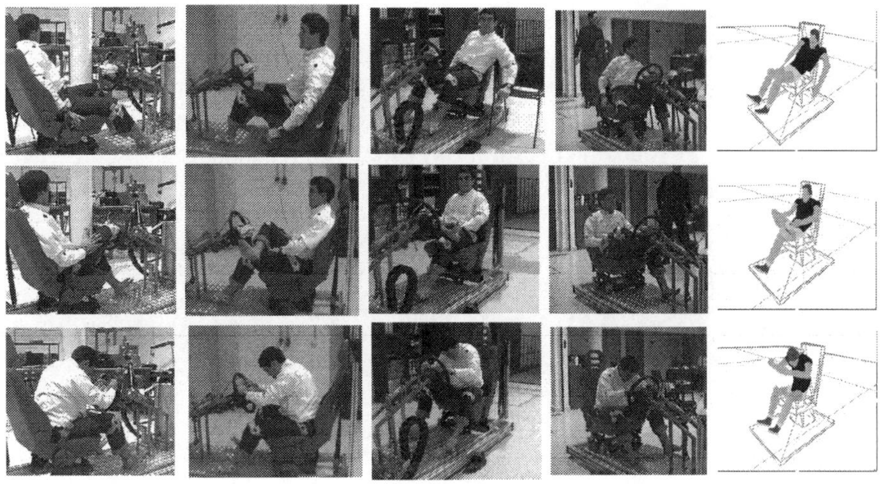

(a) (b) (c) (d) (e)

Figure 11. (a)-(d) Out-of-position occupants as viewed by the cameras and (e) spatial reconstructions.

With the setup described in Figure 9 and the vehicle seat presented in Figure 10, a seated occupant is asked to adopt seated positions similar to those that would be used when riding a car in different situations. Among the positions videotaped and reconstructed, the positions presented in Figure 11 are selected and used.

5. VEHICLE ROLLOVER

In order to appraise different aspects that have to be accounted for in the use of multibody models of the human body or of ATDs, the biomechanical models described in this work are used in the simulation of a vehicle rollover. For the purpose, a complete vehicle is modeled including the suspension systems and wheels [**33**]. This setup and the simulation outcome is compared with that of two experimental tests of the vehicle with three Hybrid III dummies that have been carried at the Transportation Research Center of Ohio (1985, 1986). An overview of the footage obtained in one of the experimental tests is shown in Figure 12.

Three occupants, with a 50%tile, are modeled and integrated with the vehicle. The two occupants in the front of the vehicle have shoulder and lap seat belts, while the occupant seated in the back of the vehicle has no seatbelt. The reconstruction techniques described in section 4 allow to define realistic postures for the biomechanical models of the occupants and to position them correctly inside the vehicle, even when this is simulated in scenarios different from those for which the positions have been obtained. The initial positions of the occupants correspond to a normally seated driver, a front passenger bent to check out the 'glove compartment', and a rear occupant with a 'relaxed' position, according to the positions reconstructed and shown in Figure 11.

Figure 12. View of the experimental test for the truck rollover.

The vehicle and occupants are simulated here in a rollover situation described in Figure 13. The initial conditions of the simulations correspond to experimental test conditions where the vehicle moves on a cart with a lateral velocity of 13.41 m/s until the impact with a water-filled decelerator system occurs. The vehicle is ejected with a roll angle of 23 degrees. The initial velocity of the vehicle, when ejected, is approximately 11.75 m/s in the Y direction while the angular roll velocity is 1.5 rad/s.

Figure 13. Initial position of the vehicle and occupants for the rollover.

The results of this simulation are shown in Figure 14, where several frames of the animation of the vehicle rollover with occupants are presented, as observed from two different points of view. It is noticeable in these sequences that the vehicle first impacts the ground with its left tires. At this point the rear occupant is ejected. The rollover motion of the vehicle proceeds with an increasing angular velocity, mainly due to the ground - tire contact friction forces. The occupants in the front of the vehicle are hold in place by the seat belts. Upon continuing its roll motion, the vehicle impacts the ground with its rollbar cage, while the ejection of the rear occupant is complete.

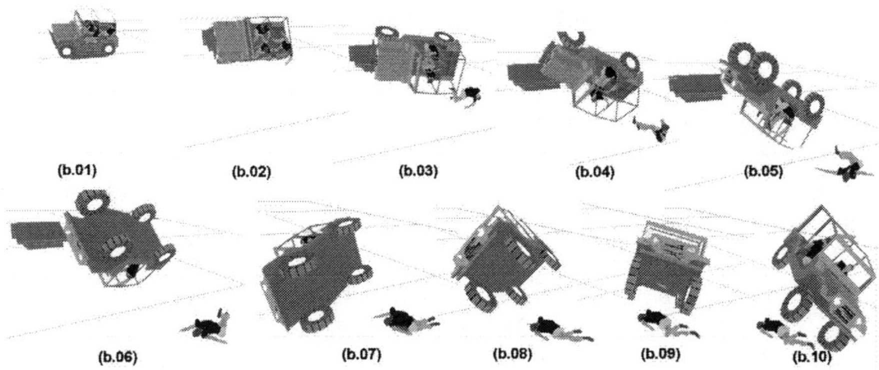

Figure 14. View of the outcome of the rollover simulation of a vehicle with three occupants.

The HICs for all occupants largely exceed 1000. The Severity Index, in Figure 15, indicates a very high probability of fatal injuries for the occupants under the conditions simulated. Notice that the model has rigid seats, interior trimming for the dashboard, side and floor panels, and that the ground is also considered to be rigid. It is expected that the head accelerations are lower if some compliance is included in the vehicle interior.

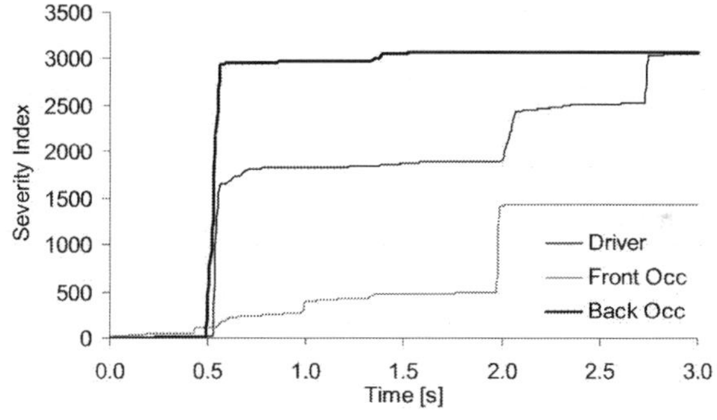

Figure 15. Severity Index for the vehicle occupants.

The kinematics of the biomechanical models of the occupants, and in particular that of the ejected occupant, are similar to the kinematics of the crash test dummies used in the experimental tests. Several simulations of the vehicle rollover with occupants seating with different postures have been performed. These simulations show that regardless of the rear occupant seating posture the ejection and post-ejection occupant kinematics remains basically unchanged.

The rollover scenario for the simulation and experimental tests described here illustrates conditions in which there is not a single suitable crash dummy available during all phases of the motion. During the rollover of the vehicle the dummy experiences side impacts, lumbar spine loading, head contact in different directions, femoral loading, eventually front contact, etc. Moreover, the vehicle occupants are subjected to multiple impacts in different directions, which are conditions not accounted for by the current injury indexes. Impact scenarios such as the one exemplified by the vehicle rollover clearly indicate the need for a more global approach to biomechanical modeling and to the definition of other measures of injury.

6. CONCLUSIONS

The methodologies used to build biomechanical models of anthropometric testing devices and humans have been reviewed here. The use of multibody dynamics formulations provides the necessary numerical tools to support the development of advanced biomechanical models with increased biofidelity.

The actual biomechanics models for whole body response neither account for the voluntary muscle activity that leads to the stiffening of the joints nor are they suitable for passenger posture studies. Some of the developments that are required in the biomechanical models to increase their biofidelity for application to general passive safety include the more detailed description of the anatomical segments, a more realistic representation of the geometrical and material properties of the body segments, improved models for the neck and trunk including the bony structures, ligaments, intervertebral discs and anatomical joints, and biofidelic muscle models that include reflexive and voluntary contraction.

The identification of the muscle reflexive actions has been addressed by several researchers, especially when applied to the head-neck muscles [**20**]. However, it is not clear how such activity will be evaluated for other muscle groups or how the voluntary muscle activity will be handled. The existence of muscle activity and the standing postures in a significant number of impact cases is a distinctive feature of the railway occupants' biomechanics. The identification of the typical postures adopted by the passengers can be achieved using standard videogrammetric techniques, such as those used for motion analysis. Due to the large volumes existing in the railway vehicle it is feasible to devise a suitable experimental program with which the kinematic and force data associated to different passenger postures and to post-impact motion can be identified.

Finally, it is suggested in this work that the human mathematical surrogates cannot replace the anthropometric testing devices as these, even if built, would not provide repeatable measures. However, their use to design more advanced dummies not only cannot be discarded but is also irreplaceable.

ACKNOWLEDGMENTS

The work reported was supported by Fundação para a Ciência e Tecnologia through the project POCTI-2001-EME-39976 *Dynamic Analysis Methodologies for Application to the Study of the Human Musculo-Skeletal Behavior.*

REFERENCES

1. Kamal, M.M., Analysis and simulation of vehicle-to-barrier impact, in *International Automobile Safety Conference Compendium*, SAE Paper No. 700414, SAE, Inc., 1970.
2. Ambrósio, J.A.C. and Pereira, M.S., Multibody dynamic tools for crashworthiness and impact, in *Crashworthiness on Transportation Systems: Structural Impact and Occupant Protection*, Ambrósio, J.A.C. and Pereira, M.S. (eds.), Kluwer Academic Publishers, Dordrecht, The Netherlands, 1997, pp. 475-521.
3. King, H.Y., Pan, H., Lasry, D. and Hoffman, R., Finite element modelling of the hybrid III dummy chest, in *Crashworthiness and Occupant Protection in Transportation Systems*, AMD-Vol. 126/BED-Vol.19, ASME, 1991.
4. Haug, E. and Ulrich, D., The PAM-CRASH code as an efficient tool for crashworthiness simulation and design, in *Proceedings of the Second European Car/trucks Simulation Symposium*, Munich, Germany, 1988.
5. Belytschko, T. and Kenedy, J.M., WHAMS-3D, An Explicit 3D Finite Element Program, KBS2 Inc., Willow Springs, Illinois, 1988.
6. Halquist, J.O., Theoretical Manual for DYNA-3D, Lawrence Livermore Laboratory, 1982.
7. Fleck, J.T. and Butler, F.E., Validation of the Crash Victim Simulator, Vol. 1: Engineering Manual, Part I: Analytical Formulations, NTIS, No. Pb86-21243-8, 1981.
8. MADYMO Theory Manual version 6.2, TNO MADYMO BV, Delft, The Netherlands, 2004.
9. Wismans, J., Design tools: human body modeling, in *Crashworthiness: Energy Management and Occupant Protection*, Ambrósio, J. (ed.), Springer, Wien, Austria, 2001, pp. 333-366.
10. Viano, D.C. and Lau, I.V., A viscous tolerance criterion for soft tissue injury assessment, *Journal of Biomechanics*, **21**, 1988, 387-399.
11. Silva, M. and Ambrósio, J., Out-of-position occupant models in a multibody integrated simulation environment, in *Proceedings of the IRCOBI2002 International IRCOBI Conference on the Biomechanics of Impact*, Munich, Germany, September 18-20, 2002.
12. Silva, M. and Ambrósio, J., Solution of the redundant muscle forces in human locomotion with multibody dynamics and optimization tools, *Mechanics Based Design of Structures and Mechanisms*, **31**(3), 2003, 381-411.
13. Seireg, A. and Arvikar, R., *Biomechanical Analysis of the Muscoloskeletal Structure for Medicine and Sports*, Hemisphere Pub. Corp., New York, 1989.
14. Jalon, J.G. and Bayo, E., *Kinematic and Dynamic Simulation of Mechanical Systems – The Real-Time Challenge*, Springer-Verlag, Berlin, Germany, 1994
15. Silva, M., Ambrósio, J. and Pereira, M., A multibody approach to the vehicle and occupant integrated simulation, *International Journal of Crashworthiness*, **2**(1), 1997, 73-90.

16. Hertz, H., *Gesammelte Werk*, Leipzig, Germany, 1895
17. Lankarani, H.M. and Nikravesh, P.E., Continuous contact force models for impact analysis in multibody systems, *Nonlinear Dynamics*, **5**, 1994, 193-207.
18. Kapandji, I.A., *The Physiology of the Joints: Vol. 3*, Churchill Livingstone, Edinburgh, United Kingdom, 1974.
19. Murial, W., *3D Modeling of the Human Upper Limb Including the Biomechanics of Joints, Muscles and Soft Tissues*, Ph.D. Thesis, Ecole Polytechnique Fed. de Lausanne, Lausanne, Switzerland, 1999.
20. Jager, De M., Sauren, A., Thunnissen, J. and Wismans, J.S.H.M., A three-dimensional head-neck model: Validation for frontal and lateral impacts, in *Proceedings of the 38th Stapp Car Crash Conference*, SAE Paper No. 942211, 1994.
21. O'Riordain, K., Thomas, P.M., Philips, J.P. and Gilchrist, M.D., Modelling and accident reconstruction of real world head injury accidents resulting from falls using multibody dynamics, *Journal of Clinical Biomechanics*, **18**, 2003, 590-600.
22. Wismans, J., Veldpaus, F. and Janssen, J., A three-dimensional mathematical model of the knee-joint, *Journal of Biomechanics*, **13**, 1980, 677-685.
23. Hirokawa, S., Biomechanics of the knee joint: a critical review, *Critical Reviews in Biomechanical Engineering*, **21**(2), 1993, 79-135.
24. Leardini, A., O'Connor, J.J., Catani, F. and Giannini, S., A geometric model of the human ankle joint, *Journal of Biomechanics*, **32**, 1999, 585-591.
25. Van der Helm, F. and Pronk, G.M., Three-dimensional recording and description of motions of the shoulder mechanism, *Journal of Biomechanical Engineering*, **117**, 1995, 27-40.
26. Hogfors, C., Sigholm, G. and Herberts, B., Biomechanical model of human shoulder joint – I. Elements, *Journal of Biomechanics*, **20**, 1987, 157-166.
27. Stronge, W.J., *Impact Mechanics*, Cambridge University Press, Cambridge, England, 2000.
28. Yamada, H., *Strength of Biological Materials*, The Williams and Wilkins Company, Baltimore, Maryland, 1970.
29. Allard, P., Stokes, I. and Blanchi, J., *Three-Dimensional Analysis of Human Movement*, Human Kinetics, Urbana-Champaign, Illinois, 1995.
30. Nigg, B. and Herzog, W., *Biomechanics of the Musculo-Skeletal System*, John Wiley & Sons, New York, 1999.
31. Addel-Aziz, Y. and Karara, H., Direct linear transformation from comparator coordinates into object space coordinates in close-range photogrammetry, in *Proceedings of the Symposium on Close-range Photogrammetry*, Falls Church, Virginia, 1971, 1-18.
32. Silva, P., and Ambrósio, J. Kinematic data consistency in the inverse dynamic analysis of biomechanical systems, *Multibody System Dynamics*, **8**, 2002, 219-239.
33. Ambrósio, J.A.C., Nikravesh, P.E. and Pereira, M.S., Crashworthiness analysis of a truck, *Journal of Mathematical Computer Modelling*, **14**, 1990, 959-964.

RECONSTRUCTING REAL LIFE ACCIDENTS TOWARDS ESTABLISHING CRITERIA FOR TRAUMATIC HEAD IMPACT INJURIES

M.C. Doorly[1], J.P. Phillips[2] and M.D. Gilchrist[1]
[1]*Department of Mechanical Engineering, UCD, Belfield, Dublin 4, Ireland*
[2]*Department of Neurosurgery, Beaumont Hospital, Dublin 9, Ireland*

Abstract. Brain injury is the leading cause of death in those aged under 45 years in both Europe and the United States. The objective of this research is to reconstruct and analyse real world cases of accidental head injury, thereby providing accurate data which can be used subsequently to develop clinical tolerance levels associated with particular traumatic injuries and brain lesions. In this study, MADYMO pedestrian models are used to analyse a well-defined set of non-fatal accidents involving simple falls. The effect of varying the initial conditions is systematically examined and the predicted MADYMO results are compared against literature data.

Key words: impact biomechanics, falls, accident reconstruction, head injury, multibody dynamics.

1. INTRODUCTION

Mechanical impact is the leading cause of injury, death and disability in people aged under 45 in the USA, Europe and increasingly so in Third World countries [1]. Costs of hospitalisation, care and rehabilitation of head injured people are estimated to be as high as $33 billion per year in the USA [2]. Much recent research has focussed on the biomechanics of traumatic head injury, an objective of which is to correlate clinical dysfunction with mechanical impact conditions, with a view to reducing or eliminating the mechanisms that cause damage.

Michael D. Gilchrist (ed.), IUTAM Proceedings on Impact Biomechanics: From Fundamental Insights to Applications, 81–90.

Road traffic accidents (RTAs), falls and assaults are the most frequently cited causes of head injury. While RTAs tend to be the leading cause of injury related death, falls tend to be the leading cause of non-fatal hospitalisation [3, 4]. In Ireland, falls are the single greatest cause of hospital admissions for both males and females across most age groups, with head injuries occurring in approximately a quarter of all admissions [4]. Falls are selected as the accidents of interest in this present study due to their lower levels of uncertainty regarding initial conditions, and due to the tendency of falls to result in focal, as opposed to diffuse, head trauma.

The aim of this research is to reconstruct a number of real life falling accidents using numerical techniques in order to provide detailed data which can be used to establish injury criteria for specific types of brain lesion. The accidents are modeled using multibody dynamics software to recreate the overall movement of the body during the accident. Once this simulation represents the accident described, the output, in the form of velocities, accelerations and forces, is subsequently used as input for a 3D finite element model of the head, which has been previously compared with experimental results [5]. This finite element model is able to simulate the effect of the overall head movement on the cranial contents, so the local deformation parameters within the brain tissue can be examined and compared to the observed clinical results. This present paper only presents the results of the multibody dynamics simulations and compares the results against those reported by other researchers.

2. METHODOLOGY

Real life cases of falls resulting in head injury are selected by Ireland's National Department of Neurosurgery at Beaumont Hospital, Dublin. Cases are screened to narrow the selection to relatively simple falls, in order to facilitate modelling of the accidents. Clinical assessments of each case are provided by the hospital, together with CT scans. The accident site is examined to determine the layout of the environment, the height of the fall, and the type of surface onto which the person fell. Informed consent is obtained from patients and witnesses, with the approval of the Ethics Committee of Beaumont Hospital.

Accident reconstruction is carried out using MADYMO (MAthematical DYnamic MOdels) [6] multibody dynamics software. MADYMO has a database of dummy models, which makes it very suitable for reconstructing accidents involving humans. For these analyses the pedestrian models, which have been validated extensively against full body pedestrian tests, were used, with altered head contact characteristics. It was found that the values for the forces and accelerations experienced by the head of the pedestrian model were very high in comparison to values cited in the literature. From previous related research [7] it was determined that the head response curve determined by Yoganandan *et al.* [8] was the most suitable for this analysis since it was independent of the head impact location.

In order to reconstruct these accidents certain initial conditions need to be applied to the model. Due to the fact that all data regarding the accidents were collected from the field, there is necessarily a degree of uncertainty regarding the precise conditions under which each accident took place. Without instrumentation attached to the person involved, it is impossible to know exactly the velocities of the people during the fall. The cases presented here have at best an eyewitness account of the accident, which is of course useful and necessary, but not scientifically rigorous. Initial conditions in MADYMO are defined by specifying the X, Y and Z components of both linear and angular velocity, and initial joint rotations and positions. For each case reconstructed here, an estimate was made of these components of the initial velocities and positions, based on available information regarding the accident. These were used to run an initial test simulation. If the graphical representation of the simulation appeared unrealistic, slight alterations were made to the initial velocities until the kinematics of the impact appeared physically realistic and correct. This method still leaves some uncertainty as to the validity of these initial conditions. In order to systematically consider the effect of these values on the results, the initial conditions were then varied by ±10% and ±50%. Figure 1 shows a sequence of events taken from one of the accident simulations. The relevant results obtained from these simulations are in the form of linear and angular accelerations $(a, \ddot{\vartheta})$, velocities $(v, \dot{\vartheta})$ and forces experienced by the head.

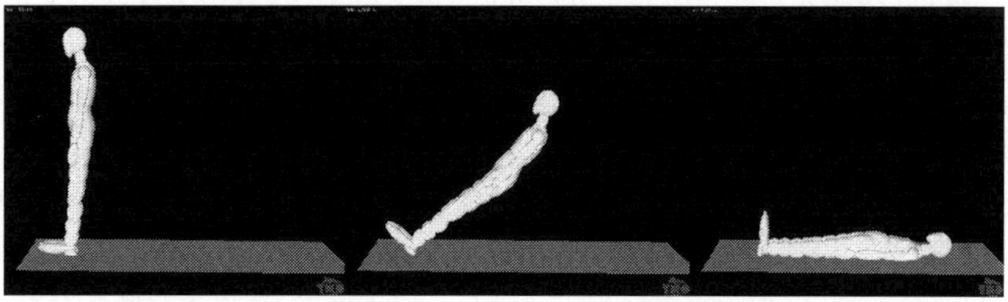

Figure 1. Images taken from the simulation of a boy falling at a water fountain.

3. RESULTS

In this study a total of ten real-life accident cases were analysed. A summary of these cases is given in Table 1. For each of the cases a sensitivity analysis was performed. This was done by varying both the initial joint positions and the initial joint velocities, both on their own and combined, by ±10% and ±50%. Some of the simulations resulting from these changes in initial conditions did not accurately represent the kinematics of the accident described, and so these results were omitted from the subsequent analysis. In some cases where there was no initial joint velocity, a certain amount of velocity was applied to the body in the direction of the movement of the fall. For example, in Cases 1 and 2, the people fell directly backwards from stationary positions. These cases were modelled by tilting the body backwards and allowing it to move under the influence of gravity. The sensitivity analysis looked at the effect of applying a small amount of backward rotational velocity to the body to see how this affected the results. In some of the cases there was very good agreement in the results despite changing the input by up to 50%. This was particularly evident in the cases where the fall was in one plane only, i.e., the person fell straight forwards or straight backwards. In the cases where there was out-of-plane motion, greater differences could be seen in the results. However, if the initial conditions were changed by too much for these accidents it was more likely that they would no longer represent the kinematics of the real-life accident accurately. Table 2 shows the range of results obtained from the simulations representing the accidents. In general there is good agreement among the results. The case with the largest differences is Case 9. This involved a man falling from a gate at a height of 138cm above standing height, giving more time for voluntary reactions by the person. The simulations do not take any voluntary reactions into account, therefore there is likely to be a higher degree of error in these results, and this is evident by the wider range of results observed for the same accident description.

The cases where the results of the sensitivity analysis are in closest agreement are those for which the accident occurred in one plane only (i.e., Cases 1-4), resulting in either frontal or occipital impact. For two of these simulations Figure 2 shows the envelope of maximum and minimum values for the range of results. It can be seen that the difference between these curves is modest, with the greatest difference occurring for the accelerations of Case 1 (± 21%). In this case the lady's head impacts the wall behind her, and not the ground, so slight differences in head position are likely to lead to larger differences in acceleration values. Figure 3 represents the accidents where the impact was fronto-parietal. There is a lot more variation in these results than for the planar impacts (greatest difference is ± 46%).

Table 1. Case summaries, providing description of accident and head injury sustained.

Case	Sex	Age yrs	Height cm	Weight kg	Brief description of accident	Head injury
Case 1	F	76	160	60	Lost her balance and fell directly backwards. Incurred occipital impact of head against concrete wall.	Small left frontal lobe contusion. Large right temporal parenchymal haematoma.
Case 2	M	11	152	37	Fell directly backwards. Incurred occipital impact of head on concrete pavement.	Right frontal lobe contusion. Traumatic subarachnoid haemorrhage.
Case 3	M	85	163	70	Fell directly forwards after losing his balance onto concrete footpath.	Left sided chronic subdural haematoma. Right sided acute subdural haematoma. Midline shift to the left.
Case 4	F	84	163	63.5	Fell directly forwards after losing her balance while walking downhill on concrete footpath.	Left sided subdural haematoma with midline shift. Dilated right ventricle.
Case 5	F	84	163	57	Tripped on a crack causing her to fall forward and to her right, hitting head off concrete footpath.	Right sided acute and chronic subdural haemorrhage with midline shift and subfalcine herniation.
Case 6	F	71	163	63.5	Lost her balance and fell forwards hitting the front left of her head off the concrete ground.	Left fronto-parietal subdural haematoma with midline shift and asymmetrical ventricles.
Case 7	M	76	170	66.7	Tripped causing him to fall forward on his right side, breaking his shoulder and hitting right side of his face.	Right chronic subdural haematoma. Left acute subdural haematoma.
Case 8	F	87	157	51	Slipped on a ramp and fell forward and hitting the right of her head off a railing.	Left sided subdural haematoma. Lateral ventricle shifted to the right side.

Table 1. (Continued).

Case 9	M	24	169	80	Fell backwards and twisted to the left while balancing on a gate, pulling a rope which broke. Incurred left lateral impact of head on tarmac.	Left temporo-parietal linear skull fracture. Left temporo-parietal extradural haematoma.
Case 10	F	24	169	55	Standing on chair, twisted sharply and fell forwards and to the right. Incurred right lateral impact of head on ceramic tiled floor.	Right frontal linear skull fracture. Right frontal extradural haematoma. Left posterior temporal contusion.

Table 2. Summary of the peak results for the sensitivity analyses.

Case	v (m/s)	$\dot{\vartheta}$ (rad/s)	a (g)	$\ddot{\vartheta}$ (krad/s^2)	Force (kN)	HIC
1	4.83 - 5.75	32.14 - 38.9	195 - 300	30.0-44.5	6.9-10.38	511.83-1200
2	6.34 - 7.2	44.05 - 49.44	333 - 403	33.9 – 42.0	12.67 – 14.7	2930 - 4308
3	5.08 - 5.64	21.1 - 24.16	321 - 365	13.2 - 16.1	15.43 - 16.88	2491 - 3255
4	4.5 – 4.77	16.22 – 20.38	343 – 364	19.9 – 27.8	11.79 – 13.03	2387 - 2831
5	3.51 – 4.21	13.62 – 19.21	242 – 298	14.9 – 17.9	8.75 – 10.96	1026 - 1786
6	3.61 – 3.67	17.62 – 19.43	302 – 307	23.8 – 24.4	11.19 – 11.31	1269 – 1312
7	4.73 – 5.11	12.12 – 17.56	303 – 330	7.4 – 15.1	9.85 – 11.84	1942 - 2408
8	5.41 – 6.11	58.83 – 72.91	265 – 351	41.9 – 49.2	11.25 – 12.64	1532 - 2914
9	6.9 – 8.19	15.52 – 26.78	189 – 456	8.1 – 22.1	8.8 – 22.11	674 – 5951
10	5.74 - 6.6	21.06 - 35.29	318 - 342	23.7 - 30.2	10.88 - 11.65	2237 - 2639

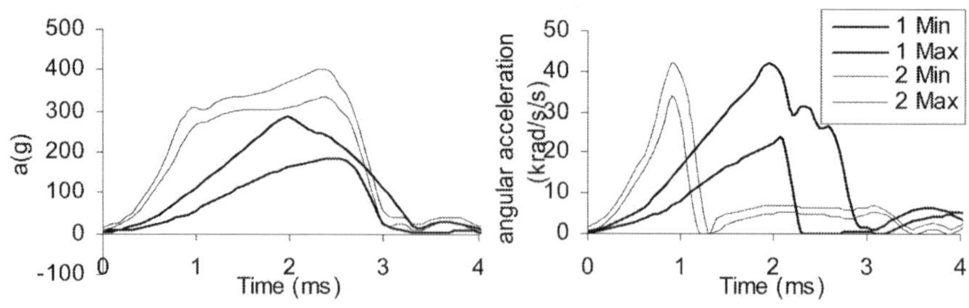

Figure 2. Maximum and minimum values obtained from the range of simulations for the falls that occurred in one plane only (Cases 1 and 2).

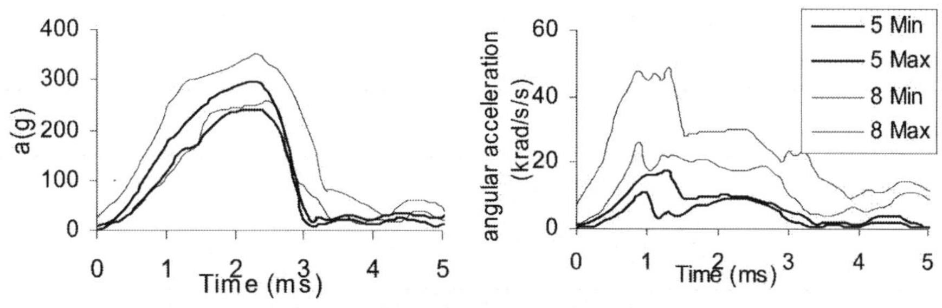

Figure 3. Envelopes of maximum and minimum values predicted for the falls that occurred in more than one plane (Cases 5 and 8).

Examining the simulations where the fall is out of plane (Cases 5-10), it can be seen that there is greater variability in the results. In these cases it is more likely that by changing the initial conditions significantly the kinematics of the simulation will change to such an extent that it no longer matches the accident description. In some of these cases the results are omitted since even a slight change in kinematics can lead to quite a large change in the resulting velocities, accelerations and forces.

4. DISCUSSION

In this set of accident cases there were two occipital impacts (Cases 1 and 2). In both of these cases contre-coup contusion was observed on the frontal lobe of the brain, with no evidence of coup contusion present. One other case involving lateral impact also presented with contusion. In all three cases where contusion was observed the linear acceleration falls above the lower tolerance curve for contusion proposed by Auer *et al* [9] indicating that contusion is a possibility. In fact, all cases observed in this study have linear accelerations above this level. However there is no upper tolerance curve provided for contusion, so there is no level given above which contusion will always be observed. In some of the other cases there may be contusion present, but it may be masked by the presence of a subdural haematoma.

In Case 1 parenchymal haemorrhage was also observed. Very little research has been carried out on the biomechanics of this specific type of lesion. One could argue that a minimum level of acceleration is necessary for parenchymal haemorrhage to occur. It is likely that it is associated with higher accelerations than contusion, as it is a more severe injury. This case has very high angular accelerations, due to the oblique nature of the impact, and it is possible that this contributes to the impact of the brain against the sphenoid ridge, which may contribute to the occurrence of both contusion and parenchymal haemorrhage.

Subarachnoid haemorrhage can be seen in Case 2 as bleeding in the right Sylvian fissure. In this case there is very high linear acceleration and reasonably high linear velocity. Bleeding in the Sylvian fissure may have been caused by damage to the blood vessels in the middle fossa region of the brain arising from this very high linear acceleration. It must be noted that this case involves an 11 year old boy. The skull does not become fully calcified until adulthood and Mohan *et al.* [10] suggest that at age 13, skull stiffness is only 90% that of an adult. If this factor were taken into account in the force penetration curves for head contact, lower forces would be predicted on the head.

There are also two cases involving lateral impact in this study (Cases 9 and 10). Both of these accidents involve a fall from a height higher than standing height. Both patients were 24 years of age at the time of the accident, and both sustained linear skull fracture and extradural haematoma. Fracture occurred at the location of impact with underlying extradural haematoma. Yoganandan *et al.* [8] found that the force necessary to fracture cadaver skulls ranged between 8.8 kN and 14.1 kN, with an average of 11.9 kN. A similar study by Allsop *et al.* [11] for temporo-parietal bone found an average of 12.4 kN. Peak forces in these cases were sufficient to cause skull fracture. Extradural haematoma involves the damage of blood vessels lying between the dura and the skull. Blumbergs [12] cites studies suggesting that skull fracture is associated with extradural haematoma in up to 97% of cases. The skull fracture and extradural haematoma can be attributed directly to contact effects (i.e., force and linear acceleration), which were quite high in these two cases.

The majority of the cases in this study involved frontal or fronto-parietal impacts. In all these cases subdural haematoma was observed, often accompanied by shifting of the midline and/or the ventricles. The location of the subdural haematoma had no apparent correlation with where the impact occurred, but was observed most often on the contrecoup side. No other injuries were observed in these cases, however there may be contusion underlying the SDH in some cases. Auer *et al.* [9] produced upper and lower tolerance curves correlating subdural haematoma with linear acceleration. All of the present cases had linear accelerations that were between these lower and upper tolerance curves, which indicates that subdural haematoma is a possibility but not a certainty. Other authors [13] suggest that high angular acceleration together with a high strain rate (i.e., rate of acceleration onset) are the important factors in ASDH causation. They suggest that falls resulting in a peak head acceleration of over 200g and a duration of 3.5ms or less would produce the conditions necessary for the occurrence of ASDH due to bridging vein rupture. These conditions are met in all cases presented here with high accelerations and very short duration impacts (in the region of

3 ms). The likelihood of subdural haematoma due to bridging vein rupture is also thought to increase with age, and in all cases presenting with this type of injury the age of the patient is quite high.

Looking at the simulations overall there is no obvious effect of increasing or reducing initial joint velocities or positions. Each case reacts differently. In general, changing the initial velocity seems to have more of an effect on the results than changing initial joint positions and rotations. In many cases changing the initial joint rotations alters the simulations to such an extent that it no longer corresponds to the kinematics of the accident described.

5. CONCLUSION

This analysis has shown that multibody modelling is a useful tool for reconstructing real-life accidents. From the results presented here it can be seen that the accuracy is appreciably greater when simulating simple accidents than more complex accidents. Falls from standing height generally gave good agreement in the results once the kinematics of the simulations represented the accident well. However, in the case where the man fell from the gate there was greater variation in the results. It is quite likely that this difference is partly due to the fact that there is more time and space for the person to react when they are falling, and voluntary reactions cannot be taken into account by the MADYMO software. This will inevitably lead to errors in the results and a wider envelope of results for a given range of input conditions. The more details that are known about the accident the more accurate the simulations will be and the less scatter will be associated with the results.

The main disadvantage of using real-life accidents is that most of the inputs rely on eyewitness reports which are often not accurate. However, in this study it can be seen that there is a limited range of input conditions that will result in a kinematically realistic simulation of the accident, and the outputs of these simulations generally agree quite well.

The main advantage of using real-life accidents is that the injuries are known. If the initial conditions are accurately reported and the injuries are known, it should be possible to see which kinematic inputs lead to particular types of brain lesions. Ongoing work in this study involves the use of finite element modelling procedures. The results from the multibody modelling, in the form of velocities, accelerations and forces, are being used as input for a 3 dimensional finite element model of the head. This gives output in the form of brain tissue deformation resulting from the head impact. This information will be compared in a subsequent study against clinical data in order to establish quantifiable mechanical thresholds for the occurrence of different types of trauma.

REFERENCES

1. Jennett, B., Epidemiology of head injury, *J. Neurol. Neurosurg. Psychiatry,* **60**, 1996, 362-369.
2. Ommaya, A.K., Thibault, L. and Bandak, F.A., Mechanisms of impact head injur', *Int. J. Impact Eng.*, **15**, 1994, 535-560.
3. Watson, W.L. and Ozanne-Smith, J., Injury surveillance in Victoria, Australia: Developing comprehensive injury estimates, *Accid. Anal. Prev.,* **32,** 2000, 277-286.
4. Scallan, E., Staines, A., Fitzpatrick, P., Laffoy, M. and Kelly, A., Injury in Ireland, *Report of the Department of Public Health Medicine and Epidemiology*, University College Dublin, 2001.
5. Horgan, T. and Gilchrist, M., The creation of three-dimensional finite element models for simulating head impact biomechanics, *Int. J. Crashworthiness,* **8**(4)**,** 2003, 353-366.
6. MADYMO, Version 6.0, TNO Automotive, The Hague, NL, 1999.
7. O'Riordain, K., Reconstruction of real world head injury accidents resulting from falls using multibody dynamics modelling, MEngSc Thesis, Department of Mechanical Engineering, University College Dublin, 2002.
8. Yoganandan, N., Pintar, F.A., Sances Jr., A., Walsh, P.R., Ewing, C.L., Thomas, D.J. and Snyder, R.G., Biomechanics of skull fractures, *J. Neurotrauma*, **12,** 1995, 659-668.
9. Auer, C., Schonpflug, M., Beier, G. and Eisenmenger, W., An analysis of brain injuries in real world pedestrian traffic accidents by computer simulation reconstruction, in *Proc. Int. Soc. Biomechanics XVIIIth Congress*, Zurich, 2001.
10. Mohan, D., Bowman, B.M., Snyder, R.G. and Foust, D.R., A biomechanical analysis of head impact injuries to children, *J. Biomech. Eng.,* **101,** 1979, 250-260.
11. Allsop, D.L., Perl, T.R. and Warner, C.Y., Force/deflection and fracture characteristics of the temporo-parietal region of the human head, in *Proc. 35th Stapp Car Crash Conf.*, 1991, 269-278.
12. Blumbergs, P.C., Pathology, in *Head Injury*, Reilly, P. and Bullock, R. (eds), Chapman & Hall, London, 1997.
13. Gennarelli, T.A. and Thibault, L.E., Biomechanics of acute subdural haematoma, *J. Trauma*, **22,** 1982, 680-686.

COMBINED MULTI-BODY DYNAMIC AND FE MODELS OF HUMAN HEAD AND NECK

Volkan Esat, David W. van Lopik and Memis Acar
Mechanical and Manufacturing Engineering, Loughborough University LE11 3TU, UK
m.acar@lboro.ac.uk

Abstract. A complete three-dimensional multi-body dynamic computational model of the human head and neck has been developed and validated using human volunteer experimental data. The complete head-neck model has been used to simulate 15g frontal and 8.5g rear-end impacts with the resulting motion compared against response corridors derived from sled acceleration tests using human volunteers. This paper reports an original work, a further development of the model that incorporates a finite element analysis of the intervertebral discs subjected to the loading conditions determined by the multi-body dynamic model of the head and neck complex.

Key words: cervical spine, multi-body model, Finite Element Method (FEM), frontal impact, rear-end impact.

1. INTRODUCTION

Recreational activities, degenerative diseases and vehicle accidents are the main causes for cervical spine disorders such as whiplash, which poses a threat for public health, resulting in a huge economic burden of medical and insurance costs and loss of work force. The cervical spine, therefore gathers utmost attention in bioengineering discipline, not only to investigate the head-and-neck to determine the biomechanical limits of its components for a better evaluation of the injury risk, but also to have an insight for the common injuries it is subjected to.

The cervical spine is a structure, composed of several components with varying anatomical and mechanical features, such as the cervical vertebrae, the spinal cord, ligaments, muscles, and the intervertebral discs. Whilst *in*

Michael D. Gilchrist (ed.), IUTAM Proceedings on Impact Biomechanics: From Fundamental Insights to Applications, 91–100.

vitro and *in vivo* clinical studies continue to provide vital but quite limited information, computational techniques such as multi-body and finite element methods are widely used to model and simulate the cervical spine in order to have a better understanding of its kinetics, kinematics, and clinical aspects. An interesting series of studies [**1-6**] charting the development of a three-dimensional multi-body model of the head and neck has shown that discrete parameter models are capable of describing global motions of the head and neck and also the local kinematics of the individual vertebrae. FE models [**7-15**] are more commonly used in the investigations of the mechanics of the cervical spine subjected to impact loading.

Whilst the multi-body models treat each element as a rigid, non-deformable body, limiting their capability to investigate stresses, FE models tend to be more complicated and very demanding on computational time. This study, therefore, proposes a novel approach, where multi-body model of the cervical spine is used in conjunction with the FE technique, combining the best features of both techniques, creating more realistic simulations and reducing the computational time significantly. In that the multi-body method is used to investigate the kinetics and kinematics of the cervical spine under specific loading scenarios, determining loads on spinal elements, which in turn is used in the FE model to determine the stress conditions of the intervertebral discs.

2. METHOD

In this study, two different computational methods in the form of a detailed biofidelic multi-body model of the cervical spine and an FE model of the cervical spine intervertebral discs are used in conjunction with each other in order to investigate the effects of two severe impact loadings on the neck, namely 15g frontal and 8.5g rear-end impacts. The loading conditions occurring at each intervertebral disc are gathered from the analyses of the multi-body model and used as dynamic loading boundary conditions for the FE model of the discs. This novel approach not only provides a detailed loading history of the impact on each disc but also provides information on how the disc is affected during the loading, yielding the exact time and location of the high stresses and strains, which may potentially be a hazard for the healthy disc.

2.1 Multi-Body Model Development and Validation

The multi-body model of the head and cervical spine provides a biofidelic response of an occupant's head and neck during a motor vehicle accident, while keeping simulation time low compared to a finite element model of the head-neck system. The model reproduces the head and neck of an adult in an upright sitting position with the arrangement of the cervical vertebrae representing the natural lordosis of the neck with mid-sagittal symmetry assumed [**16**]. The three-dimensional geometry of the vertebrae and skull were defined using Solid-Edge, version 9, CAD software from EDS (Electronic Data Systems Corporation). The body geometries were then imported into, and the model constructed, using rigid-body dynamics package visualNastran 4D 2001 R2, version 6.4, from MSC Software. The model comprises nine rigid bodies with detailed geometry representing the head, seven cervical vertebrae, and the first thoracic vertebrae. The depiction of the upper cervical vertebrae (C1 and C2) differs from the middle and lower bodies. The dens process of the axis is included as well as the concave-convex interaction of the atlanto-odontoid and atlanto-occipital joints. The inertial properties of the neck are lumped into the rigid vertebrae and represent the inertial characteristics of a slice through the neck at each vertebral level containing all surrounding soft tissues. The properties used are those derived by de Jager [**4**]. The rigid bodies are interconnected by spring and damper constraints representing the soft-tissues of the neck. (Non)-linear viscoelastic 'bushing' constraints connect adjacent vertebrae representing the cervical intervertebral discs, non-linear viscoelastic spring-damper elements are used to describe the cervical ligaments and frictionless rigid-body contact idealizes facet joint behaviour. The material properties of the various soft tissues are based on the most recent experimental results reported in the literature [**14, 17, 18**] and on decisions made by other researches [**4, 6**]. 19 muscle groups of the head and neck are incorporated in the model. Muscles with broad areas of attachment are subdivided into a number of individual muscle elements resulting in 138 individual muscle segments. Each muscle element is represented by a series of connected actuators spanning origin to insertion allowing the muscles to curve around the vertebrae during neck bending. Muscle mechanics is handled by an external application called Virtual Muscle vs. 3.1.5, developed at the Alfred E. Mann Institute at the University of Southern California that runs within Matlab and Simulink, providing both passive and active muscle behaviour [**19**]. Muscle morphometry and fibre type composition of the various muscles of the neck are based on values reported in the literature [**20-24**] and on choices made by other researches [**5, 6, 25**]. Virtual Muscle calculates the force generated by each muscle at each integration step of Visual Nastran based on the musculotendon path length (outputted from the muscle elements of the model at each time-step) and activation level of the muscle. The resulting muscle force is applied at both the origin and insertion of the muscle element in the tangential direction to the muscle curvature. Figure 1 shows the final neck model from occiput to T1 with all the soft tissue elements and musculature attached.

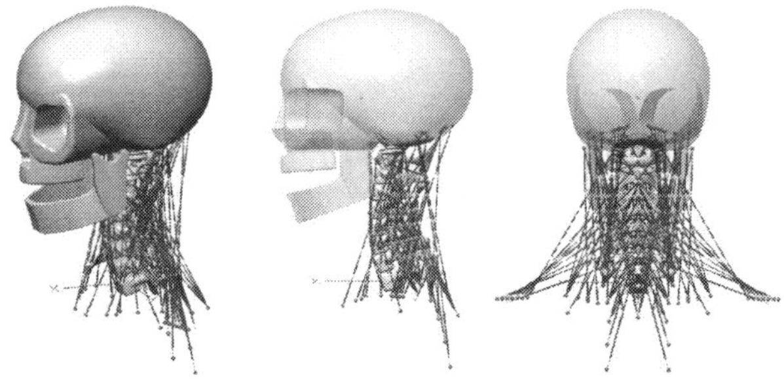

Figure 1. The multi-body model of the head and neck.

The head-neck model has been evaluated to check the accuracy of the individual components, motion segments and the model as a whole in response to different loading conditions. The completed model has been rigorously validated against experimental results, ranging from the individual motion segment response to the dynamic response of the whole head neck model to frontal, lateral and rear end impacts. The response of the motions segments to small and large static loading was found to be in good agreement with experimental results in all directions of loading. The coupling characteristics of the cervical spine were shown to be accurately reproduced and the moment generating capacity of the muscle elements was found to be realistic. The model has been used to simulate the frontal and lateral sled acceleration tests performed at the Naval Biodynamics Laboratory (NBDL) using human volunteers. Response corridors based on these sled tests have been used to evaluate the model and investigate the effect of muscle activation on the head-neck motion. These corridors have also been used by other researchers to validate mathematical and mechanical models of the head and neck [**4-6**, THOR, 2001]. The corridors specify the response that a valid model of the human head and neck should meet. The effect of passive and fully active muscle behaviour was investigated and was shown that for both impact directions the inclusion of active muscle tensioning results in closest agreement with the experimental data. Good agreement was seen for both impact directions [**26**]. The model has also been implemented without musculature to simulate bench-top trauma experiments using cadaveric isolated cervical spine specimens. The model successfully reproduced the characteristic 'whiplash' motion and resulting head and vertebral rotations and displacements seen in the experimental results for rear impact accelerations. Following this the same experiments have been simulated with the inclusion of active and passive musculature and their effects on the forces in the neck studied [**27**].

2.2 Finite Element Model Development and Validation

In order to investigate the effects of the dynamic impact loading on the discs, a 3-D biofidelic FE model of the 6 discs (C23, C34, C45, C56, C67, and C7T1) in the cervical spine has been developed by using FEA software, MSC.Marc/Mentat (Figure 2). The dimensions, positions and the orientations of the discs were taken from the quantitative anatomy of the cervical spine [**28, 29**]. Intervertebral discs were modelled as 8 node brick elements, the material properties of which were adapted from literature [**9, 14, 30**] (Table 1). Each disc model comprises 1815 elements and 938 nodes. The intervertebral disc is mainly composed of two parts: nucleus and annulus fibrosus. In the FE model, all annulus fibrosus bands were modelled with a fibre orientation of 30° with respect to the lateral plane. The fibres are in opposite alternating directions in two adjacent bands.

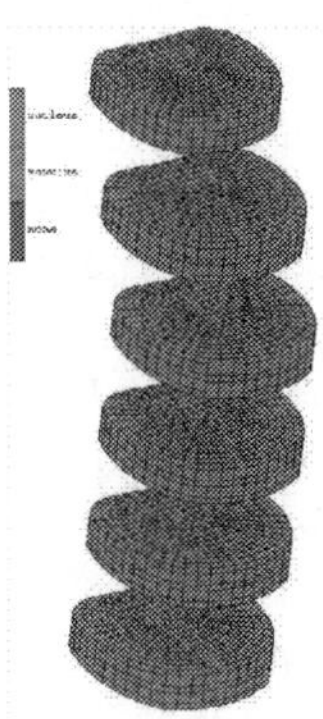

Figure 2. The FE model of the discs.

Table 1. Material properties of the intervertebral discs.

Component	Young's Modulus, E (MPa)	Poisson's Ratio, ν
Cortical Shell	12000	0.30
Cancellous Core	100	0.20
Nucleus Pulposus	3.4	0.49
Annulus Matrix	4.2	0.45
Annulus Fibers	450	0.30
Endplate	600	0.30

The FE model was validated against published experimental measurements, in compression and flexion/extension only, as the multi-body model yields results in sagittal plane. Firstly, C4-C6 model was built in order to comply with the experimental set-up. Therefore, for this purpose only, vertebral bodies were built with the endplates, which surround the discs C45 and C56. Then, the model was subjected to two different loadings: (a) 1 mm axial compression and (b) 1600 Nmm flexion and extension together with a 73.6 N axial compressive preload. The results from the 1 mm axial compression loading are given in Figure 3. For the latter loading case, the FE model yielded 6.23° for flexion and 6.60° for extension moments, while the experimental results show 7.02° and 4.80° with 2.23° and 1.41° standard deviations, respectively, resulting in a good agreement with the experimental data.

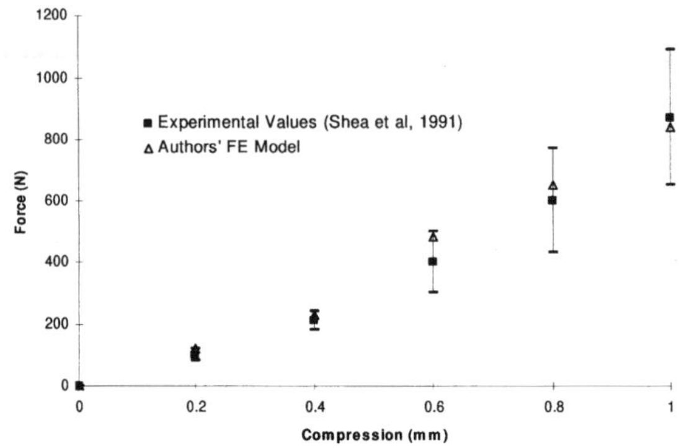

Figure 3. Validation of the FE disc.

3. RESULTS

The multi-body head and neck model was used to simulate 15g frontal and 8.5g rear-end impacts with the resulting motion compared against response corridors derived from sled acceleration tests using human volunteers [**31**]. The intervertebral disc loads from the 15g frontal and 8.5g rear-end impact simulations for the first 200 ms period (Figure 4) are used as force boundary conditions for the FE model of the discs. The disc forces F_x and F_z are shown on the left axis and moment M_y on the right axis. The predicted results of maximum von Mises stresses in the annulus and the intradiscal pressure in the nucleus of each disc of the FE model are depicted in Figure 5. The intradiscal pressures occurring in the nuclei of the discs have almost the same pattern with the von Mises stress distributions with respect to time increments as in Figure 5, but possessing less magnitudes; a max of 0.5 MPa for the frontal and a max of 0.1 MPa for the rear-end case.

The von Mises stress distribution in the C56 intervertebral disc for the last stage of frontal impact at 200 ms can be seen in Figure 6.

4. DISCUSSION & CONCLUSION

Due to the nature of a direct frontal impact no forces were developed in lateral shear and the moments were experienced about the y-axis only. A peak in anterior shear and tension of the discs at all levels can be seen at around 100ms at maximum neck excursion. In the early stages of impact, anterior shear is the dominant force of the C7-T1 disc. Following maximum neck excursion, compression in the discs develops reaching a maximum at all levels at around 180-190ms. Torque in the discs increases with flexion of the vertebrae and reaches a maximum at all levels at around 160 ms in conjunction with maximum neck rotation.

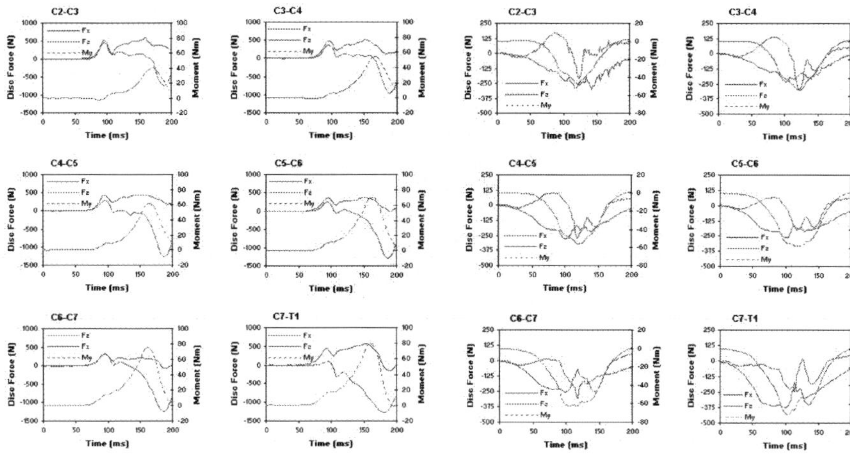

(a) (b)

Figure 4. Intervertebral disc loadings: (a) frontal impact, and (b) rear-end impact.

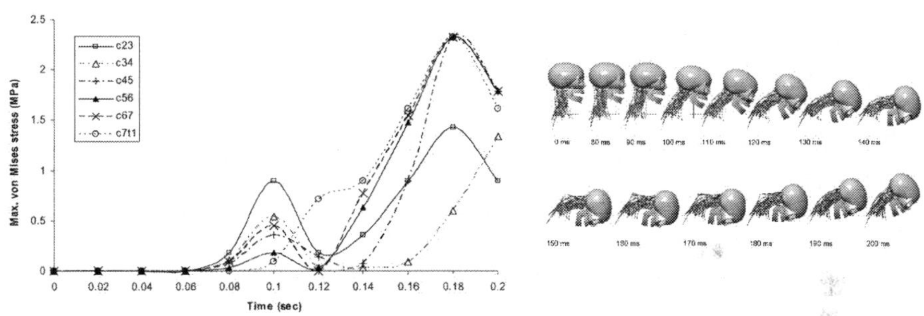

(a)

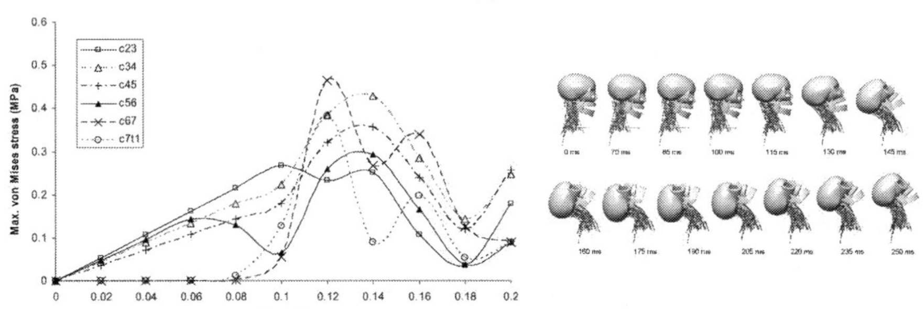

(b)

Figure 5. FE results: (a) Max. von Mises stresses in the annulus for frontal impact with multi-body head and neck response illustration, and (b) max. von Mises stresses in the annulus for rear-end impact with multi-body head and neck response illustration

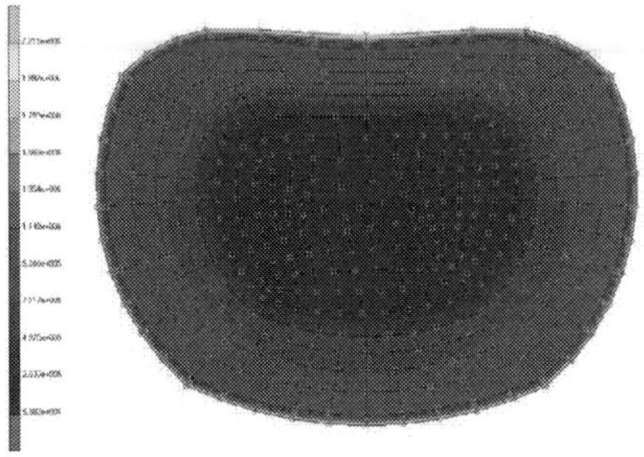

Figure 6. Von Mises stress distribution of C56 disc at 200 ms for frontal impact.

In rear-end impact, as was the case with the frontal impact, no forces were developed in lateral shear and moments were experienced about the y-axis only. The discs experience increasing levels of posterior shear up to around 100ms followed by a sharp decrease after the T1 acceleration ends and the facet joints become loaded as the head and neck start to return towards their initial position. Axially the discs experience a small compressive force at the start of the acceleration pulse before being pulled in tension as the head and neck extend, however the C7-T1 disc remains in compression for the first 100ms. Peak tensile force is reached just prior to the end of the acceleration pulse following which the discs are compressed as the head continues to displace vertically.

The results from the FE models show that the responses of the annulus and the nucleus are similar but different in magnitude due to different material properties. The annulus possesses much higher stresses when compared to the nucleus in both loading cases. In frontal impact case, the stresses reach a peak between 80 ms and 120 ms and the highest peak at about 180 ms. This is due to the high concentration of the loads, especially axial force and the moment, at these intervals. In rear impact case, the peaks occur about 120-140 ms, where the head and neck almost reach their most extended posture.

This study shows that the proposed novel approach that combines the multi-body and FE models have the potential to provide a powerful, cost-effective and versatile platform to investigate the kinetics and the kinematics of the whole cervical spine and its components and the mechanical response of the intervertebral discs under complex dynamic loading histories.

ACKNOWLEDGMENTS

Authors gratefully acknowledge the research studentships from the UK Engineering and Physical Sciences Research Council and Wolfson School of Mechanical and Manufacturing Engineering, Loughborough University, UK.

REFERENCES

1. Reber, J.G. and Goldsmith, W., Analysis of large head-neck motions, *Journal of Biomechanics*, **12**, 1979, 211-222.
2. Merrill, T., Goldsmith, W. and Deng, Y-C., Three dimensional response of a lumped parameter head–neck model due to impact and impulsive loading, *Journal of Biomechanics*, **17(2)**, 1984, 81-95.
3. Deng, Y.C. and Goldsmith, W., Response of a human head/neck/upper-torso replica to dynamic loading-II. Analytical/numerical model, *Journal of Biomechanics*, **20(5)**, 1987, 471-486.
4. De Jager, M.K.J., *Mathematical Head-Neck Models for Acceleration Impacts*, PhD Thesis, Technical University of Eindhoven, 1996.
5. Van der Horst, M.J., Thunnsissen, J.G.M., Happee, R., Van Haaster, R.M.H.P. and Wismans, J.S.H.M., The influence of muscle activity on the head-neck response during impact, *SAE Conference Proceedings,* **315**, 1997, pp. 487-508.
6. Van der Horst, M.J., *Human Head Neck Response in Frontal, Lateral and Rear End Impact Loading - modelling and validation*, PhD Thesis, Technical University of Eindhoven, 2002.
7. Dauvilliers, F., Bendjellal, F., Weiss, M., Lavaste, F. and Tarriere, C., Development of a finite element model of the neck, in *Proceedings of the 38th Stapp Car Crash Conference*, Society of Automotive Engineers, SAE Paper No. 942210, 1994, pp. 77-91.
8. Nitsche, S., Krabbel, G., Appel, H. and Haug, E., Validation of a finite-element-model of the human neck, in *International Conference of the Biomechanics of Impact*, Dublin, Sept. 1996, pp.107-122.
9. Yang, K.H., Zhu, F., Luan, F., Zhao, L. and Begeman, P.C., Development of a Finite Element Model of the Human Neck, in *Proceedings of the 42nd Stapp Car Crash Conference*, Society of Automotive Engineers, SAE Paper No. 983157, 1998, pp. 195-205.
10. Kumaresan, S., Yoganandan, N., Pintar, F.A. and Maiman, D.J., Finite element modeling of the cervical spine: Role of intervertebral disc under axial and eccentric loads, *Medical Engineering & Physics*, **21**, 1999, 689-700.
11. Kumaresan, S., Yoganandan, N. and Pintar, F.A., Finite element analysis of the cervical spine: A material property sensitivity study, *Clinical Biomechanics*, **14**, 1999, 41-53.
12. Kumaresan, S., Yoganandan, N., Pintar, F.A., Maiman, D.J. and Goel, V.K., Contribution of disc degeneration to osteophyte formation in the cervical spine: A biomechanical investigation, *Journal of Orthopaedic Research*, **19**, 2001, 977-984.
13. Ng, H.W., Teo, E.C. and Lee, V.S., Statistical factorial analysis on the material property sensitivity of the mechanical responses of the C4-C6 under compression, anterior and posterior shear, *Journal of Biomechanics*, **37**, 2004, 771-777.
14. Yoganandan, N., Kumaresan, S. and Pintar, F.A., Biomechanics of the Cervical Spine Part 2: Cervical spine soft tissue responses and biomechanical modelling, *Clinical Biomechanics*, **16**, 2001, 1-27.

15. Schmitt, K-U, Muser, M.H., Walz, F.H. and Niederer, P.F., Fluid-structure interaction in the biomechanics of soft tissue neck injuries: A finite element study to analyze blood vessel pressure effects in the venous plexus, *Traffic Injury Prevention*, **3**, 2002, 65-73.
16. Van Lopik, D.W., *A Computational Model of the Human Head and Cervical Spine for Dynamic Impact Simulation*, PhD Thesis, Loughborough University, UK, 2004.
17. Camacho, D.L., Nightingale, R.W., Robinette, J.J., Vanguri S.K., Coates D.J. and Myers B.S., Experimental flexibility measurements for the development of a computational head-neck model validated for near-vertex head impact, in *Proceedings of the 41st Stapp Car Crash Conference*, Society of Automotive Engineers, SAE Paper No. 973345, 1997, pp. 473-486.
18. Yoganandan, N., Kumaresan, S. and Pintar, F.A., Geometric and mechanical properties of human cervical spine ligaments, *Journal of Biomechanical Engineering*, **122**, 2000, 623-629.
19. Cheng, E.J., Brown, I.E. and Loeb, G.E., Virtual muscle: A computational approach to understanding the effects of muscle properties on motor control, *Journal of Neuroscience Methods*, **101**, 2000, 117-130.
20. Winters, J.M. and Woo, S.L-Y., *Multiple Muscle Systems: Biomechanics and Movement Organization*, Springer-Verlag, 1990.
21. Johnson, G., Bogduk, N., Nowitzke, A. and House, D., Anatomy and actions of the trapezius muscle, *Clinical Biomechanics*, **9**, 1994, 44-50.
22. Kamibayashi L.K. and Richmond F.J.R., Morphometry of human neck muscles, *Spine*, **23(12)**, 1998, 1314-1323.
23. Richmond, F.J.R., Singh, K. and Corneil, B.D., Neck muscles in the rhesus monkey I. Muscle morphometry and histochemistry, *Journal of Neuroscience,* **86**, 2001, 1717-1728.
24. Boyd-Clarke, L.C., Briggs, C.A. and Galea, M.P., Comparative histochemical composition of muscle fibres in a pre- and a postvertebral muscle of the cervical spine, *Journal of Anatomy*, **199**, 2001, 709-716.
25. Vasavada, A.N., Li, S. and Delp, S.L., Influence of muscle morphometry and moment arms on the moment-generating capacity of human neck muscles, *Spine*, **23 (4)**, 1998, 412-422.
26. Van Lopik, D.W. and Acar, M., A computational model of the human head and neck for frontal and lateral impacts, in *ESDA 2004, 8th Biennial Conference on Engineering Systems and Design Analysis,* 2004.
27. Van Lopik, D.W. and Acar, M., A computational model of the human head and neck system for the analysis of whiplash motion, *International Journal of Crashworthiness*, **9(5)**, 2004, 465-473.
28. Panjabi, M.M., Duranceau, J., Goel, V., Oxland, T. and Takata, K., Cervical human vertebrae. Quantitative three-dimensional anatomy of the middle and lower regions, *Spine*, **16(8)**, 1992, 861-869.
29. Nissan, M. and Gilad, I., The cervical and lumbar vertebrae-and anthropometric model, *Engineering in Medicine*, **13(3)**, 1984, 111-114.
30. Teo, E.C. and Ng, H.W., Evaluation of the role of ligaments, facets and disc nucleus in lower cervical spine under compression and sagittal moments using finite element method, *Medical Engineering & Physics*, **23**, 2001, 155-164.
31. Wismans, J., van Oorshot, E. and Woltring, H.J., Omni-directional human head-neck response, in *30th Stapp Car Crash Proceedings*, Society of Automotive Engineers, SAE Paper No. 861893, 1986, pp. 313-331.

THREE YEARS OLD CHILD NECK FINITE ELEMENT MODELISATION

Frontal Validation

Raphaël Dupuis, Frank Meyer and Rémy Willinger
Université Louis Pasteur, Institut de Mécanique des Fluides et des Solides,
2, rue Boussingault, 67000 Strasbourg, France
Phone: (33).3.90.24.29.45, Fax: (33).3.88.63.41.00, Email: dupuis@imfs.u-strasbg.fr

Abstract. Despite of recent progresses in occupant safety, the protection of children is still not optimal. To offer a better comprehension of child injury mechanisms, we developed a human-like finite element of a three years old child's neck. The subject was scanned with a medical scanner. The images were first semi-automatically segmented in order to extract the soft tissues and the bones. In the second step, we separate the different bones slice by slice on the geometry previously reconstructed. The anatomic structures are identified and each vertebra is reconstructed independently with special attention for the articular process. In a second step, we have generated an original meshing on the previous geometry to obtain a finite element model of the child's neck. The anatomical structures incorporated are the head, the seven cervical vertebrae (C1–C7), the first thoracic vertebra (T1), the intervertebral discs and the principle ligaments which are modelled using non-linear shock-absorbing spring elements. The stiffness values used are taken from literature, and scaled down using scale factors from Irwin (1997). This model incorporates 7340 shell elements to model the eight vertebrae, the head and 1068 solid 8-node elements to model the intervertebral discs. Contact between the articular surfaces is represented by interfaces permitting frictionless movement. Since this study does not aim to reproduce bone fractures, we have modelled the cervical vertebrae as rigid bodies.
Given that validation data were not available, we realized some Q3 dummy component sled tests. The accelerometric responses of the head model were similar to those recorded experimentally with a Q3 dummy neck in frontal impact direction.

Key words: children, Finite Element Method (FEM), neck, crash.

Michael D. Gilchrist (ed.), IUTAM Proceedings on Impact Biomechanics: From Fundamental Insights to Applications, 101–109.

1. INTRODUCTION

Each year, more than 700 children are killed on European roads and 80,000 are injured [1]. The EC project CHILD (Child Injury Led Design) aims to improve the protection offered to children in cars by increasing the understanding about the injuries sustained and providing innovative tools and methods for improvement of Child Restraint Systems (CRS) in cars.

One of the tools developed is a three year old child head and neck finite element model. While some models are existing in the literature like Van Ratingen's [2] or Yoganandan's [3] model, they differ largely in term of purpose and methodology.

1.1 Multi-body Child Neck Model

Child multi-body neck finite element models are mainly models developed under MADYMO. Thus, TNO developed 3, 4, 5, 6, 8, 10 and 12 year old child models for use in automotive crash test reconstruction. The models are carried out by the assembly of cylinders, ellipsoids, parallelepipeds connected to each other by joints with one or more degrees of freedom and different stiffness according to mobility.

The three year old child MADYMO model is the most recent child model. It was developed by TNO in parallel of the Q3 dummy that was validated against National BioDynamic Laboratory (NBDL) tests [4] after scaling [2, 5]. Its validation was conducted within the framework of the EC CREST project. The Q3 model is directly issued from the dummy CAD. The head/neck elements were similar to those of the dummy. The validation was carried out by reproducing tests on the model similarly as previously realized on Q3 in frontal, rearward, lateral direction and of pendulum test. The stiffness and the damping coefficients of the various articulations were then tuned to adjust the dummy response.

1.2 Detailed Finite Element Models

Only two child human like cervical spine models were found in the literature.

The first was that developed by Kumaresan and Yonganandan [3]. They developed three finite element models for three different ages: 1, 3 and 6 years. These models were limited to the cervical C4-C6 segment and resulted directly from the adult model [6]. It has to be noticed that this adult model was developed in order to realize static simulations. Three types of model construction were adopted: first a pure geometric scaling, then the introduction of anatomic specificities without any scaling, and finally a method where the two preceding approaches were combined. While no experimentation on children is available to validate these models, the tendencies observed seemed to be in conformity with the experimental results on animals [7].

The other three year old child finite element model found was that developed by Mizuno [8] by scaling from a Total Human Model for Safety (THUMS) AM50human finite element model to investigate the potential injury risks from restraints. The geometric scaling factors were chosen so that λ_x, λ_y, λ_z have values as similar as possible, and the material properties scaling factors were determined in the literature [9, 10]. The model has been validated for thorax impact according to Hybrid III 3YO dummy requirements. No information was available on neck validation.

2. MATERIAL AND METHODS

2.1 Geometrical Reference

A three year old male child was scanned with an ELSCINT Helix 3.0 (Elscint Ltd., Ma'alot, Israel) scanner, in order to realize a medical exam. The slice thickness was 1.1 mm with a table feed of 1 mm (pitch 0.9). After ensuring that no abnormality was detected, and after depersonalising the exam, the images were semi-automatically segmented in order to extract skin and bones. This stage was conducted at IRCAD from software developed in partnership with the Epidaure project of INRIA for the automatic 3D patient reconstruction [11, 12], and provided a rapid and precise result [13] (see Figure 1).

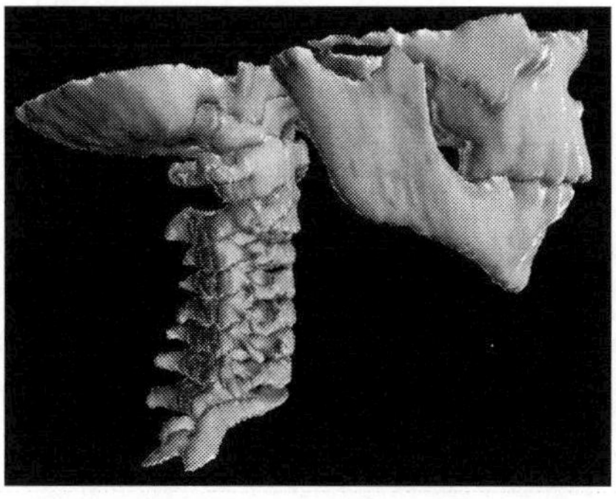

Figure 1. Complete reconstruction of the cervical spine of a three year old child.

2.2 Finite Element Modeling

In order to stick to our geometrical reference, we deformed and remeshed the geometrical meshing of an existing adult model [14]. Each finite element vertebra was superposed on the 3D reconstruction and the different area of the meshing was deformed by proportional transformation in order to fit to the main vertebral dimensions. We were then obliged to remesh the vertebral body and the articular apophyses that are not proportional to the adult's. The anatomical structures incorporated and illustrated in Figures 2 and 3 are the head, the seven cervical vertebrae (C1–C7), the first thoracic vertebra (T1), the intervertebral discs and the principle ligaments, including the anterior longitudinal ligament (ALL), anterior-atlanto occipital membrane (AA-OM), posterior-atlanto occipital membrane (PA-OM), techtorial membrane (TM), posterior longitudinal ligament (PLL), flavum ligament (LF), supraspinous ligament, interspinous ligament (ISL), transverse ligament (TL), alar ligament (AL), capsular ligaments (CL) and the apical ligament (APL).

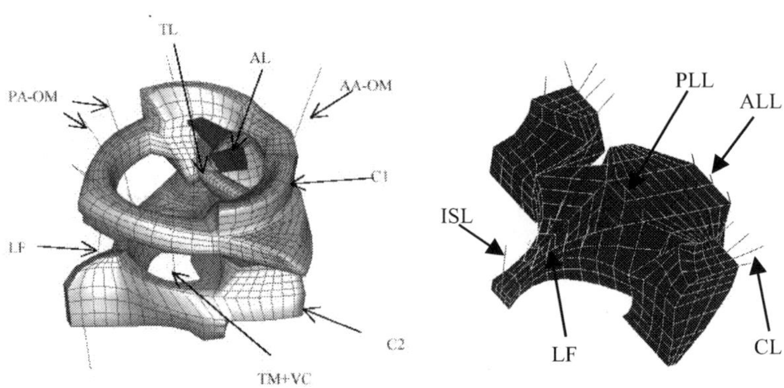

Figure 2. Ligamentary system of the upper cervical spine (C1-C2) and of the lowercervical spine (C3).

These are modeled using non-linear shock-absorbing spring elements. The behaviour laws of each ligament in both the lower and upper cervical spines, are defined by referencing to three complementary studies: Myklebust [15], Chazal et al. [16] and Yoganandan et al. [17]. The Chazal et al. study [16] highlights the non-linear viscoelastic behavior of ligaments whereas Yoganandan et al. [17] gives information on their failure properties.

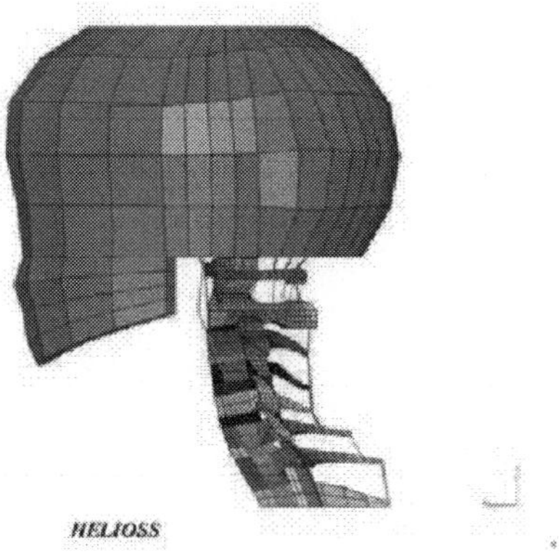

Figure 3. Complete finite element model of the head and neck complex of a three year old child.

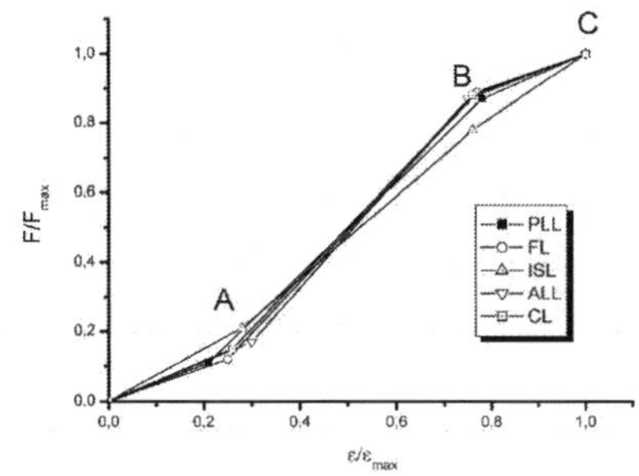

Figure 4. Behaviour laws of the anterior longitudinal ligament (ALL C2-C5), posterior longitudinal ligament (PLL C2-C5), flaval ligament (FL C2-C5), interspinous ligament (ISL C2-C5), capsular ligament (CL C2-C5).

The total height of the model is 17.3 cm and its weight is 4.57 kg. This model incorporates 7340 shell elements to model the eight vertebrae, the head and 1068 solid 8-node elements to model the intervertebral discs. Contact between the articular surfaces is represented by interfaces permitting frictionless movement. Since this study does not aim to reproduce bone fractures, we have modelled the cervical vertebrae as rigid bodies, taking their inertial moments and masses from Deng [18] and scaled down using scale factors from Irwin [9].

Most models use an elastic law for the intervertebral discs and a wide range of Young's modulus values has been observed, varying from 3.4 MPa in Yoganandan's [17] model to 4.3 MPa for that of Golinski [19] and 200 MPa for that of Dauvilliers [20]. A scalling factor of 0.705 given by Yoganandan [21] for the 3 year old child intervertebral disc is supposed, this values conduced to adopt a disc modulus of the order of 100 MPa.

2.3 Model Validation

Given that validation data in term of acceleration were not available and that sled tests were only realized on full complete dummy, we were obliged to realize some Q3 dummy component sled tests (see Figure 5).

Figure 5. Q3 dummy component sled test on head and neck.

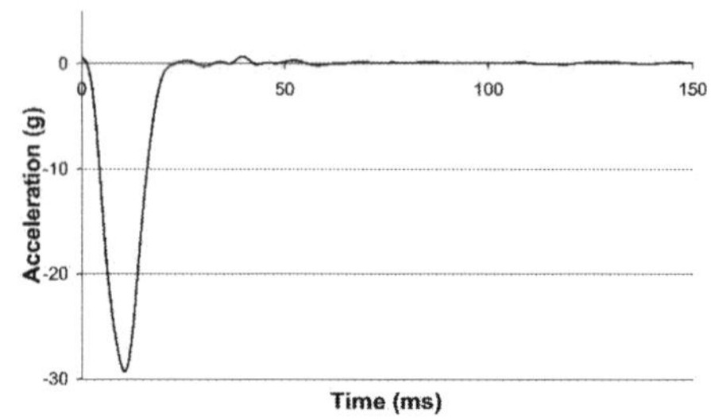

Figure 6. Sled acceleration in frontal impact.

Therefore, the base of the Q3 dummy neck was fixed on the sled. A set of three accelerometers was attached to the dummy head to measure linear acceleration. The sled is accelerated in frontal direction. In order to reproduce the experimentation with the numerical model, the model was controlled in terms of first thoracic vertebra speed (see Figure 6). Head linear acceleration values were computed and compared with those recorded experimentally.

3. RESULTS

The parameters of the model has been tuned in order to fit to the experimental results as shown in Figures 7 and 8 for frontal impact. The main parameters tuned were the ligament damping on which no literature reference is available, and on the vertebral inertia.

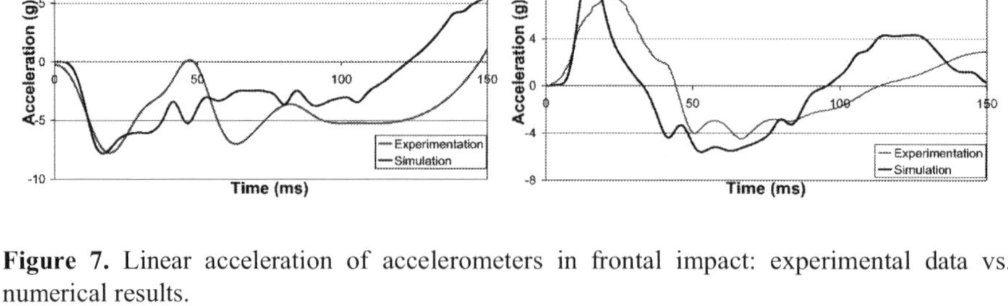

Figure 7. Linear acceleration of accelerometers in frontal impact: experimental data vs. numerical results.

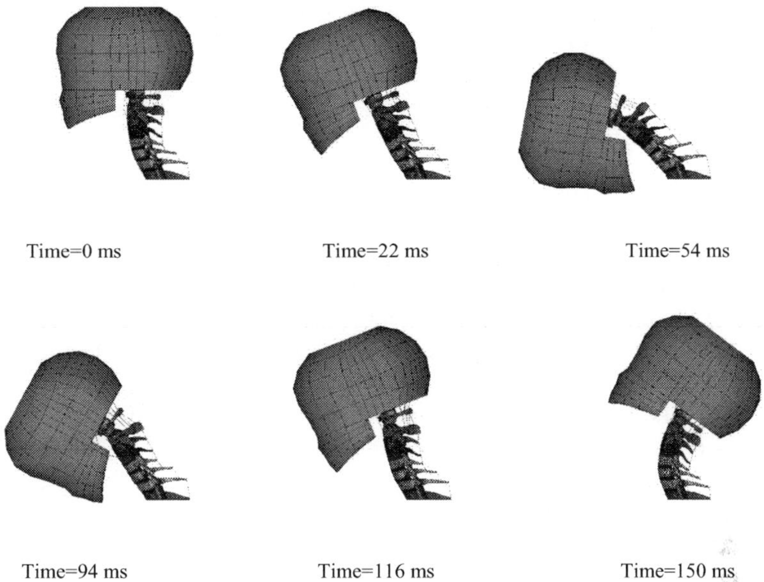

Figure 8. Model configurations in frontal impact.

4. DISCUSSION

The main discussion concerning the method is the validation of the model with regards to a dummy. In fact, no in-vitro or in-vivo experimentation on child's neck has even been conducted. The data usable for infant finite element model validation are never "first hand" data, but only obtained by scaling adult results. The scaling coefficient of mechanical properties is based on three in-vitro tests (2 newborns and a 6 year old child) on parietal bone. With the inter-individual variation on human mechanical properties, we can wonder about the credibility of these scaling factors. That is why we decided to use the Q3 dummy as reference, because it proved its reliability in accident reconstructions that offer good correlation with injuries (EC CREST and CHILD program). We can notice that the methodology used by Mizuno [8] to validate their three year old child model is the same, even if it was on the torso.

The choice of the geometrical reference to realize our finite element model meshing can also be discussed. The medical scanner was realized on a three year old child, and because of anonymisation procedure, we didn't have information about its corpulence (weight, height...). However, it appeared that it was a better solution to remesh a geometrical reference, than to apply a pure scaling on an adult model, moreover with similar scaling factor in the three directions as made Mizuno [8].

5. CONCLUSION

A three year old child human like neck finite element model was developed, based on a three year old child medical scanner. The model includes intervertebral discs and almost all intervertebral ligaments. It was compared with Q3 dummy neck that was validated with regards to scaled NDBL corridors. The three year old child neck finite element model validation was performed in frontal impact. The model will be used for accident reconstruction in order to evaluate local injuries and to provide a basis for injury criteria.

ACKNOWLEDGEMENTS

The authors would like to acknowledge the efforts of INRETS for their contribution in the validation of the three year old child neck finite element model.

The development and the validation of the model were funded by the European Community under the G3RD-CT-2002-00791 CHILD project.

REFERENCES

1. Jager, K. de, Waagmeester, K. and Ratingen, M.R. van, 2004, Development and validation of a newborn child dummy: Q0, Proc. JSAE.
2. Ratingen, M.R. van, Twisk, D., Schrooten, M., Beusenberg, M.C., Barnes, A., and Platten, G., 1997, Biomechanically based design and performance targets for a 3-year old child crash dummy for frontal and side impact, Proc. 26th STAPP Car Crash Conf., 243–260.
3. Kumaresan, S., Yoganandan, N. and Pintar, F.A., 1997, Age specific pediatric cervical spine biomechanical responses: three dimensional nonlinear finite element models, Proc. 26th STAPP Car Crash Conf., 31–61.
4. Mertz, H.J. and Patrick, L.M., 1971, Strength and response of the human neck, SAE710955.
5. Lange, R. de, Made, R. van der, Feustel, J.R., Subbian, T. and Hoof, J. van, 2001, Development and evaluation of MADYMO child occupant dummy models, 4th North American MADYMO User's Meeting.
6. Yoganandan, N., Kumaresan, S., Voo, L. and Pintar, F.A., 1997, Finite element model of the human lower cervical spine, Journal of Biomechanical Engineering, 119(1): 87–92.

7. Mertz, H.J., Driscoll G.D., Lenox, J.B., Nyquist, G.W. and Weber D.A., 1982, Responses of animals exposed to deployment of various passenger inflatable restraint system concepts for a variety of collision severities and animal positions, Proc 9th International Technical Conference on Experimental Safety Vehicles, Kyoto, Japan.
8. Mizuno, K., Deguchi, T., Furukawa, K. and Miki, K., 2004, Development of three-year-old child human FE model, IRCOBI Conference, Graz, September, 335–336.
9. Irwin, A. and Mertz, H.J., 1997, Biomechanical basis for the CRABI and Hybrid III Child Dummies, Proc. 26th STAPP Car Crash Conf., 261–272.
10. Curry, J.D. and Butler, G., 1975, The mechanical properties of bone tissues in children, J. Bone and Joint Surg. 57A, 6: 810–814.
11. Ayache, N., 1995, Medical computer vision, virtual reality and robotics, Image and Vision Computing, 13(4): 295–313.
12. Cotin, S., Delingette, H. Bro-Nielsen, M., 1996, Geometric and physical representation for a simulator of hepatic surgery, in: Weghorst, S.J., Sieburg, H.B. and Morgan K.S. (eds), Health Care in the Information Age. Future Tools for Transforming Medicine. Medicine Meets Virtual Reality: 4, Studies of Health Technology and Informatics 29, Amsterdam/Oxford/Tokyo/Washington: IOS Press, 139–151.
13. Soler, L., Delingette, H., Malandain, G., et al., 2001, Fully automatic anatomical, pathological, and functional segmentation from CT scans for hepatic surgery, Computer Aided Surgery, 6(3): 131–142.
14. Meyer, F., Willinger, R. and Legall, F., 2004, The importance of modal validation for biomechanical models, demonstrated by application to the cervical spine, Finite Elements in Analysis and Design, 40: 1835–1855.
15. Myklebust, J., Pintar, F., Yoganandan, N., Cusick, J., Maiman, D., Myers, T., and Sances, A., 1988, Tensile strength of spinal ligaments, Spine, 13(5): 526–531.
16. Chazal, J., Tanguy, A., Bourges, M., Gaurel, G., Escande, G., Guillot, M. and Vanneuville, G., 1985, Biomechanical properties of spinal ligaments and a histological study of the supraspinal ligament in traction, Journal of Biomechanics, 18(3): 167–176.
17. Yoganandan, N., Kumaresan, S. and Pintar, F., 2000, Biomechanics of the cervical spine Part 2. Cervical spine soft tissue responses and biomechanical modeling, Clinical Biomechanics, 16: 1–27.
18. Deng, Y.-C.; Goldsmith, W. 1987. "Response of a human head/neck/upper-torso replica to dynamic loading-II. Analytical/numerical model." J. Biomech. 20 (5): 487–497.
19. Golinski, W.Z. 2000 "3D-dynamic modelling of the human cervical spine in whiplash situations." Ph.D. Thesis, Notthingham University.
20. Dauvilliers, F. 1994. "Modelisation Tridimensionnelle et Dynamique du Rachis Cervical." Ph.D. Thesis, LBM ENSAM, Paris.
21. Yoganandan, N.; Pintar, F. A.; Kumaresan, S.; Gennarelli, T. A. 2000. "Pediatric and small female neck injuryscale factor and tolerance based on human spine biomechanical characteristic", IRCOBI Conference, Monpellier, September 21-23.

RESPONSE ANALYSIS OF LUMBAR VERTEBRA MODEL UNDER AXIAL IMPACT LOADING

J. Fang[1], J. Cao[1], J. Zhang[1], Q.G.Rong[1] and D.G. Zhou[2]
[1]*Department of Mechanics and Engineering Science, Peking University, Beijing 100871;* [2]*Department of Orthopaedics, People's Hospital, Peking University, Beijing 100081, P.R. China*

Abstract. Both experimental analysis of dynamic photoelasticity and numerical simulations of finite element computation have been carried out to study the responses of the lower lumbar spine (L_3-L_5) model to the axial impact load. In the photoelastic tests, the impulsive forces were loaded on the anterior and middle columns of the spine, to study the stress wave propagations with different forms including the stress focusing in the vertebra bodies which might cause bone fracture. In the FEM computations, the responses of the intervertebral discs were studied associated with von Mises stress movement in the vertebras, to show the stress history in those bodies with different phases and magnitudes during the impact energy transferring.

Key words: lower lumbar spine, vertebral columns, intervertebral discs, impact loads, transient stress responses.

1. INTRODUCTION

Connected by the intervertebral discs as the tissues to absorb shock and stabilize the segments, the vertebral bodies in the lumbar spine support the human trunk and upper extremities to resist both the intrinsic weight and the external loads. Consisting of five vertebral bodies from cephalad to caudad, or numbered by L_1 to L_5 from the upper to lower, the vertebrae in the lumbar spine become larger and larger to carry the majority axial load (about 80%) by providing more surface areas and connecting with the sacrum as the support foundation. As the major elements of the posterior column of the

Michael D. Gilchrist (ed.), IUTAM Proceedings on Impact Biomechanics: From Fundamental Insights to Applications, 111–118.

spine, the facet joints not only enclose and protect the spinal cord and nerve roots but also sustain the remaining axial force, together with the tissues like ligaments to subject to torsion and shear to keep the spine stability.

Trauma to the spine may cause instability of the spine by injury to the bone elements and/or disruption of the ligamentous tissues, which implies capability reduction of the spinal structure to sustain normal loads. When the lumbar bodies are impulsively forced by accident fall or jumps, bone fracture may happen for the vertebral bodies under impact loads [1]. Relying on the loading position of the external forces, the injury may appear on the anterior, middle and posterior columns that were classified by Denis [2]. Even though some classifications of thoracolumbar fracture have been developed for static loads [3], investigations are needed to reveal the mechanisms to cause spinal failure under impact forces.

In this work, dynamic photoelastic tests of axial impact were performed on a model of L_3-L_4-L_5 vertebral bodies to study response behavior of the lower lumbar assembly under impulsive loads acted on different columns. The experimental purpose is to visualize the stress wave propagation in these vertebral bodies under various impact locations of the columns. Numerical analysis of finite element computation is then carried out to quantitatively study the histories of the transient stresses in the L_3-L_4-L_5 vertebral bodies and the deformation of the intervertebral discs in the energy transferring, to analyze their characteristic variations during the stress wave movement.

2. EXPERIMENT OF PHOTOELASTICITY

The bone models of L_3, L_4 and L_5 made of polycarbonate were connected by the intervertebral disc models with adhesion of silicon rubber. Soft rubber was used to simulate the contribution of the facet capsular elements to the force transferring. As the specimen was impulsively loaded by a falling weight, sixteen sparks of light flashes were sequentially initiated to offer strong light beams to illuminate the photoelastic specimen. During the impact, sixteen individual cameras were used to receive the corresponding rays from the circularly polarized light field, to record the isochromatic fringe patterns with time interval of 20 μs . The impact load was concentrated in the anterior and posterior parts of the anterior column and the anterior part of the middle column of the L_3 body, respectively, based on Denis' classification [2]. The whole assembly was fixed at the bottom of L_5 to simulate the lower foundation from the sacrum. Experimental results present interesting phenomena with the dynamic fringe patterns to show the behaviors of stress wave propagation in the spinal structure.

As example, Figure 1 gives the isochromatic fringe patterns of the impact acted in the posterior part of the anterior column. After the contact of the falling rod with the specimen top of the L_3 body, transient stress fringes were initiated from the loading point where a dark shadow of caustics appeared to show the action of the concentrated force. The stress fringes moved downwards in the vertebra and the impact energy was then transferred to the lower bodies. It was shown that the isochromatic fringes in the L_4 and L_5 bodies were evenly distributed to show a relatively uniform action of the dynamic stress. This might represent the function of the intervertebral discs that absorbed part of the energy and uniformly transferred the force to the neighboring vertebral bodies. The dynamic fringes in L_4 and L_5, in fact, were increased after the stress waves had been reflected from the bottom foundation of the sacrum body. The superposition of the reflected wave upwards and the sequentially propagating wave downwards produced strong stress focusing in the lower two bodies, which might cause fracture failure with body disruption. During this time, the transient stress distribution in L_4 and L_5 varied strongly, including the stress concentrations in the corners connecting with the facet elements.

The impact loaded in the anterior part of the anterior column, however, had different wave propagation from that in the posterior part. After reaching the lower edge of L_3, the stress fringes directly passed through the disc below to propagate in the L_4 body with a relatively nonuniform wavefront. During this period, in fact, a large part of impact energy still remained in L_3 due to the resistance of the intervertebral disc so that the stress fringes moved to the posterior facet joint. With the momentum movement in the assembly, the impact energy was transferred to the L_4-L_5 bodies with the increase of fringe numbers and variations, accompanying the focusing of the stress waves by superimposing and the stress concentrations on the anterior surface of the vertebral bodies.

When the impulsive load was acted in the middle column of the spine, the downward process of the wave propagation mainly happened in this local region, which represented that the column was the main structure to carry the axial loads. After the stress waves were reflected from the bottom foundation of the L_5 body, however, the fringes moved upwards in the whole bodies of the assembly, to meet those coming from the upper bodies and produce stronger variations of stress patterns than those in the former two cases. This resulted from the location of the impact load that was in the inner part of the lower lumbar spinal curvature, i.e., the combination of the impact load in this column of L_3 with the reaction force from the sacrum foundation produced an eccentric compressive force including the bending component, to generate higher stress variations together with the concentrations at the joints with the facet elements.

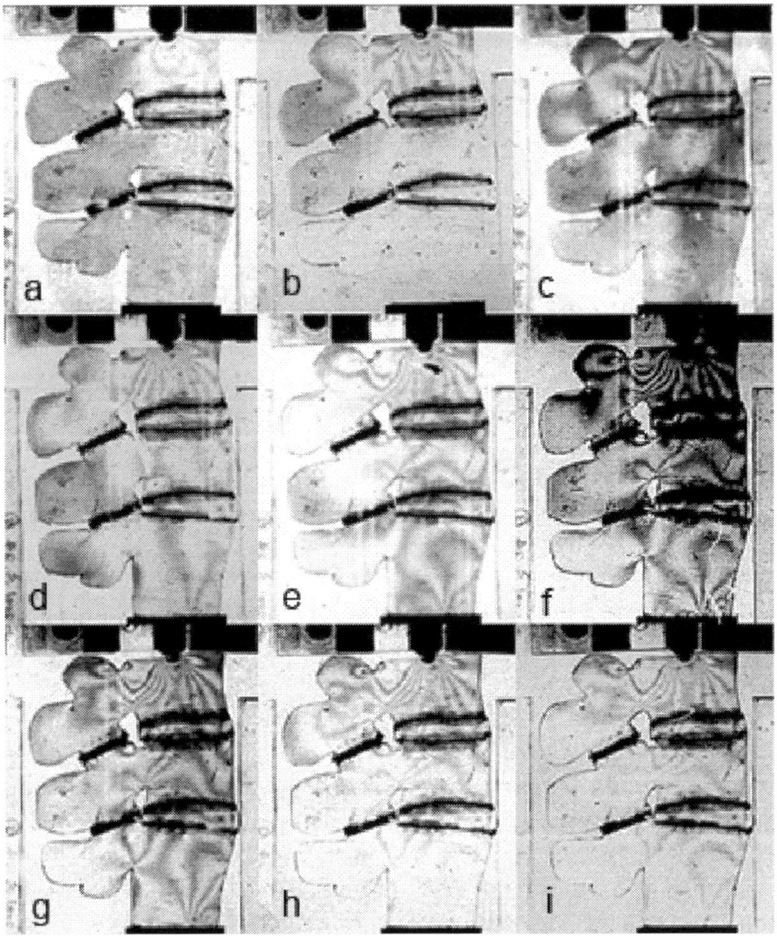

Figure 1. Pictures of photoelastic fringe patterns showing stress wave propagation in the vertebras of the lower spine loaded near the middle column.

3. NUMERICAL ANALYSIS BY DYNAMIC FEM

To understand the interactive response of the vertebras and the intervertebral discs in the lower lumbar spine, a dynamic analysis of FEM computation was carried out to study the histories of the transient stresses responding to the impulsive pressure applied on the top of the L_3 body.

The geometrical model of the lower spine was formed by ten-node tetrahedral elements (SOLID92, ANSYS). The model was transformed from Smit's shared model [4] by extracting the basic characteristics (points, lines and surfaces) and using a CAD software to reconstruct the solid elements suitable for the ANSYS computation. Table 2 summarized the material properties of different part of the lumbar spine. The data came from the resources referred by [5-9]. Meanwhile, the supraspinous ligaments were simulated by using a 3-D cable element (LINK10, ANSYS) with Young's modulus of 10Mpa, to subject the tensions on the tissues.

To reflect the interaction of the neighboring bodies in the L_3-L_4-L_5 assembly, two pads with similar properties to the intervertebral discs were put on the top of L_3 body and under the bottom of L_5 body, respectively, to relate the action of the upper vertebra (L_2) and the sacrum body where the L_5 vertebra was fixed in all directions. A compressive impact pressure was applied on the top of the superior pad by reaching a peak value of 100, 000 Pa in a short time of 3E-6s. That constant impulsive load was kept until at 3E-4s and then reduced to only 100Pa in a short period of 5E-6s, which was then remained till the end time of the computation at 1.9E-3s.

Figure 2 shows the stress propagation in the spinal assembly presented by the von Mises stress contours. The dynamic stresses moved downwards to pass through L_3 and then L_4, respectively. When the L_4 body was about fully loaded (Figure 2b), the intervertebral disc between L_3 and L_4 was extremely compressed to expand to a big size in the radial direction. Sequentially, similar phenomenon appeared in the disc between L_4 and L_5 after the L_5 body was acted by the impact stresses (Figure 2c). However, that intervertebral disc reached the maximum size in the lateral direction at *t‡E 3s* when the stress wave was reflected from the bottom foundation and then acted again on the disc together with the sequentially downward stress, during that time the stresses in the L_5 segment went also to higher values than those in the other two bodies by the stress focusing (Figure 2d). With the movement of the reflected stress upwards passing through L_4 and then L_3, the intervertebral disc between them expanded again laterally (Figure 2e), due to the coming of the focusing region to the upper part of the spine.

Table 1. Material properties of the FE models.

Material Type	E (MPa)	ν	ρ (kg/m^3)
Cortical bone	12000	0.3	1900
Cancellous bone	100	0.2	700
Posterior elements.	3500	0.3	1410
Annulus substance	4.2	0.45	1065
Nucleus pulpous	1	0.499	1000

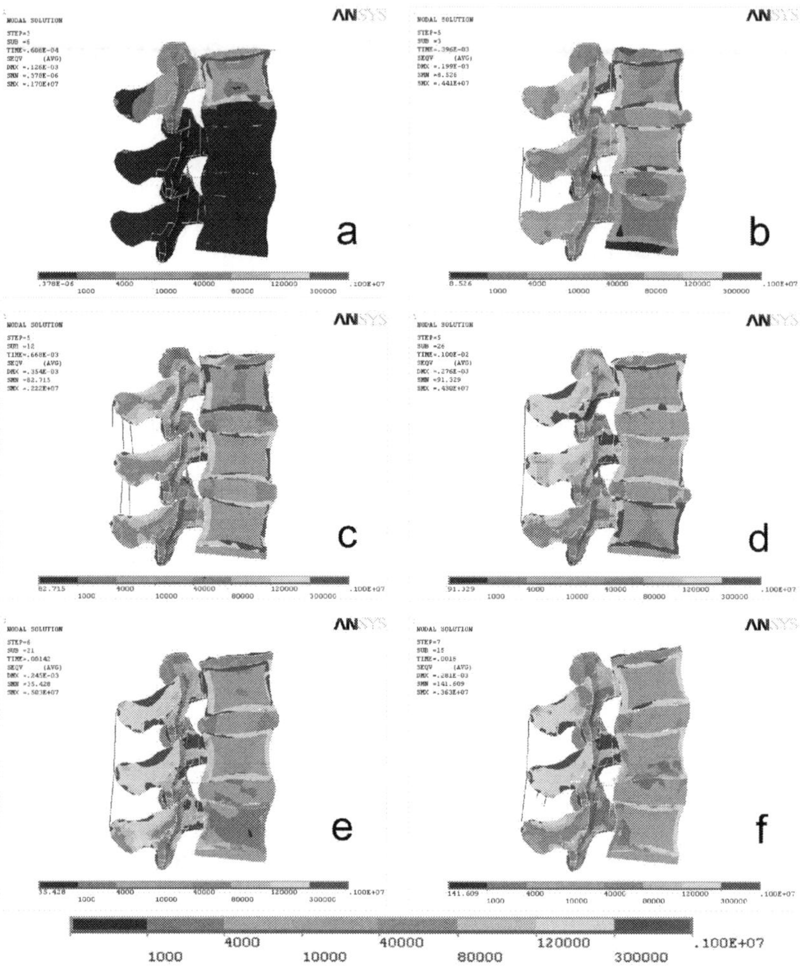

Figure 2. The von Mises stress propagation in the lower spine, at the instants of t= .0606E-3s, 0.396E-3s, 0.668E-3s, 1.00E-3s, 1.42E-3s, 1.80E-3s, respectively.

These processes present not only the function of the intervertebral discs to absorb the impact energy with their deformation, but also the transient stresses distributed in each vertebral body. It could be observed that the maximum von Mises stresses were distributed in the outer shell of each body where the cortical bone occupied, which coincided with the anatomic architecture of the vertebra by the thin layer of the hard bone to bear the major forces from either static load or impact action. Based on the experimental results presented in the last section and the transient analysis of the stress distributions, we focused on the middle point on the anterior edge of each vertebral body where the high stresses might appear under the impact load to cause the bone fracture. Figure 3 demonstrates the variations of the von Mises stress at three points (P1, P2 and P3) on the anterior surface of L_3, L_4 and L_5 bodies, respectively. We could observe that in the beginning stage the stress oscillations appeared in the L_3 and L_4 bodies, respectively, with phase delay and frequency reduction of the transient stresses. The effective stress in the L_5 body, however, became much smoothened due to shock absorption of the intervertebral discs. At instant *t#Es* , the stress in the L_5 body reached its peak that was about the double of the mean levels of stress variations in other bodies, resulting from a direct focusing of the stress waves propagating in the opposite directions. The focusing stresses appeared also later in the L_4 and L_3 bodies, respectively, but with less amplitudes due to energy attenuation and still oscillated due to the wave reflection /diffraction from the multiple interfaces within the intervertebral discs.

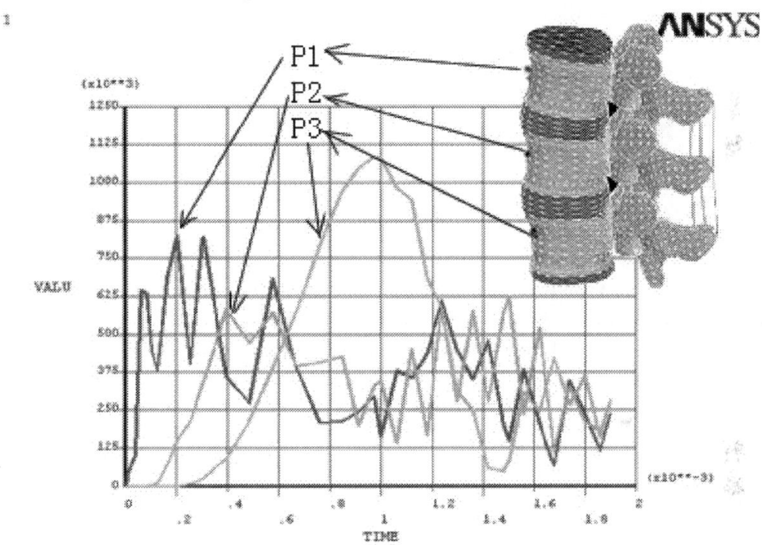

Figure 3. The von Mises stress variation at three points P1, P2 and P3, on the anterior surface of the vertebral bodies.

4. CONCLUSIONS

By using the experimental techniques of dynamic photoelasticity and the computational technique of finite element method, the transient stress distributions were analyzed to study the stress wave propagations in the vertebral bodies and the intervertebral discs of the lower lumbar spine. The patterns of isochromatic fringes visualized the wave movement in the specimen with transmission through the intervertebral disc models and the reflection from the bottom foundation of the spine. Different dynamic behaviors were presented in the vertebral bodies as the impact was loaded in the middle column, the anterior part and posterior part of the anterior column of the L_3-L_4-L_5 assembly, respectively. The FE simulation results revealed the propagation of the transient von Mises stress in the vertebral bodies together with the deformation of the intervertebral discs, showing the energy absorption of the discs during the wave movement and the stress focusing of the waves propagating in the opposite directions. The stress histories in the cortical shells of the vertebras presented high levels of the effective stress that might cause the fracture of the spinal bone.

ACKNOWLEDGEMENTS

The research was support by NNSFC grant of No. 10125211 and by the BioMed-X Research Center of PKU.

REFERENCES

1. Harrington, K.D., Vertebral compression fractures: differentiation between benign and malignant causes, *Iowa Orthop J* , **13,** 1993, 85–96.
2. Denis, F., The three-column spine and its significance in the classifications of acute thoracolumbar spine injuries, *Spine*, **8,** 1983, 817.
3. Gertzbein, S.D., Spine update: Classification of thoracic and lumbar fracture, *Spine,* **19,** 1994, 626.
4. Smit, Th.H., The mechanical significance of the trabecular bone architecture in a human vertebra, Ph.D,Thesis Technical University Hamburg, Shaker Verlag, Aachen., 1996.
5. Chosa, E., Totoribe, K. and Tajima, N., A biomechanical study of lumbar spondylolysis based on a three-dimensional finite element method, *J. Orthop. Res.,* **22**, 2004, 158–163.
6. Noailly J., Lacroix D. and Planell A., The mechanical significance of the lumbar spine components – A finite element stress analysis, Technical University of Catalonia, Barcelona, 2000.
7. Pitzen, T., Geisler, F., Matthis, D., Müller-Storz, H., Barbier, D., Steudel, W. and Feldges, A., A finite element model for predicting the biomechanical behaviour of the human lumbar spine, *Control Eng. Pract.,* **10**, 2002, 83–90.
8. Ezquerro, F., Simon, A, Prado, M. and Pérez, A., Combination of finite element modeling and optimization for the study of lumbar spine biomechanics considering the 3D thorax-pelvis orientation, *Med Eng Phys.,* **26,** 2004, 11–22.
9. Lee, K.K. and Teo, E.C., Effects of laminectomy and facetectomy on the stability of the lumbar motion segment, *Med. Eng. Phys.*, **26**, 2004, 183–192.

SESSION 3

HEAD IMPACT BIOMECHANICS IN SPORT

Implications to Head and Neck Injury Tolerances

David C. Viano

[1]National Football Leaugue, Mild Traumatic Brain Injury Committee, 280 Park Avenue, New York, NY 10017, USA; [2]Sport Biomechanics Laboratory, Wayne State University, 818 W. Hancock, Detroit, MI 48201, USA; [3]ProBiomechanics LLC, 265 Warrington Rd., Bloomfield Hills, MI 48304, USA; phone: 248.645.5832, fax: 248.645.5598, dviano@comcast.net

Abstract. Head impacts in sport are analyzed, including 1) concussive impacts in professional football simulated in laboratory tests to determine injury mechanisms 2) neck compression forces with head down impacts by striking players and 3) straight punches to the frangible face of the Hybrid III dummy by US Olympic boxers to determine punch force and head dynamics. Concussed NFL players experienced impacts at 9.3 ± 1.9 m/s velocity with 7.2 ± 1.8 m/s head ΔV. Peak acceleration was 98 ± 28 g over 15 ms. Concussion correlated with translational acceleration. The nominal tolerance for concussion was HIC = 250. Impact force was 7191 ± 2352 N with 56.1 ± 22.1 g head acceleration in the striking player and 4221 ± 1885 N neck compression force. Nij was greater than tolerance in $1/3^{rd}$ of the striking players who had neck compression force of 6614 ± 1006 N and Nij of 1.37 ± 0.25. In boxing, the punch force was 3427 ± 811 N, hand velocity 9.14 ± 2.06 m/s and effective punch mass 2.9 ± 2.0 kg. Punch force was higher for the heavier weight classes, due primarily to a greater effective punch mass. Jaw load was 876 ± 288 N and translational acceleration 58 ± 13 g. In the NFL, concussion occurs with considerable impact velocity and head ΔV due to translational acceleration. Olympic boxers deliver straight punches with similar impact velocity, but the punch mass is much lower, so the head ΔV is considerably less than in football concussions. The role of rotational acceleration in concussion is unclear from these studies. Helmets and padding need to reduce HIC to lower the risk of concussion. Neck compression forces often exceed Nij tolerances in professional football. This sports-related biomechanics data is relevant to impact conditions in automotive crashes and the assessment of safety systems.

Key words: human tolerance, concussion, head-neck injury, biomechanics, sport.

Michael D. Gilchrist (ed.), IUTAM Proceedings on Impact Biomechanics: From Fundamental Insights to Applications, 121–130.

1. INTRODUCTION

In 1973, NOCSAE established standards for the impact performance of football helmets by limiting the Severity Index (SI) of head acceleration [1]. By 1980, there was a significant reduction in serious head injury with improved helmets, including a 51% reduction in fatal head injuries and 65% reduction in skull fractures [2]. Using SIs, head injury risks were reduced 12-55% with helmets meeting NOCSAE standards. While continued improvements were made through the 1990s, attention increased on mild head injuries. Concussion emerged as a concern in football [3]. Recent studies of concussion in the NFL used instrumented Hybrid III dummies to determine head dynamics with player injury [4, 5].

Similarly in the 1970s, there was concern for head down tackling (spearing) that resulted in catastrophic neck injuries in striking players. This was related to head impacts into the tackled player's torso, where the mass of the struck player's body increased the load in the striking player's neck and led to compression-flexion or other compression-bending injuries. Quadriplegia and death were the most serious consequences. Video of spearing injuries showed that cervical fracture-dislocations were due to axial loading. This resulted in rule changes in high school, college and professional football banning deliberate spearing or the use of the top of the helmet as the initial contact in a tackle. There was a significant reduction in cervical injuries by the late 1970s with a continued decline until today [6]. However, there is another concern with head down tackling in helmet-to-helmet impacts. This relates to the risk of concussion in the struck player hit oblique on the facemask or on the side of the helmet [7].

Boxing exposes athletes to head impacts and brain injury risks, often due to repeat impacts. Punch force has been measured in a world ranked heavyweight boxer using an instrumented target suspended as a ballistic pendulum [8]. The boxer's fist reached 8.9 m/s impact velocity with a peak force of 4096 N. Another study found punch forces of 4800 ± 601 N for elite, 3722 ± 375 N for intermediate and 2381 ± 328 N for novice English boxers [9]. The punch of 3 amateur boxers showed a peak translational acceleration of 21.5 ± 4.6 g for a left jab to 44.6 ± 15.6 g for a left hook and peak rotational acceleration of 293 ± 72 r/s^2 for a left jab to 676 ± 231 r/s^2 for a left hook. Based on tolerances of 200 g for linear acceleration and 4500 r/s^2 for angular acceleration, the researchers concluded that neither translational nor rotational acceleration reached a level that was injurious to the boxer for a single blow [10].

Currently, the head impact tolerance is HIC = 700 for a 15 ms time window and the neck compression tolerance is 4000 N to avoid serious injury risks. This study addresses concussion (mild traumatic brain injury) with padded head impacts that occur in professional football and the role of translational and rotational acceleration in boxing and football head injuries. Neck compression was also studied in striking NFL players to provide insight on tolerances to neck compression force and Nij. The research uses sport impact biomechanics to study human tolerances.

2. METHODS

Reconstruction of Head Impacts in Professional Football: Details of game film selection and analysis have been published by Pellman et al. [4, 5]. Network tape was obtained from the NFL on 182 game impacts causing concussion or severe head impacts (Figure 1). The 3D impact velocity, orientation and helmet kinematics were determined for cases with at least two clear views. This gave the impact speed and helmet kinematics. Laboratory reconstruction of 31 impacts was made with instrumented, Hybrid III dummies. A helmeted, head-neck representing the struck player was attached to a 7.1 kg mass simulating the struck player's torso and guided in freefall from a height to match the impact velocity determined from video analysis of the game collision. The Hybrid III head and neck weighed 4.38 kg with instrumentation. The helmet and facemask weighted 1.92 kg so the falling mass was 15.1 kg. Impact was against another helmeted head-neck attached to the torso and pelvis of the Hybrid III dummy. This dummy weighed 46.4 kg without arms and legs, and it was suspended by flexible cables.

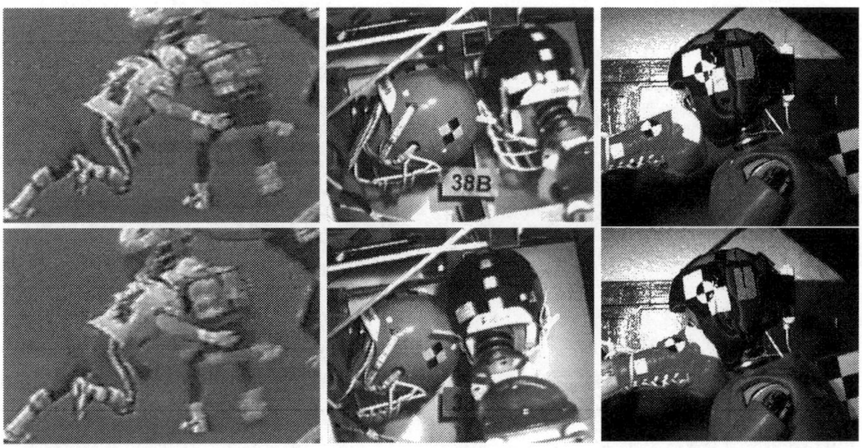

Figure 1.. NFL concussion of a struck player (left), reconstruction in a laboratory test with dummies (center) and Olympic boxer straight punch to the Hybrid III jaw (right).

Each Hybrid III head was equipped with standard accelerometers at the head cg and nine linear accelerometers in the "3-2-2-2 configuration" to determine rotational acceleration using a correction for centripetal and Coriolis acceleration [4, 5]. The striking player had a 6-axis, upper neck transducer. High speed video recorded head kinematics in the reconstruction. The camera was positioned similarly to a view from the game video. More than half of the impacts involved the facemask or area where the facemask attaches to the helmet shell of the struck player, whereas virtually all striking players involved the front crown or top portion of the helmet as their head was down and the axis of impact was through their neck and torso [7]. The top of the helmet is substantially stiffer than the side or facemask region.

Collision Biomechanics: Impact force from the striking player was determined by adding the head inertia and neck compression force: $F = m_{Striking} a_{Striking} + F_N$ [7], where $a_{Striking}$ is the resultant acceleration of the striking player's head, F_N is the resultant neck compression force and $m_{Striking}$ is the mass of the striking player's head above the neck load cell. The mass was $m_{Striking}$ = 5.90 kg and included the Hybrid III head (3.64 kg), load cell above the sensing element (0.34 kg) and helmet with facemask (1.92 kg).

The impact force is equilibrated by the struck player. Their head, neck, helmet and a portion of the torso are involved, so $F = m_{Struck} a_{Struck}$, where a_{Struck} is the resultant acceleration of the struck player's head. The mass of the struck player is m_{Struck} = 8.40 kg and includes the head (4.38 kg), neck (1.06 kg), helmet and facemask (1.92 kg) and portion of the torso mass (1.04 kg). The difference in the Hybrid III head mass between the striking and struck players reflects the full weight of a bracket that is used in place of the neck load cell.

Head acceleration of the striking player is lower than that of the struck player, so the effective mass of the striking player is greater than that of the struck player. The neck load cell measures the contribution from the torso mass in the collision, which adds to the impact force. The effective mass of the striking player was $m_{Eff.\ Striking}$ = 14.0 kg, indicating a mass ratio of $m_{Eff.\ Striking}/m_{Struck}$ = 1.67 or a 67% greater effective mass of the striking player than that of the struck player during peak force [7].

Head Injury Tolerances: The resultant head acceleration is used to calculate two head injury criteria. The NOCSAE football helmet standard uses the Severity Index: $SI = \int_T a(t)^{2.5} dt$, where a(t) is the resultant translational acceleration and T is the duration of impact. NHTSA uses a variation of SI to assess head injury risks. The Head Injury Criterion is: $HIC = \{(t_2 - t_1)[\int_{t_1}^{t_2} a(t)dt/(t_2 - t_1)]^{2.5}\}_{max}$, where t_1 and t_2 are determined to maximize HIC using a 15 ms limit.

Neck Injury Tolerances: The historic neck tolerance to axial compression is 4000 N. This was determined from a Hybrid III dummy when struck by a tackling block that had produced serious head and neck injuries in football players [11]. More recently, a neck injury criterion Nij was developed, where the "ij" represent four injury mechanisms: tension-extension (Nte), tension-flexion (Ntf), compression extension (Nce) and compression-flexion (Ncf). Nij is calculated as a function of time by normalizing My and Fz with four intercepts [12]. The normalized flexion-extension moments are added to the normalized axial loads to give $Nij=(Fz/Fzc)+(My/Myc)$, where Fzc is the critical intercept for axial neck loading and Myc the critical intercept for flexion-extension bending moment at the occipital condyles. The critical intercepts are Fzc = 6806 N for tension, Fzc = 6160 N for compression, Myc = 310 Nm for flexion and Myc = 135 Nm for extension. The neck extension intercept is substantially higher than the earlier 57 Nm tolerance.

Measurement of Olympic Boxer Straight Punches to the Jaw: Ten US Olympic boxers, weighing 48-109 kg and representing four weight classes, delivered three punches each to the jaw of an instrumented Hybrid III dummy. Details of the punch evaluation and head dynamics have been presented by Walilko et al. [13]. The tests involved the frangible Hybrid III face (www.ftss.com), which provided realistic impact force and transfer of head acceleration.

A biaxial accelerometer was placed in the boxer's hand, and five axial accelerometers in the Hybrid III dummy head to measure sagittal plane translational and rotational acceleration using an in-line technique [14]. A six-axis loadcell measured upper neck loads and moments. Tekscan pressure sensor was inserted between the frangible foam and the head skin to measure pressure distribution on the jaw. High speed video captured the impact at 1000 frames per second.

Punch force was determined in the axis of impact by: $Fp_x - Fn_x = mA_x$, where Fp_x is the punch force, Fn_x is the neck shear force, m is the mass of head and A_x is translation acceleration in the x direction. The effective mass of the boxer's punch was determined using conservation of momentum: $m_h V_p = (m_h + m) V_h$, where m_h is the effective mass of the puncher's hand, V_p is the punch velocity, V_h is the Hybrid III head velocity and m is the mass of the Hybrid III head [13].

3. RESULTS

Concussion in Struck NFL Players: Helmet impacts involve a player running toward another who is generally unaware of the closing angle. With concussion, the average impact speed was 9.3 ± 1.9 m/s. The striking players line up their head, neck and torso and impact the struck player. The strike is oblique on the facemask or facemask attachment to the helmet usually below the head cg, or on the side of the helmet above the head cg. Since the struck player is unaware, only their head is initially involved in the impact. This gives a higher effective mass for the striking player, and more momentum is transferred to the struck player causing a rapid change in head velocity of 7.2 ± 1.8 m/s, which is statistically higher ($t = 2.9$, $p < 0.005$) than the 5.0 ± 1.1 m/s in uninjured players. The striking players experience a head ΔV of only 4.0 ± 1.2 m/s.

Figure 2 shows peak accelerations. Head acceleration was 98 ± 28 g with 15 ms duration for concussion and was significantly greater than 60 ± 24 g for uninjured struck players ($t = 3.1$, $p < 0.005$). The strongest correlation with concussion was HIC ($t = 3.2$, $p < 0.005$) and SI ($t = 3.0$, $p < 0.005$). The conventional measures of head injury risk were effective in assessing concussion risks. Nominal tolerances for concussion are HIC = 250 and SI = 300. Concussion occurred with head rotational acceleration of 6432 ± 1813 r/s^2 and rotational velocity of 34.8 ± 15.2 r/s.

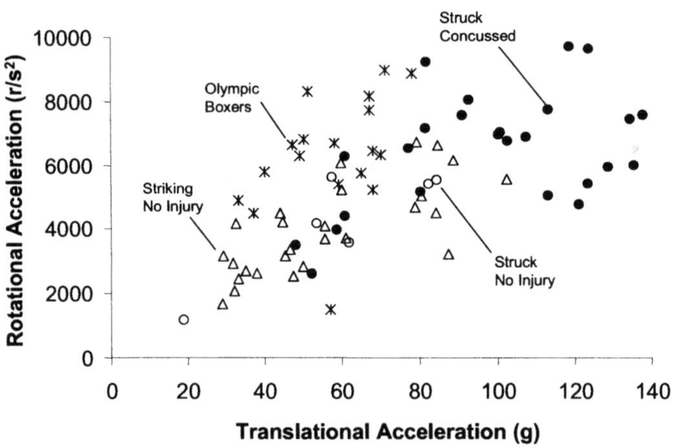

Figure 2. Peak translational and rotational acceleration with NFL concussions and severe impacts (closed circles are concussions, open symbols no injury) and responses for Olympic boxer's straight punches to the jaw of the Hybrid III dummy ("x" symbol).

Striking NFL Players: More momentum was transferred to the struck player than the striking player in the collisions. The mass ratio based on ΔV was $m_{Eff.\ Striking}/m_{Struck} = 1.67$ (6.8/4.1). HIC was 117 ± 101. The average peak neck compression force was 4221 ± 1885 N for the striking players. The average fore-aft shear was 707 ± 323 N and lateral shear was 494 ± 238 N irrespective of direction. The peak neck bending moment was 51 ± 29

Nm in extension and 38 ± 19 Nm in flexion. The average peak neck moment in flexion-extension was My = 44 ± 25 Nm and the lateral bending was Mx = 32 ± 17 Nm irrespective of direction. Nij was 0.86 ± 0.38. All striking players experienced neck compression without injury.

Figure 3 shows the neck compression force and flexion-extension bending moment at peak Nij for the NFL reconstructions. Superimposed on the plot is the NHTSA neck tolerance criterion with Nij=1.0 and 4000 N limit on neck compression force. These values are NHTSA's current human tolerance limits for neck loading. The NFL reconstructions had 9 players outside the tolerable limit. In all 9 cases, the neck compression force exceeded the tolerance limit of 4000 N and in a few cases it exceeded by almost a factor of two. For these cases, the average neck compression force was 6614 ± 1006 N (range 5056 – 8194 N) and Nij was 1.37 ± 0.25 for those above 1.0. There was a moderate level of neck flexion-extension moment. The primary impact response was neck compression.

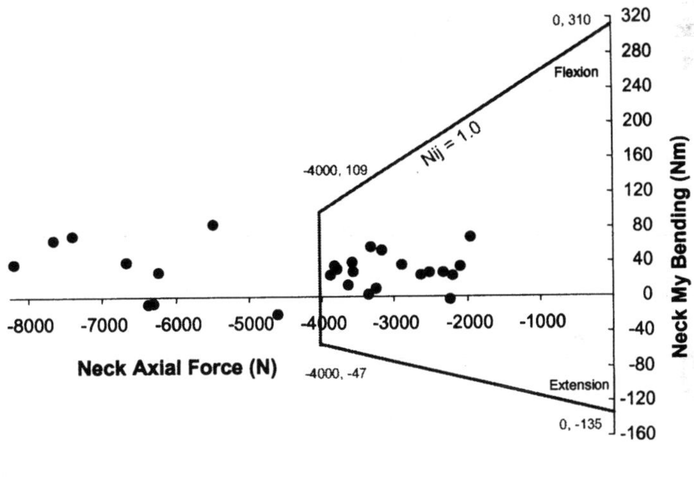

Figure 3. Neck compression force and bending moment at peak Nij for striking NFL players, including lines of Nij=1.0 and tolerance for neck compression force. (from [7]).

The striking player's head acceleration was 56.1 ± 22.1 g in helmet-to-helmet impacts [7]. The struck player's head acceleration reached 89.4 ± 27.5 g. The peak impact force was 7191 ± 2352 N acting on the struck player. The peak neck compression force was 4221 ± 1885 N contributing over half of the impact load. Double integration of the struck player's head acceleration indicated 48 mm displacement at peak force. The biomechanical analysis showed a 67% greater mass of the striking player using a head down tackle than the struck player at peak force.

Impact Conditions with Concussion: Peak head acceleration was 94.3 ± 27.5 g with concussion and 67.9 ± 14.5 g without (t=3.02, df=12, p=0.005) in helmet-to-helmet impacts [7]. The peak impact force averaged 7642 ± 2259 N with concussion and 5209 ± 1774 N without injury (t=2.62, df=7, p=0.017). The head ΔV was 7.08 ± 1.88 m/s with concussion and 5.38 ± 0.48 m/s without (t=3.75, df=24, p=0.0005). The striking player experienced 4.26 ± 1.23 m/s head ΔV with concussion and 3.44 ± 0.90 m/s without injury (t=1.69, df=8, p=0.064). The average peak neck compression force was 4539 ± 1931 N in the striking player causing concussion compared to 2823 ± 725 without injury (t=3.27, df=18, p=0.002).

Olympic Boxing Punches to the Jaw: The peak punch force was 1990-4741 N. The average force was 3427 ± 811 N. There was a significant difference (p=0.021) in punch force between the various weight classes [13]. Tekscan measured the portion of the impact force delivered to the jaw region. The average force on the mandible was 876 ± 288 N. The average neck shear force was 994 ± 318 N and Nij averaged 0.27 ± 0.07. Hand velocity was determined by integrating the resultant hand acceleration up to face contact. The average hand velocity was 9.14 ± 2.06 m/s at impact. The average effective mass for all punches was 2.86 ± 2.03 kg. The average head cg acceleration was 58 ± 13 g with 11.4 ms duration. The average head ΔV was 2.97 ± 0.81 m/s. The average rotational acceleration was 6343 ± 1789 r/s^2. HIC was 71 ± 49, which varied with weight class (p = 0.002). The energy transfer averaged 17.2 ± 8.9 J and the punch power was 6574 ± 3453 W.

4. DISCUSSION

Figure 4 summarizes the existing body of biomechanical data on human tolerance to head impact. The NFL data provides unique and new scientific information on human tolerance. Automotive crashes typically have impact durations below 6 ms for head impact on vehicle rails, pillars and structures. Head impacts into airbags with seat belt restraints have durations above 40 ms. The NFL results provide new information on tolerances in the 15 ms range, where there has been an essential absence of scientific data. Historically, the acceleration tolerance for 15 ms duration head impacts has been estimated at 42-80 g based on extrapolation of studies using military volunteers. The NFL reconstructions indicate 70-75 g for concussion in padded impacts, which is at the high end of the earlier tolerance levels, but consistent with the Wayne State University Concussion Tolerance Curve.

Concussion in NFL football involves exceptionally high impact severities, averaging 9.3 m/s impact velocity, 7.2 m/s ΔV and 98 g. Since the mass of the Hybrid III head is 4.55 kg, the peak force on the head is 4.4 ± 1.2 kN with concussion. Since this is within the experimental range for skull fracture for short duration impacts, the helmet shell and padding are doing a good job of distributing load and lowering risks for serious brain injury and skull fracture in NFL collisions.

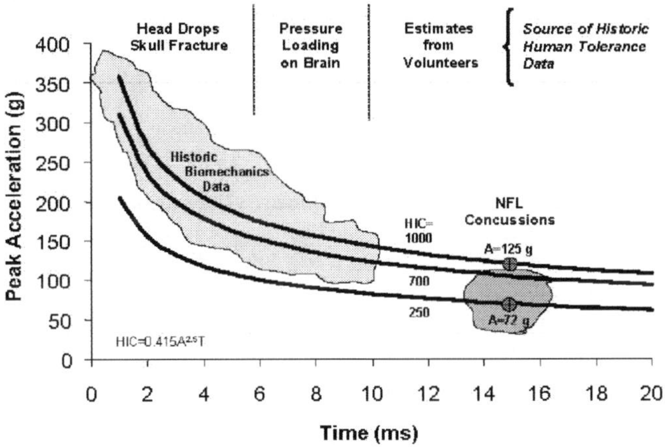

Figure 4. Peak head acceleration and duration for various HICs and the types of historic data used to develop the SI and HIC criteria. The NFL data is from [4].

There has been much speculation that concussion is related to rotational acceleration of the head. The helmet impacts demonstrate a stronger correlation of concussion with translational acceleration indicating it is the primary measure to assess the performance of helmet protection systems. Moreover, there is a strong correlation between peak translational and rotational acceleration so reducing translational acceleration (SI or HIC) proportionately reduces rotational acceleration [4].

Even though there were high compression forces in the neck of striking players, no NFL player experienced serious neck injury or concussion. In 9 out of the 27 NFL collisions, the compressive neck force exceeded Nij tolerances for serious neck injury. The avoidance of neck injury is primarily by maintaining an axial alignment of the impact-force vector through the neck and torso, by minimizing neck flexion and lateral bending that increase the risk of injury, and by engaging the helmet of the struck play. This delivers more momentum to the struck player. The primary means of reducing NFL concussions in helmet-to-helmet collisions is to enforce head up tackling techniques [7].

NFL concussions occurred at peak rotational accelerations averaging 6600 r/s^2. These levels are consistent with the rotational accelerations found in Olympic boxing at 6300 r/s^2. While rotational acceleration may be a factor in head injury, these sport studies are unclear in proving a relationship. The strongest correlation with concussion was with HIC and translational acceleration. No concussion occurred with average accelerations of 68 ± 15 g and HIC of 143 ± 37, while NFL concussions occurred with 94 ± 28 g and 345 ± 181 HIC. These levels are well above what the Olympic boxer delivered to the Hybrid III dummy. Using injury risk functions [4], concussion risks were 13% ± 10% for HIC and 20% ± 2% for translational acceleration with US Olympic boxers.

REFERENCES

1. Hodgson VR. National Operating Committee on Standards for Athletic Equipment football helmet certification program. Med Sci Sports. Fall;7(3), 1975, 225-32.
2. Hodgson VR, Thomas LM. Mechanisms of Cervical Spine Injury during Impact to the Protected Head. 24th Stapp Car Crash Conference, Society of Automotive Engineers, Warrendale PA, SAE 801300, 1980.
3. Vastag B. Football brain injuries draw increased scrutiny. JAMA. Jan 23-30;287(4), 2002, 437-9.
4. Pellman EJ, Viano DC, Tucker AM, Casson IR, Waeckerle JF. Concussion in Professional Football: Reconstruction of Game Impacts and Injuries. Neurosurgery 53(4), 2003, 799-814.
5. Pellman EJ, Viano DC, Tucker AM, Casson IR, Waeckerle JF. Concussion in Professional Football: Location and Direction of Helmet Impacts - Part 2. Neurosurgery 53, 2003, 1328-1341.
6. Heck JF, Clarke KS, Peterson TR, Torg JS, Weis MP. National Athletic Trainers' Association Position Statement: Head-Down Contact and Spearing in Tackle Football. J Athl Train. 39(1), 2004, 101-111.
7. Viano DC, Pellman EJ. Concussion in Professional Football: Biomechanics of the Striking Player – Part 8. Neurosurgery 56, 2005, 266-280.
8. Atha J, Yeadon MR, Sandover J, Parsons KC. The Damaging Punch. British Medical Journal. Dec, 1985, 1756-57.
9. Smith MS, Dyson RJ, Hale T, Janaway L. Development of a boxing dynamometer and its punch force discrimination efficacy. Journal of Sports Sciences 18, 2000, 445-450.
10. Smith TA, Bishop PJ, Wells RP. Three dimensional analysis of linear and angular accelerations of the head experienced in boxing. IRCOBI, 1986, 271-285.
11. Mertz HJ, Hodgson VR, Thomas LM, Nyquist GW. An Assessment of Compressive Neck Loads Under Injury Producing Conditions. The Physician and Sportsmedicine, 6(11), 1978, 95-106.
12. Kleinberger M, Sun E, Eppinger R, Kuppa S, Saul R. Development of Improved Injury Criteria for the Assessment of Advanced Automotive Restraint Systems. NHTSA Docket No 1998-4405-9, September, 1998.
13. Walilko TJ, Viano DC, Bir CA. Biomechanics of Olympic Boxer Punches to the Face. in print, British Journal of Sport Medicine, 2005.
14. Shee TR, Viano DC. Computing Body Segment Trajectories in the Hybrid III Dummy Using Linear Accelerometer Data. Journal of Biomechanical Engineering, 116(1), 1994, 37-43.

BIOFIDELITY OF DUMMY AND FEM NECKS IN THE FREQUENCY DOMAIN

N. Bourdet, R. Fischer, F. Meyer and R. Willinger
UMR 7507 ULP-CNRS, University of Strasbourg, 2 rue Boussingault, 67000 Strasbourg, France

Abstract. Modal analysis technique is used in order to characterize the human head-neck system *in vivo*. The extracted modal characteristics consist of a first natural frequency at 1.3±0.1 Hz associated with head-neck extension motion and a second mode at 8±0.7 Hz associated with head translation. By recording experimentally the apparent mass of dummies head-neck system under the same experimental condition as the volunteer subjects, it was possible to compare the human and the dummies frequency response functions and to evaluate their bio-fidelity. The evaluation methodology based on validation parameters extracted in the frequency domain is tested on six dummy necks and on a FEM of head-neck system. Only BioRID is presented in this paper. The results of the two models are similar and present natural frequencies close to the volunteers. Beside dummy evaluation this study also gives new insight into injury mechanisms given that a given natural frequency can be related to a specific neck deformation.

Key words: dummy necks, modal analysis, natural frequency.

1. INTRODUCTION

Neck injuries in motor vehicle accidents continue to be a serious and costly societal problem. The cervical spine is one of the most complex structures in the human skeletal system and its behavior during impact is still poorly understood. Most injury prevention strategies are based on results from anthropomorphic test dummies and computational models.

Typically dummy necks are validated against volunteers or post mortem human subjects in the temporal domain and in term of corridor. In the past

Michael D. Gilchrist (ed.), IUTAM Proceedings on Impact Biomechanics: From Fundamental Insights to Applications, 131–138.

under rear end impact condition, the human neck has been characterized experimentally by decelerating a human subject seated in a car seat with or without a headrest (Eichberger *et al.* 1996, Ono *et al.* 1997 [1, 2]). More recently Ono *et al.* 2001 [3] suggested characterizing the human cervical spine by directly loading the head with a force close to 150 N (over 50 ms) applied horizontally or vertically to the chin.

In the present study, contrary to the works carried out in the time domain, we propose to identify the head-neck system in the frequency domain. This will enable us to extract the intrinsic parameters from the head-neck system such as stiffness and damping. Some similar inputs are used for the experimental tests by loading directly the forehead (150 N; 3Gx; 50 ms) and recording its kinematics. The originality of our research is to proceed with a frequency analysis of the head-neck response rather than a temporal one, followed by the extraction of the system's modal characteristics which are inherent to the system whatever the loading. The main advantages of investigating in the frequency domain are the extraction of noise from the recorded signals. However an important restriction of the method is its limitation to the linear domain. This hypothesis is acceptable in case of low speed rear impact conditions as explained later in the discussion.

The experimental modal analysis of the human head neck system *in vivo* will provide us with natural frequencies and mode shapes which must be reproduced by the dummy head-neck system. In a first section we present the technique applied to a single human volunteer and we show how this experiment constitute the biomechanical background of the proposed dummy and numerical evaluation methodology. Experimental and numerical modal analysis is then applied to several existing rear impact dummy and a FEM under similar loading conditions as for the volunteer. The results of dummy are presented on this paper.

2. BIOMECHANICAL BACKGROUND AND METHODOLOGY

2.1 Modal Analysis of the Head-Neck System In Vivo

As outlined briefly in the introduction, the head-neck system is firstly characterized in vivo. The experimental impact device is represented in Figure 1. It consists of a simple pendulum (4 kg; 0.6 m) which slightly impacts frontally the volunteer's forehead. The volunteer is seated on a rigid seat without a headrest. The subject is asked to close his eyes and to remain totally relaxed in order to prevent as much as possible any active muscle contribution. It is hypothesized that the head motion remains in the sagittal plane and that the head motion amplitude remains sufficiently small (a few degrees) so that the two recorded responses i.e. the applied frontal force and the linear head acceleration can be assumed as unidirectional in the antero-posterior direction.

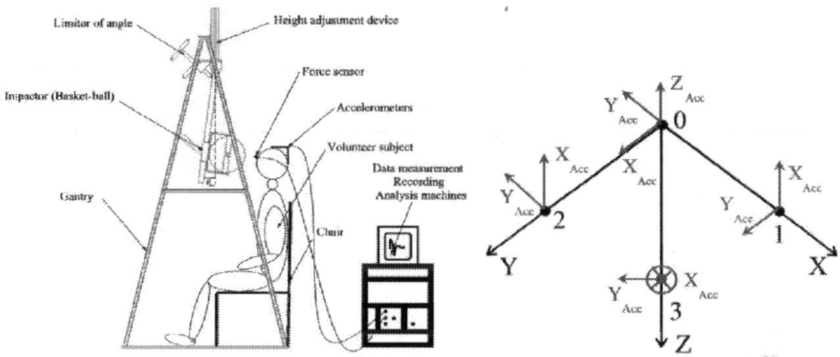

Figure 1. Experimental test device for the modal analysis of the head-neck system and the 3-2-2-2 configuration of nine accelerometers of the head.

The head acceleration is measured using nine accelerometers (Entran EGA ± 10 g) arranged in the well-known 3-2-2-2 configuration as illustrated in Figure 1 in order to calculate the linear component of the head acceleration at any point. The impulsive force is recorded using a force sensor (PCB 208A02 11.432 mV/N).

After impact, the transfer function between force and acceleration is estimated in terms of apparent mass. Special attention is paid to noise management with the coherence function.

A total of six human male volunteers of very different sizes and masses were tested and lead to two natural frequencies. f_1 =1.3 Hz (± 0.1 Hz) and f_2 =8 Hz (± 0.7 Hz). This is a typical result for a system with two degrees of freedom.

The imaginary parts of three transfer functions (at vertex S, at atlanto-occipital point O_H, and the zero transfer function at T1 O_N) are given in Figure 3, in order to extract the system's deformed mode shapes. It is finally a three dimensional representation of this parameter obtained by including the spatial dimension between O_N, O_H and S which permits to draw the deformed mode shapes of the neck at respectively 1.3 and 8 Hz. Figure 3 represents the two shapes schematically, by plotting the imaginary part of the selected geometrical point transfer function for the two natural frequencies: the first mode is an extension motion and the second an retraction motion.

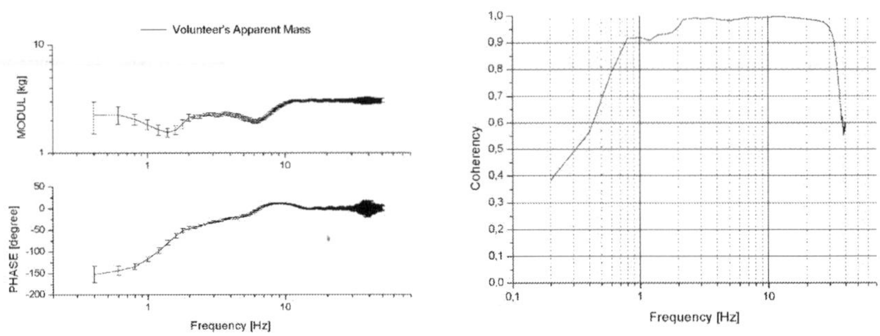

Figure 2. Experimental transfer function of the head-neck system in terms of Apparent Mass and its coherence function.

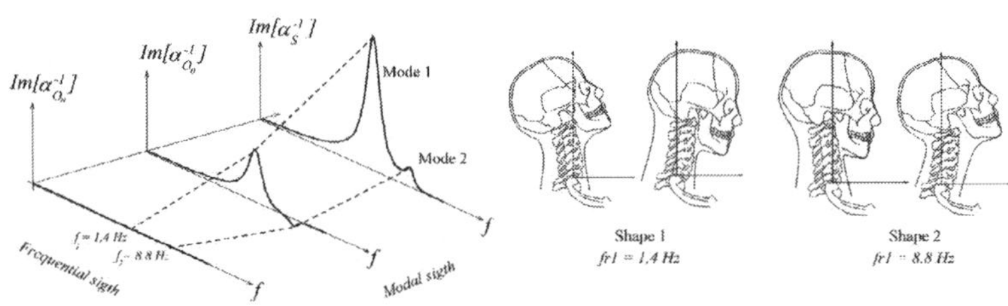

Figure 3. Three-dimensional representation of the imaginary part of the transfer function; Representation of the two shapes.

2.2 Methodology for Dummy and FE Necks Evaluation

Dummy evaluation is conduced in accordance with the previous experimental modal analysis of the human head-neck in vivo. Six dummies are tested strictly under the same conditions as the volunteers. Firmly fixed on a rigid seat without headrest, their forehead is impacted with the pendulum. Impact force and head acceleration are determined at the vertex (point S) and center of gravity (point O_H) of the dummy head.

The above described experimental modal analysis was performed numerically by simulating the experience with the head-neck FE model (Meyer and Bourdet, 2004 [4]) followed by a results analysis in the frequency domain as illustrated in Figure 4. The experimental results presented in this modal analysis correspond to the response of the subject modeled. We assumed that the first thoracic vertebra was fixed and reproduced the impact load in its amplitude and duration. To simulate the force of the impact between the ball and the head a Young's modulus of

E=0.3 MPa (ν=0.49) was implemented for the modeling of the ball and an initial horizontal speed of 0.6 m/s. The acceleration at the vertex was computed over 5 s.

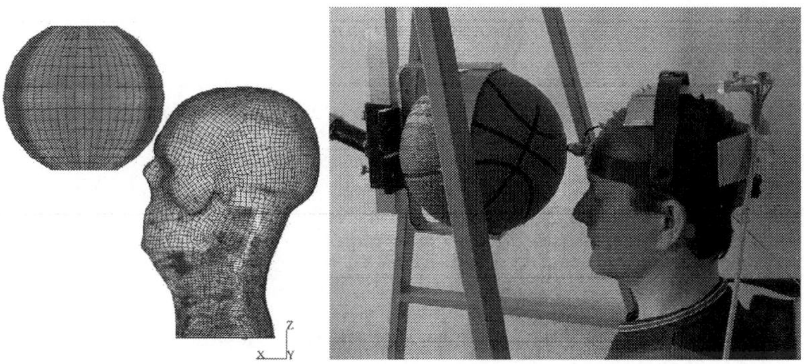

Figure 4. Experimental configuration of the experimental modal analysis (at left), and numerical replication with the finite element model of the neck (at right).

3. BIOFIDELITY EVALUATION OF DUMMY AND FEM NECKS

In 1992 Svensson and Lövsund [5] developed a new dummy neck (the RID neck). This neck presented improved kinematics of the dummy response in terms of head rotation but further modifications were needed in order to also reproduce head translation. Today, after several dummy versions, a flexible vertebral column has been added to the neck in order to reproduce T1 rotation accurately. The new name of this rear impact dummy prototype is BIORID II (Davidsson 1999 [6]).

BIORID II's experimental apparent mass in Figure 5 shows an important improvement of neck flexibility as its first natural frequency is decreased to 2.5 Hz. Compared to the human neck, it can be concluded that this dummy neck is still too stiff and that its motion damping is too low. Concerning the second mode it is interesting to notice that the retraction mode exists but at a frequency of 10 Hz against 6 Hz for the human being.

In spite of a difference between the model and the volunteer, the frequency response of the model presents resonance frequencies very close to the volunteer, namely 1.6 Hz for the first mode as against 1.3 Hz and 8 Hz for the second as for the volunteer. This result demonstrates the importance of the mechanical properties of the ligaments and in particular their damping coefficients, which have a major influence on the resonance frequencies as well as on the volunteer's apparent mass amplitude. As the signals recorded during the temporal validation are too complex, it is quasi-impossible to observe the influence of these parameters on the retraction phenomenon. In the contrary, the proposed modal analysis allowed us to adjust these parameters in a more realistic way.

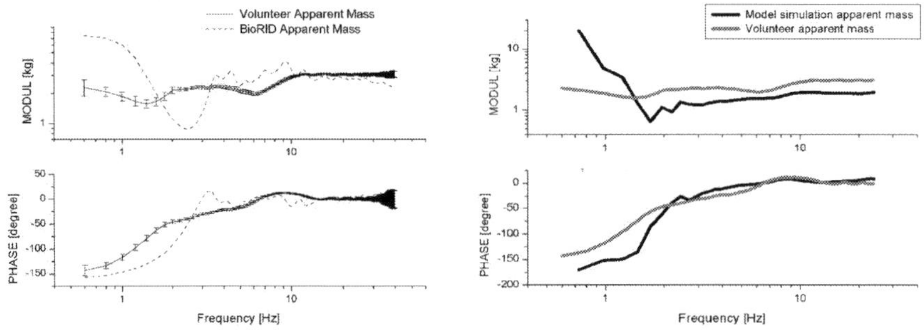

Figure 5. Superimposition of the experimental apparent mass recorded on BioRID II and FEM of the head-neck system with the human volunteer response.

4. DISCUSSION

This new neck models investigation approach is interesting in two different ways: first the pertinence of the new validation parameters acquired *in vivo* and second the evaluation of rear impact neck model.

The extracted modal characteristics of the human head-neck system *in vivo* can be compared to observations made in the time frame.

The new issue at this level is that the first mode at 1.3±0.1 Hz is associated with the neck extension (C shape) and the second mode at 8±0.7 Hz to translation or retraction (S shape) obtained on six very different human male subjects. These deformation modes have often been observed in the literature and it is generally mentioned that the "S mode" appears before the neck extension in the time domain (Deng *et al.* 1987, Kleinberger *et al.* 1993, Walz *et al.* 1995, Ono *et al.* 1997 [2, 7-9]). The present research shows that S shape mode is only excited if energy is introduced in the system around 8 Hz, and this is the case only if the impact duration is short enough or if there are high-loading ramps within the loading function.

The hypothesis of linearity is checked in the transfer function with the coherence function and is in relation to the low impact energy involved in the experimental impact. It is therefore important to remember that the methodology is well designed to describe the human neck and to evaluate dummy neck behavior for low energy impact, or before non-linearity due to saturation (hyper-elongation of ligaments, bone contact, muscle activity) occurs. Under these restrictions, how can modal analysis techniques, inform us about the complex properties of the human and dummy neck?

The main limitation of the present study is its focus on mechanical behavior of the neck and possible neck injury mechanisms, excluding any investigation of tolerance limits. It must be pointed out also that this study is restricted to impulse to the forehead, focusing therefore on rear motion of the head as it occurs in rear impact configuration. Finally, only adult males have been considered and muscle action has not been taken into account, so further research is needed on human volunteers.

5. CONCLUSION

An experimental and theoretical modal analysis of the human head-neck system under frontal head impact, has been successfully conducted and lead to original results : For the human head-neck system *in vivo,* the extracted modal characteristics consist of a first natural frequency at 1.3±0.1 Hz associated to neck extension and a second mode at 8±0.7 Hz associated with head translation or neck retraction. For five very different volunteer male subjects similar results were obtained. This set of data constitutes new validation parameters in the frequency domain suitable for dummy evaluation under moderate rear impact.

By recording experimentally the apparent mass of the dummy and using a FEM of head-neck systems under the same experimental condition as the volunteer subjects, it was possible to compare the human and the models frequency response functions and to evaluate their bio-fidelity against validation parameters in the frequency domain. Further conclusions could be drawn from this study: BioRID presents two natural frequencies close to the volunteer's ones. Improvements are however needed in order to reproduce closer the extension mode and to set the retraction mode at a realistic natural frequency. The FEM of the head-neck system is able to reproduce the mode shapes at the same natural frequencies than the experimental. Its frequency behavior is close to the BioRID dummy.

To the authors' knowledge it is the first time that modal characteristics of the human head-neck system are extracted. The results lead to new models evaluation methodology and give new insight into injury mechanisms given that if a natural frequency and its mode shape are excited, the related injury mechanism is potentially present. Impact characteristics in the frequency domain or neck loading rate should consequently be able to avoid a given neck deformation mode. This opens up new possibilities for protective system evaluation and optimization. Consequently, it could be suggested that we might well suppress or reduce transmissibility of seat to occupant around 8 Hz in order to avoid the neck retraction mode.

ACKNOWLEDGEMENTS

The authors wish to thank the French government for his support in framework of PREDIT program n° 00A008601. Acknowledged are also UTAC and TNO for their collaboration.

REFERENCES

1. Eichberger, A., Geigl, B., Moser, A. et al. (1996) Comparison of different car seats regarding head-neck kinematics of volunteers during rear end impact, Paper presented at the IRCOBI Conference, Dublin, Ireland.
2. Ono, K., Kaneoka, K., Wittek, A. and Kajzer, J. (1997) Cervical injury mechanism based on the analysis of human cervical vetebral motion and head-neck-torso kinematics during low and speed rear impacts, 339-356.
3. Ono, K., Kaneoka, K., Sun, E. A., Takhounts, E. G. and Eppinger, R. H. (2001) Biomechanical response of human cervical spine to direct loading of the head, Paper presented at the IRCOBI conference, Isle of Man, UK.
4. Meyer, F., Bourdet, N., Deck, C., Willinger, R. and Raul, J. S. (2004) Human neck finite element model development and validation against original experimental data, Stapp Car Crash Journal, Vol. 48, pp. 177-206.
5. Svensson, M. and Lövsund, P. (1992) A dummy for rear-end collisions - Development and validation of a new dummy-neck, Paper presented at the IRCOBI Conference, Verona, Italy.
6. Davidsson, J. (1999) BioRID II final report (Göteborg, Sweden, Crash Safety Division, Department of Machine and Vehicle Design, Chalmers University of Technology).
7. Deng, Y.-C. and Goldsmith, W. (1987) Response of a human head/neck/upper-torso replica to dynamic loading-II. Analytical/numerical model, Journal of Biomechanics, 20, 487-497.
8. Kleinberger, M. (1993) Application of finite element techniques to study of cervical spine mechanics, Paper presented at the 37th Stapp Car Crash Conference.
9. Walz, F. and Muser, M. (1995) Biomechanical aspects of cervical spine injuries, Paper presented at the 39th Stapp Car Crash Conference.

INFLUENCE OF HELMET INERTIAL PROPERTIES ON NECK INJURY RISK DURING EJECTION: A NUMERICAL HUMAN MODEL APPROACH

S. Laporte[1,3], E. Chavary[1,2,3], W. Skalli[1,3] and A. Guillaume[2]

[1]*Laboratoire de Biomécanique, ENSAM – UMR CNRS 8005, 151 Boulevard de l'Hôpital, 75013 Paris, France; Phone: +33 (0)1.44.24.63.64; Fax: +33 (0)1.44.24.63.66, E-mail: {sebastien.laporte, estelle.chavary, wafa.skalli}@paris.ensam.fr*

[2]*IMASSA, BP 73, 91220 Brétigny sur Orge, France; Phone: +33 (0)1.69.23.76.53; Fax: +33 (0)1.69.23.70.02; E-mail: {echavary, aguillaume}@imassa.fr*

[3]*GDR CNRS 2610 "Biomécanique des chocs", France*

Abstract. Survival rate after jet ejection is about 94% , but 50% of major injuries are due to neck injuries . Mechanical parameters responsible for those injuries are not yet fully understood. The purpose of the present study is to use a existing finite element model to investigate helmet mass and inertial properties influence on neck injuries that may occur during ejection.

The model position represented the head-neck in a specific MK16© ejection seat, with a functional 30° oblique inclination. Vertical acceleration was imposed in order to simulate the first 350ms ejection event. Then the case of bare head and three alternative existing helmets were considered.

All the quantified parameters are under failure level and these results could be explained by a helmet weight less than 2Kg. Furthermore, no shock due to the parachute opening or windblast effect is modelled. Indeed, the simulation represents the first 350ms of an ejection, and there was no relative speed during the experiment.

Key words: Finite Element Method (FEM), added mass, ejection, neck injury, numerical simulation.

Michael D. Gilchrist (ed.), IUTAM Proceedings on Impact Biomechanics: From Fundament Insights to Applications, 139–146.

1. INTRODUCTION

Due to aircraft speed, escape from military jet is assisted by pyrotechnic devices. However, this rescued maneuver is stressful for the pilot and the crew, particularly for the cervical spine. Although the survival rate after jet ejection is about 94%, 36% of the survival pilots have neck injuries and 50% of the severe injuries concern neck structures. In 100% of the lethal case, neck injuries are found [1].

Moreover, in this stressful environment pilots have to use more sophisticated helmets. Head Mounted Display (HMD) or Night Vision Goggles (NVG) are then used in order to improve operation effectiveness. These systems modify inertial properties of the helmet and strains in the cervical spine. This effect then seems to increase the neck injury risk during an ejection [SHA97].

There is a lack of knowledge about cervical spine limitations to support masses and biomechanical factors responsible of neck injuries, although these data will be useful for future development of mounted head systems.

A fiftieth percentile human finite element model was previously developed and evaluated for automotive crash in order to predict injury risk on the cervical spine [5, 6]. The purpose of the presented study is then to use this finite element model to quantify mechanical stresses in anatomical neck structures for different helmet configurations cases during an ejection. Results are then compared with baselines values to determinate the more safety configuration.

2. METHODS

An existing fiftieth percentile human finite element model was used [5, 6], representing cervical spine anatomical structures (vertebra, ligaments, soft tissues and muscles) and a Hybrid III head geometry. The head-neck complex is set in ejection position with a sagittal inclination of 19 degree and a head support. These geometric boundary condition represent the specific case of a MK10© ejection seat (Figure 1). An acceleration is then imposed on C7 vertebra in order to simulate the 350 first milliseconds of an ejection, corresponding only to cockpit extraction - accelerometric data used in this study was stored during a 0/0 shot (Altitude 0m, relative velocity 0m.s-1) with a modified Hybrid III dummy. Simulations were developed with the explicit finite element code Radioss©.

Global reference system is related to the head, X-axis is positive forwards, Y-axis is positive on the left of right-hand side and Z-axis is positive upwards, parallel to gravity orientation. This skew system centre is defined at head gravity center considered at the middle of external auditory meatus (MEA) [8].

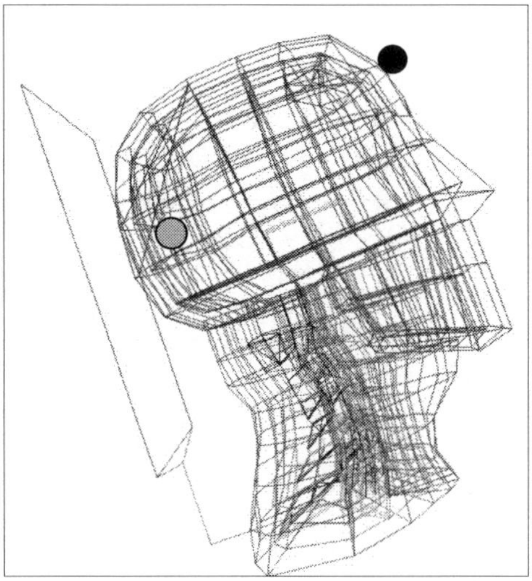

Figure 1. Finite Element model of the head-neck complex in ejection position.Grey round represents the counterweight. Black round represents the Night Vision Goggle (NVG).

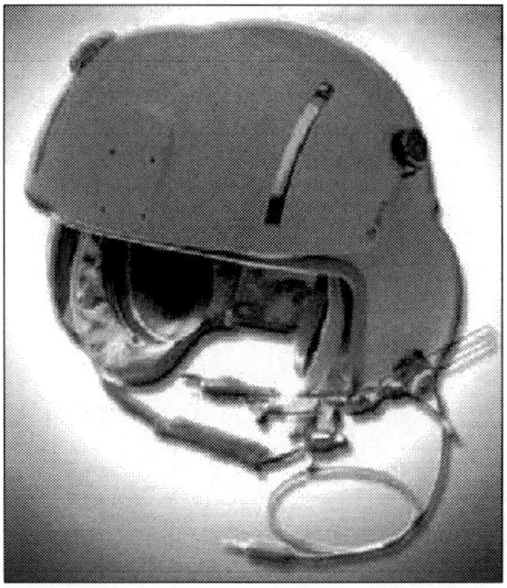

Figure 2. Helmet used in this study (*www.gentexcorp.com*).

Four simulations were considered: bare head (existing geometry without added masses), a helmet (Figure 2), NVG mounted on this helmet, and the same NVG with a counter weight placed backward on the helmet. Helmet inertial properties were defined from literature (Figure 2) [7], NVG and counterweight masses were defined following the French Army recommendations. Localizations of these different masses are defined with regard to the global skew system (Table 1).

Table 1. Localizations and values for helmet, NVG and counterweight masses.

	Mass (g)	*Localization with regard to global skew system* *X* (mm)	*Y* (mm)	*Z* (mm)
Helmet	1386	14.7	8.4	43.9
NVG	700	106.0	0	48.4
Counterweight	350	-80.9	0	-67.1

Effect of mass distribution was quantified and analyzed in terms of compressive loads and moments on each vertebra of the cervical spine and supraspinous ligament elongation.

3. RESULTS

Results are presented as maxima of compressive loads, extension/flexion moments and supraspinous ligament elongation. These values were extracted from their temporal variations.

Figure 3 presents the maxima of vertebral body compression loads along the cervical spine for the bare head and the three helmet configurations. Results show compression loads only, which is coherent with the literature. Most important amplification appears at C1 level, with 17% for the heaviest helmet configuration (Helmet plus NVG and counterweight) relatively to bare head case. Nevertheless results show that compressive loads are all under injury level.

Figure 4 presents the maxima for vertebral body extension/flexion moments along the cervical spine. Upper cervical spine is in flexion and lower cervical spine is in extension. The extension moment increases by 88% at C1 level between bare head and the helmet alone. The flexion moment increases by 53% at C7 level between bare head and the helmet alone configuration. Results show that moments are all under injury level.

Legend for the following figures

Bare Head
Helmet only
Helmet, NVG and counter weight
Helmet and NVG

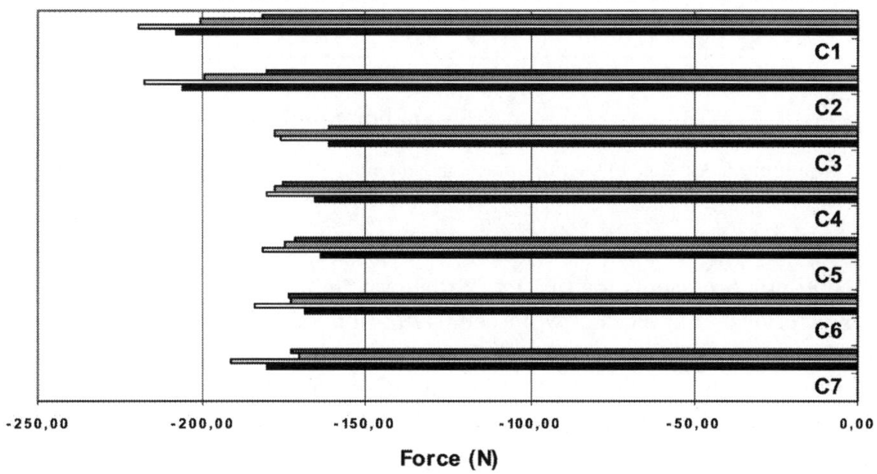

Figure 3. Maxima vertebral body compressive loads along the cervical spine for the four configurations.

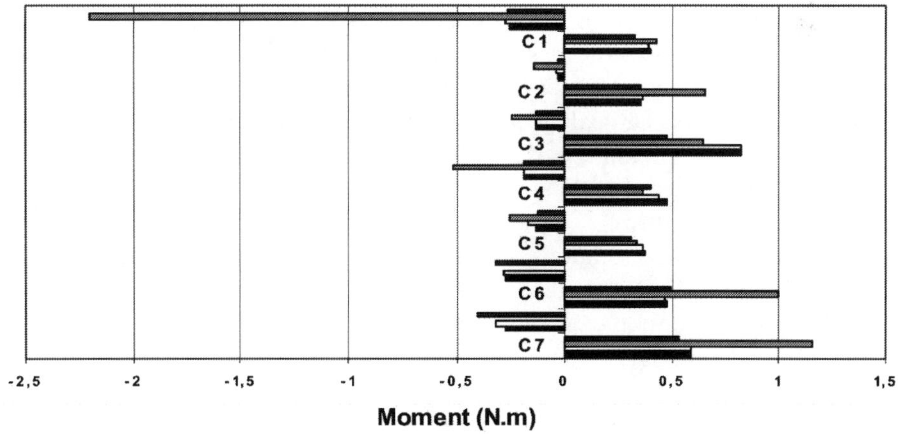

Figure 4. Maxima vertebral body extension/flexion moment among the cervical spine for the four configurations.

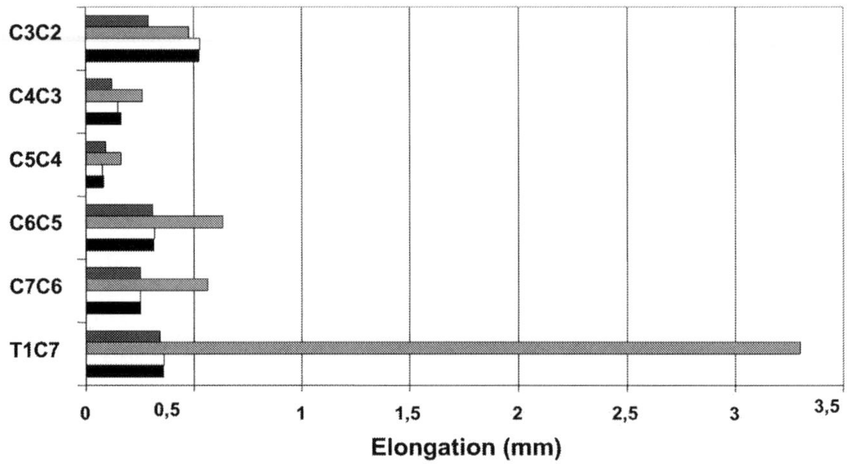

Figure 5. Maxima of supraspinous ligament elongation along the cervical spine for the four configurations.

Figure 5 presents the maxima for supraspinous ligament elongation along the cervical spine. The worst configuration is for helmet alone with an increase of 750% at C7T1 level. This result is in accordance with mechanical loading of the cervical spine. Results show that elongations are all under injury level.

4. DISCUSSION

Even if all the analysed parameters are under failure level, this study shows the global influence of mass distribution on cervical spine mechanical loading. Mass value and anterior position seem to be the main parameters of influence on mechanical loading. The under failure values could be explained by the simulation of a simple ejection.

First the acceleration data were collected from a 0/0 shot without relative speed and without sink rate or roll and yaw. However, the aerodynamic effects and complex rotations are worsening factors of the injuries neck for the ejected pilot [9].

Furthermore only the ejection the first 350 ms are simulated, representing only the catapult phase. Then no shock due to the parachute opening or landing appears.

However, the model was in an ejection correct position, cervical vertebrae alignment out of position is a fundamental factor in neck injuries apparition among ejected pilot [9].

Then, neck muscular contractions against the head box are not taken into account as it is requested to the pilot during ejection [10]. Finally, the considered configurations tested do not present an oxygen mask; mask mass is about 300g and located forward from head gravity centre. This localization could have influence on the loads sustained by neck anatomic structures.

5. CONCLUSIONS

The presented study confirmed the global influence of head mass distribution on cervical spine mechanical loading. Helmet gravity centre localization and mass distribution have a negative influence on loads applied on neck during ejection, and then could increase pilots neck injury risk.

However this model needs to be fully validated. Experimentations are under process in order to compare model global kinematics with kinematics of a manikin on an ejection tower. Another way of validation is the simulation of real reported ejections with comparison between crewmembers medical data and mechanical values calculated for the different anatomic structures of the cervical spine. Other biomechanical parameters will have to be analyzed such ligament stress or inter articular capsules deformations.

Once this ejection numerical simulation will be fully validated, it should allow to search for optimal distributions of helmet mass around head gravity centre, to minimize the risk of cervical injuries in case of egress emergency [11].

This study shows that involving finite elements simulations are useful to analyse and investigate influence of several protective helmets. Numerical simulations usually carried out for this field of research, used Articulated Total Body models. This kind of model is based on rigid bodies. Human neck model presented in this study is a predictive model; it could quantify constraints in the anatomical structures and injury criteria.

REFERENCES

1. Rowe, Brooks, Head and neck injuries in Canadian Forces ejections, Aviation, Space and Environmental Medicine, Vol. 55, 1984, 313–315.
2. Visuri, T. and Aho J., Injuries associated with the use of ejection seats in Finnish pilots, Aviation, Space and Environmental Medicine, Vol. 63, 1992, 727–730.
3. Perry, C.E., Buhrman, J.R. and Knox, F.S., Biodynamic testing of helmet mounted systems, Proceedings of the 37th Human Factors and Ergonomics Society Annual Meeting, 1993, 79–83.
4. Frechede, B., Contribution to the study of the behavior of the cervical rachis subjected to a shock, Ph.D. Dissertation, 2003, Laboratoire de BioMécanique, ENSAM.

5. Bertholon, N., Modelisation 3D of the human neck in situation of omnidirectionnal shocks: kinematics analyze and injurious aspects, Ph.D. Dissertation, 1999, Laboratoire de BioMécanique, ENSAM.
6. Frechede, B., Saillant, G., Lavaste, F. and Skalli, W., Spine Week Annual Meeting, Porto, Portugal, May 30-June 05 2004.
7. Ashrafiuon, H., Alem, M.A. and MCEntire, J.B., Effects of weight and centre of gravity location of head supported devices and neck loading, Aviation, Space and Environmental Medicine, Vol. 10 No. 68, 1997, 915–922.
8. Vital, J.M. and Senegas, J., Anatomical bases of the study of the constraints to which the cervical spine is subject in the sagittal plane: A study of the centre of gravity of the head, Surgical Radiologic Anatomy, Vol. 8, 1986, 169–173.
9. Guill, F.C. and Herd, G.R., An evaluation of proposed causal mechanisms for ejection associated neck injuries, Aviation, Space and Environmental Medicine, Vol. 60 No. 7, 1989, A26–A27.
10. Clère, J.M., L'évacuation en vol d'un aéronef, Manuel de médecine aérospatiale à l'usage des navigants, Commandement des Ecoles de l'Armée de l'Air, LeMot, 1994, pp. 111–123.
11. Perry, C.E., The effect of helmet inertial properties on male and female response during +Gz impact accelerations, Proceedings of the 35th Annual Symposium of the SAFE Association, 1997, 189–195.

QUASI-ANALYTIC ACCELERATION INJURY RISK FUNCTIONS: APPLICATION TO CAR OCCUPANT RISK IN FRONTAL COLLISIONS

Denis Wood[1], Ciaran Simms[2], Colin Glynn[3], Anders Kullgren[4] and Anders Ydenius[4]
[1]*Denis Wood Associates, Isoldes Tower, 1 Essex Quay, Dublin 7, Ireland; 00-353-1-6704566, 00-353-1-2780159, dpw@iol.ie*
[2]*Centre for Bioengineering, Trinity College Dublin*
[3]*Dept. of Mechanical and Manufacturing Engineering, Trinity College Dublin*
[4]*Folksam Research, S23, SE – 10660 Stockholm, Sweden*

Abstract. There is continuing controversy regarding the role of car mass in occupant – risk. Injury risk is influenced not only by the vehicle deformation characteristics and the occupant restraint system, but also by the size and mass of both case and partner cars. Recent research has shown that injury is better correlated with mean car acceleration than with velocity change. Injury risk relations for belted/unbelted drivers in all crashes with AIS 3+ and fatal injuries were derived from US data using a modified Joksch [7] type risk equation. These risk equations are shown to compare with recent Swedish AIS 2+ and AIS 3+ frontal collision data and with US frontal collision fatality data. The derived AIS 3+ and fatality risk functions, in conjunction with car population stress/density characteristics, were combined with US, German and Japanese car mass and collision velocity distributions to predict the variation in fatality and AIS 3+ injury risk with car mass and mass ratio using Monte Carlo simulation techniques. The resulting predictions were compared to published real world accident risk versus car mass data for each country, and a high degree of correspondence obtained. The results show that empirically derived car population structural characteristics and injury risk functions satisfactorily explain the real world fatality and injury versus mass effect.

Key words: mean acceleration injury and fatality risk functions, car occupant risk, car mass effect, frontal collisions.

Michael D. Gilchrist (ed.), IUTAM Proceedings on Impact Biomechanics: From Fundamental Insights to Applications, 147–154.

1. INTRODUCTION

The influence of car mass and size on injury causation in frontal collisions remains in dispute. Mechanics dictates that case car acceleration in two-car collisions declines with increasing case car mass, but structural properties of both vehicles determine the shape of the acceleration pulse. Further, increased case car mass increases risk for the partner car occupants.

In moderate severity collisions, the acceleration of a restrained occupant is a reflection of the vehicle acceleration [1]. In high severity collisions, intrusion influences injury. For unrestrained occupants, injury is due to contact with the car interior and vehicle "ride down" inertial loading.

Real world data indicates that, for similar severity of frontal collision, injury risk increases with vehicle structure acceleration [2, 3], despite reduced intrusion. Folksam [4, 5] used on-board collision recorder data to show that the mean vehicle acceleration ($\bar{a}$) is a better predictor of injury risk than velocity change (Δv). 50ms moving average vehicle acceleration based injury criterion is used to evaluate safety barriers [6].

This paper presents analytic injury risk functions based on mean car acceleration during impact. These functions are combined with recent real world car population structural behaviour data [8–10] to predict the variation in injury risk with car mass in frontal collisions.

2. INJURY RISK FUNCTION

Joksch [7] proposed the empirical injury risk function given by equation (1a), where $\Delta v_{critical}$ is the lower limit of Δv for which risk equals 1.0. For > $\Delta v_{critical}$, P_i=1. Following recent research [4, 5], it is now proposed that injury risk is a function of mean acceleration, see equation (1b), in which $\bar{a}_{critical}$ is the lower limit of $\bar{a}$ for which risk equals 1.0. The empirical constants $\bar{a}_{critical}$ and n are derived by regression of real world injury risk data.

$$P_i(\Delta v) = \left(\frac{\Delta v}{\Delta v_{critical}} \right)^n \quad \text{(1a)}, \qquad P_i(\bar{a}) = \left(\frac{\bar{a}}{\bar{a}_{critical}} \right)^n \quad \text{(1b)}$$

3. RISK FACTORS IN FRONTAL COLLISIONS

The independent variables in frontal collisions are:

- distribution of vehicle masses
- structural and crashworthiness characteristics of the vehicles
- interactive structural effects and configuration of the vehicles
- distribution of collision velocities
- risk versus $\bar{a}$ distributions for restraint use/ non-use.

The distribution of car masses in any car fleet is obtained from analysis of the records of the registered fleet in service.

3.1 Structural Characteristics of Car Population

Wood et al. [8, 9] analyzed real world collision recorder data from 269 single and two-car collisions (overlap > 25%, collision direction 11-1 o'clock) and showed that the mean real world behaviour of individual car types could be described by empirically derived statistically robust power regressions. Further, the scatter in the real world responses results from the distributions of overlap, vehicle crush profiles and structural interaction effects, while the mean curves reflect the mean structural responses [10]. Analysis [8, 9, 11] showed that the fundamental structural parameter for cars is the crumpling stress/density ratio, σ/ρ, which is independent of car size and mass and is equivalent to the specific energy absorption capacity, $\bar{a}L$. Power regressions for individual car types and also for a 'mean' car type of the form

$$\bar{a}.L = \sigma/\rho = C_{da}.(d/L)^{da} \quad (2a) \qquad \bar{a}.L = C_{va}.\Delta V^{va} \quad (2b)$$

show high statistical correlation, where C_{da} is the key stress/density parameter, which is independent of mass, while d_a is a function of C_{da} with a secondary mass influence [9]. These relations have been applied to successfully predict peak barrier force versus vehicle mass for 64 km/hr tests with offset deformable barriers [9].

3.2 Collision Velocity Distribution

The independent velocity parameter is collision closing speed. However, the available data is for damage-plus collisions and is in the form of velocity change, Δv, which, as shown in equation (3), is a dependent variable being a function of mass and collision closing speed.

$$\Delta v_{1,2} = \left[\frac{M_{2,1}}{M_1 + M_2} \right] V_{CCS} \quad (3)$$

In countries with in-depth accident research, the Δv and vehicle mass distributions for tow away-plus collisions are known. The log form of equation (3) yields: $\ln(\Delta v_{1,2}) = \ln(M_{2,1}/(M_1 + M_2)) + \ln(V_{CCS})$. Therefore, provided the mass and Δv distributions can be satisfactorily described by logarithmic statistical distributions the mean and variance of the collision closing speed, respectively, can be obtained,

$$[\ln(V_{CCS})]_{mean/\mathrm{var}} = [\ln(\Delta v_{1,2})]_{mean/\mathrm{var}} - [\ln(M_{2,1}/(M_1 + M_2))]_{mean/\mathrm{var}} \quad (4)$$

Analysis of Δv and mass distribution data for the US [13, 14] showed both can be represented by log distributions, thus allowing derivation of the V_{ccs} distribution.

4. INJURY PROBABILITY FUNCTIONS

4.1 Real World Data

Evans [13] detailed injury probability for belted and un-belted drivers in all car crashes. Using the US car mass [14] and V_{CCS} distributions and the car population stress/density, σ/ρ, characteristics [8, 9], mean accelerations for each risk level in [13] were found using Monte Carlo methods. The data was regressed using the form of equation (1b) for AIS 3+ injuries and fatalities for belted/unbelted drivers, see Table 1.

Table 1. Injury & Fatality Risk : Regression Parameters.

Case	$\bar{a}_{\mathrm{critical}}$ (g)	n	r^2	N
Belted Driver Fatality	19.72	6.168	0.96	13
Belted Driver AIS 3+ Injury	19.25	3.68	0.93	13
Unbelted Driver Fatality	20.05	4.9	0.96	14
Unbelted Driver AIS 3+ Injury	19.05	3	0.94	12
Folksam Belted AIS 2+ Injury	18.07	1.95	0.96	8

Figure 1 shows the data and regression lines for driver fatality risk. Regressions for AIS 3+ injury risk were determined similarly. All regressions have high correlations and the values of $\bar{a}_{\mathrm{critical}}$ are similar (from 19.05g for AIS 3+ unbelted to 20.05g for unbelted fatalities). Statistical analysis shows that there is no significant difference in $\bar{a}_{\text{critical}}$ for AIS 3+ injuries and $\bar{a}_{\text{critical}}$ for fatalities nor between belted and unbelted drivers.

Folksam Research [5] reported risk probability for AIS 2+ belted front seat occupants based on 72 frontal injury collisions, including some cases with airbags. The regression parameters are detailed in Table 1.

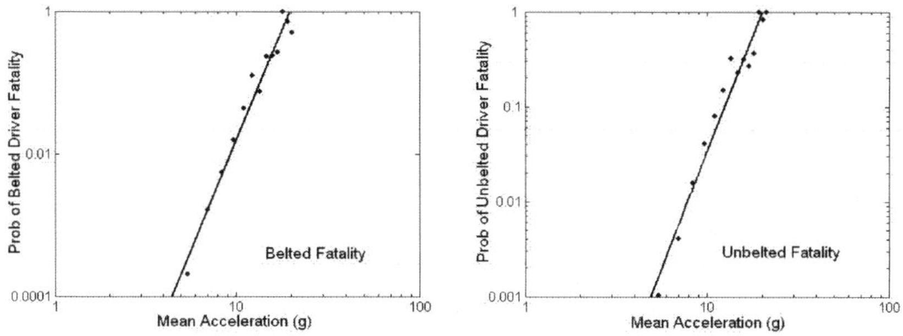

Figure 1.

4.2 Application of Risk Equations to Frontal Collisions

The suitability of the US accident data (containing some side/rear impact cases) for application in frontal collisions was statistically analysed. Comparison of $\bar{a}_{\text{critical}}$ for US AIS 3+ injuries with $\bar{a}_{\text{critical}}$ from the Folksam AIS 2 + shows no significant difference ($t = 0.54$). The regression exponent for the Folksam AIS 2 + risk is, as expected, less than that for the AIS 3+ unbelted which is less than the AIS 3+ belted, both being less than the exponents for fatalities.

Evans [15] reported fatality risk versus Δv for frontal collisions for belted and unbelted drivers between 1982 and 1991. Using the proportions of belted drivers over this period [16] and the described procedures (refer 4.1) the predicted fatality risk was computed using the regression data in Table 1. Figure 2 shows the comparison. Correlation analysis shows a coefficient of determination, $r^2 = 0.919$ with a proportionality coefficient = 1.006 and standard error = 0.09.

Figure 3 compares the derived AIS 3+ belted injury relation with recent Folksam AIS 3+ injury risk data derived from 18 AIS 3+ frontal injury collisions. Analysis shows that there is no significant difference between the risk equation and the Folksam data ($P > 5\%$).

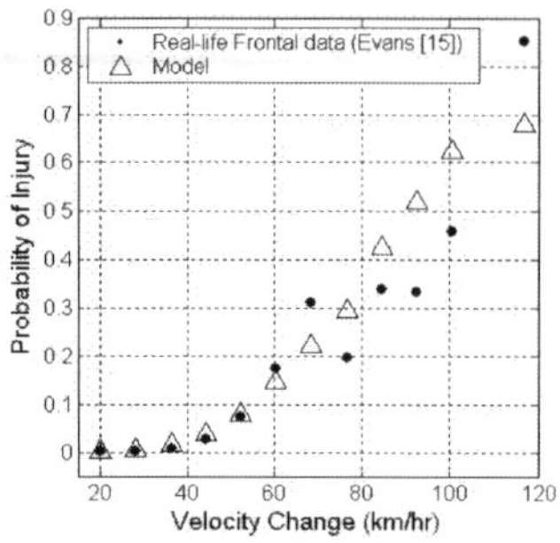

Figure 2.

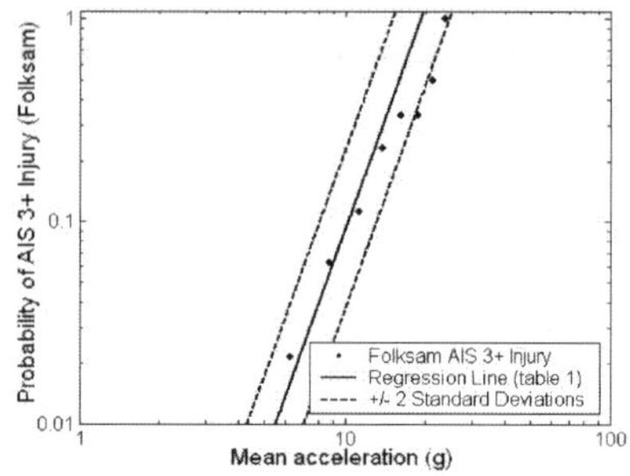

Figure 3.

5. MASS EFFECTS – DISCUSSION

In the analysis and graphs that follow, mass ratio is the defined as the ratio of heavier vehicle mass to the lighter vehicle mass.

Figure 4 compares the relative fatality risk to drivers as a function of car to car mass ratio in frontal collisions for 1991-1994 for the US [17] with the predictions using the fatality risk functions in Table 1 and the ratio of reported seatbelt use [16]. The predicted injury risk versus mass ratio and the real world data closely match (t = -1.06).

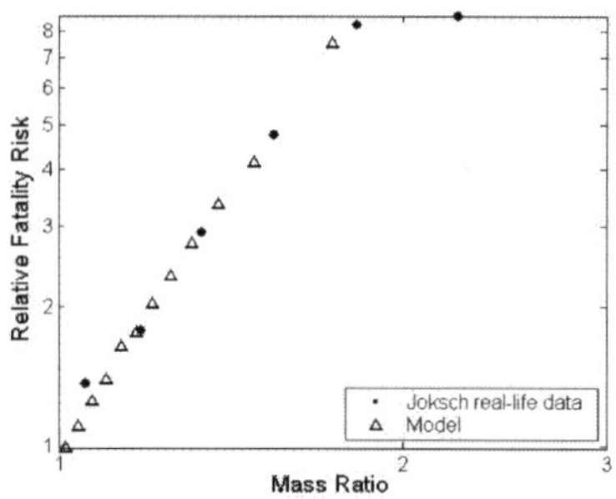

Figure 4.

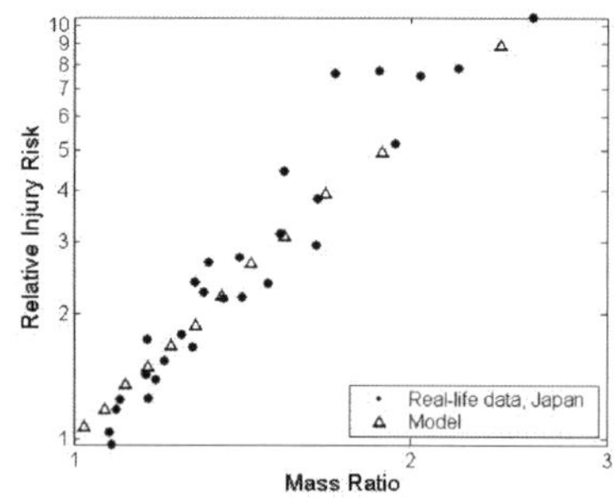

Figure 5.

Figure 5 compares the relative AIS 3 + injury risk to drivers as a function of car to car mass ratio in Japan [18] from 1992-1995 with the model predictions, showing close correspondence (t= 0.95).

Ernst et al. [19] reported on the variation in AIS 3+ injury car mass in rural two car frontal collisions in Rhine Westphalia between 1984 and 1988. The V_{CCS} distribution for Rhine Westphalia was not available, instead overall data for West Germany for the period 1973-1988 was used [20]. Figure 6 shows the real world variation in AIS 3+ risk with car mass compared with the predicted results for 50%^{ile} collision closing speeds between 51 km/hr and 62 km/hr. The absolute risk is very sensitive to the magnitude of the 50%^{ile} V_{CCS}. A 50% increase in injury risk for all cars results from a 20% increase in 50%^{ile} Vccs - from 52km/h to 62 km/h. Figure 6 shows that predicted risk varies with mass in a similar manner to the real world data and bounds the data. Figure 7 shows the close match between real world variation in risk with mass normalized for the 50%^{ile} German car mass (1050 kg). Statistical analysis of the normalised data for different 50%^{ile} closing speeds shows no significant difference between the real world data and the model predictions, see Figure 7.

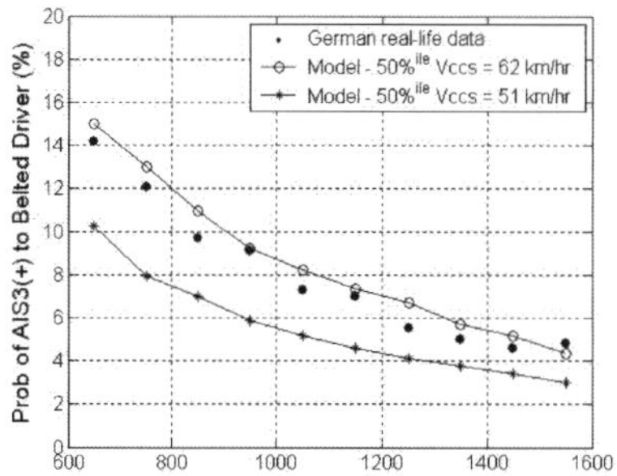

Figure 6.

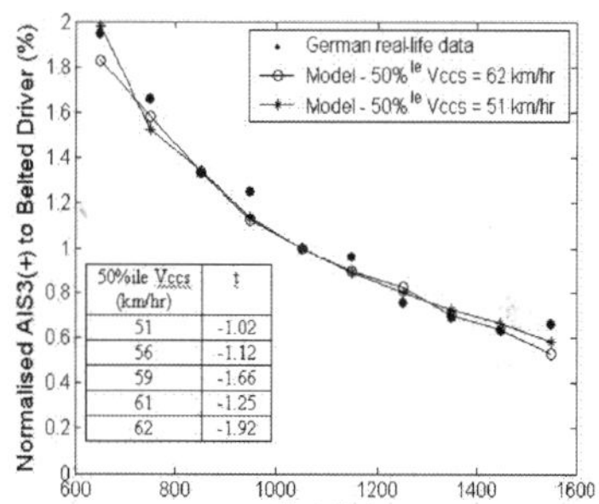

Figure 7.

6. CONCLUSIONS

The derived mean car acceleration based injury risk functions are shown to apply to frontal collisions. The risk functions in combination with car population structural characteristics derived from real world crash recorder data are shown to satisfactorily predict the variation in injury and fatality risk with car mass and mass ratio.

REFERENCES

1. Wood, D.P. and Simms, C.K., "Communication on: Edwards,M.J., Davies, H., Thompson, A. and Hobbs, A. "Development of test procedures and performance criteria to improve compatibility in car frontal collisions", Proc. IMechE 2003, Vol. 217, Part D, pp. 233-245, pub. Proc. IMechE, Vol. 218, Part D, pp. 1063-1064, 2004.
2. Prasad, P., Laituri, T.R. and Sullivan, K., "Estimation of AIS3+ Thoracic Injury Risks of Belted Drivers in NASS Frontal Crashes", Proc. IMechE 2004, Vol. 2118, Part D, pp. 591-610, 2004.

3. Prasad, P. and Smorgonsky, L., "Comparative Evaluation of Various Frontal Impact Test Procedures", Pub. Society of Automotive Engineers (SAE), 1995, paper 950646.
4. Kullgren, A., "Crash Pulse Recorder – Results from Car Acceleration Pulses in Real Life Frontal Impacts", Proc. 1996 IRCOBI Conference, Sept. 11-13, 1996, Dublin.
5. Ydenius, A. and Kullgren, A., "Pulse Shapes and Injury Risks in Frontal Impacts Based on Real Life Crashes" Proc. 2002 IRCOBI Conference Sept 18-20 2002, Munich.
6. Gabauer, D. and Gabler, H.C., "A Comparison of Roadside Crash Test Occupant Risk Criteria Using Event Data Recorder Technology", Proc. 2004 IRCOBI Conference Sept. 2004, Graz.
7. Joksch, H.C., "Velocity Change and Fatality Risk in a Crash – A Rule of Thumb", Accid. Anal. Prev., Vol. 25, pp. 103-104, 1993.
8. Wood, D.P., Ydenius, A. and Adamson, D., "Velocity Changes, Mean Accelerations and Displacements of Some Car Types in Frontal Collisions", Int. J. Crashworthiness, Vol. 8, No 6, pp. 591-603, 2003.
9. Wood, D.P., Adamson, D. and Ydenius, A., "Car Frontal Collisions: Occupant Compartment Forces, Interface Forces and Stiffnesses'', Int. J. Crashworthiness, Vol. 9, No 3, pp. 311-325, 2004.
10. Wood, D.P., Glynn, C. and Simms, C.K., "Frontal Collision Behaviour: Comparison of Onboard Collision Recorder Data with Car Population Characteristics", accepted for publication Int. J. Crashworthiness, 26 November 2004.
11. Wood, D.P. and Simms, C.K., "Car Size and Injury Risk: A Model for Injury Risk in Frontal Collisions", Accid. Anal Prev. Vol. 34, pp. 93-99, 2002.
12. Edwards, M.J., Davies, H., Thompson, A. and Hobbs, A., "Development of Test Procedures and Performance Criteria to Improve Compatibility in Car Frontal Collisions", Proc. IMechE 2003, Vol. 217, Part D, pp. 233-245, 2003.
13. Evans, L., "Driver Injury and Fatality Risk in Two-Car Crashes versus Mass Ratio Inferred Using Newtonian Mechanics", Proc. Accid Anal and Prev. Vol. 26, No. 5, pp. 609-616, 1994.
14. US Department of Transportation, National Highway Traffic Safety Administration "FARS 90 – Fatal Accident Reporting System 1990", US Department of Transportation, 1991.
15. Evans, L., "Car Size and Safety: A Review Focused on Identifying Causative Factors", Proc of the 14th ESV Conference, Munich, Germany 23-26 May 1994, Vol. 1, pp. 712-733, 1995.
16. Evans, L., Private communication, 2003.
17. Joksch, H.C., "Fatality Risks in Collisions between Cars and Light Trucks", US Department of Transport, Washington DC, 1998.
18. Mizuno, K., Umeda, T. and Yonezawa, H., "The Relationship between Car Size and Occupant Injury in Traffic Accidents in Japan", SAE Paper 970123, 1997.
19. Ernst, G., Bruhning, E., Glaeser, K.P. and Schmid, M., "Compatibility Problems of Small and Large Passenger Cars in Head-on Collisions", 13th ESV Conference, Paris, Paper S1-0-12, 1991.
20. Appel, H., Krabbel, G. and Meissner, T., "Safety of City Cars – Conflict between Ecology, Economy, Road Traffic Benefits and Safety", Forum of European Road Safety Institutes, September 1992, Berlin, Paper BER.NR.487/92.

MODELLING OF FRONTAL COLLISIONS FROM ONBOARD RECORDER AND FULL WIDTH BARRIER DATA

Ciaran Simms[1], Denis Wood[2], Colin Glynn[3], Anders Kullgren[4] and Anders Ydenius[4]
[1]*Centre for Bioengineering, Trinity College Dublin, Ireland*
email: csimms@tcd.ie ,tel: ++353 16083768, fax 6795554
[2]*Denis Wood Associates, Isoldes Tower, Essex Quay, Dublin 8, Ireland*
[3]*Dept of Mechanical & Manufacturing Engineering, Trinity College Dublin, Ireland*
[4]*Folksam Research, Sweden*

Abstract. Estimating vehicle collision severity is a key factor in biomechanics as it allows correlation of injuries with impact severity in real accidents. Key vehicle parameters for injury are the mean, peak and shape of the acceleration pulse. Real world frontal collision patterns reflect the distributions of overlap, impact angle, vehicle compliance and car-to-car frontal structural interactions. In this paper, a model relating displacement to crush as a function of overlap is combined with a full-width barrier characteristic to predict the acceleration-displacement response for given car types. Application yields predictions for 30 degree angled barrier and 45% overlap rigid barrier tests which closely match the corresponding experimental results. Separately, regression analysis of data from onboard collision recorders of real world collisions for 16 car types in 269 collisions yielded relationships between mean acceleration, velocity change and displacement for individual car types. This shows that the mean crumpling stress/density is equivalent to the normalised mean acceleration and is independent of size or mass. The geometric crush model combined with a single full-width barrier characteristic for the car population is used to model the variability of structural characteristics in frontal collisions. These are combined with overlap and crush profile distributions to predict overall collision characteristics using Monte Carlo methods. The results confirm that the collision recorders data reflects the distribution of frontal collisions. A new method of estimating the frontal response of individual cars at various overlaps from a single full-width barrier test is presented, with predictions that correlate closely with collision recorder data.

Key words: frontal collisions, injury, real world accidents.

Michael D. Gilchrist (ed.), IUTAM Proceedings on Impact Biomechanics: From Fundamental Insights to Applications, 155–162.

1. INTRODUCTION

Correlation of injuries with impact severity in frontal collisions in real life accidents is a major goal in impact biomechanics. Occupant loading is most strongly related to the acceleration of the vehicle, especially in low-medium speed impacts where intrusion does not occur and this paper focuses on the pulse shape and mean vehicle deceleration during a frontal collision.

Full-width barrier tests show that cars have repeatable crush characteristics, and these can therefore be used to relate the residual crush to the Δv in cases involving a single vehicle in a head-on strike with a wide and flat rigid barrier. However, in real world accidents occupant injuries arise from a broad range of collision configurations including single/multiple collisions, wide/narrow collision partners and different offsets, impact angles and structural compliances. Consequently, it is not generally possible to directly equate the residual crush in a real accident to crush in a barrier test.

Analysis has shown that the shapes of crushed vehicles can be categorised as either segmental or triangular [1-3]. The mechanisms resulting in segmental or triangular profiles are not fully understood, but the outcome depends on the collision overlap, orientation and severity and the structural characteristics of the colliding pair and their resulting interactions. A car type may assume either profile.

1.1 Geometric Crush Model

This paper details the application of a geometric crush model derived by Wood et al. [4, 5]. The model relates displacement of the vehicle cg (d_{cg}) to mean crush depth (d_{mean}) for varying collision overlaps (OL). It is assumed that the length of the car front (w) is inextensible, and that the non-struck side hinges at the front of the occupant compartment. For segmental crush:

$$d_{cg}' = \bar{d}' + \left(\frac{\cos\alpha}{2}\right)(1 - OL)^2, \qquad [1]$$

while for a triangular crush profile

$$d_{cg}' = \bar{d}' + (1 - OL)\frac{\cos\alpha}{2}. \qquad [2]$$

The parameter α is a non linear function of the overlap and the vehicle length. It is assumed that energy absorbed in deformation is solely due to longitudinal crush and is a unique function of the mean crush depth. Therefore, combining the geometric model with the full-width barrier velocity-crush characteristic for a given type allows a prediction of crush versus speed for different overlaps.

The model was applied to eight vehicle types where the full-width barrier response as well as the 45% offset and 30 degree angled barrier test data were available [6]. Figure 1 shows magnitude and phase of the predictions compared to the test data for one car type.

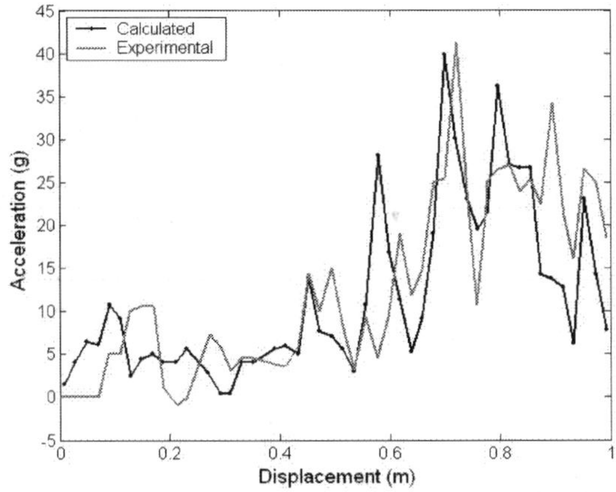

Figure 1. Comparison of geometric shape model prediction with experimental data for sample car, 30 degree angled barrier test, from [6].

1.2 Theoretical Considerations: The Stress/Density Parameter

The kinetic energy of a vehicle striking a rigid barrier is absorbed by the vehicle structure as it collapses to a crush depth (d) under a mean crumpling force ($\overline{F}$). For a plastic impact,

$$\frac{1}{2} M v^2 = \overline{F}\, d \qquad [3]$$

Using the relations between mean acceleration, force and stress ($M\bar{a}$ = $\overline{F} = \overline{\sigma} A$), and between mass and density ($M = \overline{\rho} A L$), two important equalities result:

$$\frac{1}{2} v^2 = \left(\frac{\overline{\sigma}}{\rho}\right) \frac{d}{L} \qquad [4] \qquad \text{and} \qquad \bar{a} L = \left(\frac{\overline{\sigma}}{\rho}\right) \qquad [5]$$

The stress/density parameter (σ/ρ) relating speed to crush is $\bar{a}L$.

1.3 Crash Pulse Recorder Data

Folksam Research have developed and installed crash pulse recorders in 160,000 cars and reported on crashes involving this fleet [7]. Wood *et al* [8-9] analyzed these data for 269 single and two car frontal collisions with overlap ≥25% and collision direction 11-1 o'clock. The analysis, involving twelve car types manufactured between 1985 and 2002 (lengths 3.6-4.8m), shows that linear log-log regressions adequately represented the relationships between $\bar{a}L$ and d/L, between $\bar{a}L$ and Δv and between Δv and d/L. For $\bar{a}L$ versus d/L, the form of the regression was

$$\bar{a}L = C_{da}\left(\frac{d}{L}\right)^{da} \quad [6]$$

The stress/density regression coefficient (C_{da}) is independent of car size and mass and represents the distribution of structural characteristics in the population. The exponent (da) is independent of size and mass, but correlated with C_{da}, on its own ($r^2 = 0.66$), and with C_{da} and mass combined ($r^2 = 0.95$). The mean characteristics of individual car types were statistically robust and independent of collision and configuration.

2. APPLICATION OF CRUSH MODEL

The model was applied in a Monte Carlo procedure to predict the scatter in collision responses evident from the crash recorders. Δv was randomly selected from a log normal approximation of the real world collision velocity distribution derived from the crash pulse data. A single barrier characteristic representative of the vehicle population was used to calculate the resulting mean crush depth (d_{mean}/L). Following selection of the crush profile and collision overlap using previously reported distributions [10], the resulting centre of gravity displacement was calculated. Equation 6 were used to calculate the mean acceleration response ($\bar{a}L$) as a function of normalized crush d/L. Figure 2 compares the predicted overall population responses with the onboard collision recorder data for the stress/density parameter ($\bar{a}L$) versus normalised displacement (d/L) and velocity change (Δv) versus d/L.

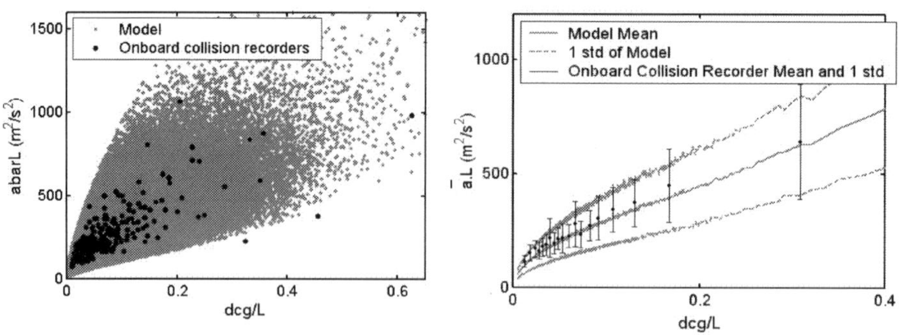

Figure 2. Monte Carlo model versus real world crush characteristics.

Figure 2 shows the model scatterplots and the mean and ±1 standard error curves for both the model predictions and the real world collision recorder data. These graphs show good matching between the model and the onboard collision recorder data, confirming that the mean collision recorder responses reflects the mean behaviour in frontal collisions.

2.1 Distribution of Responses

Scatter in the full-width barrier tests is due to variation between car types (σ_{type}). In contrast, the collision recorder data includes overlap ($\sigma_{overlap}$) and frontal collision effects ($\sigma_{frontal\ effects}$) due to car-to-car interactions and impact angle. The collision recorder data can only be separated into standard errors for car type (σ_{type}) and a combined error for overlap and frontal effects ($\sigma_{overlap+frontal\ effects}$). The full-width barrier data provides a standard error for car type as does the onboard recorder data. Table 1 shows the results: the overall standard error ($\sigma_{overall}$) for the model was computed and the standard error due to overlap was obtained. As the model response was computed from barrier data, structural interactions and collision configurations are not included. Hence, the overall and overlap standard errors for the model are less than for the collision recorder data and this allows $\sigma_{frontal\ effects}$ to be computed. Table 2 shows the mean values of the standard errors.

Table 1. Variability between model and onboard recorder data.

Standard error	Model	Collision Recorders
σ_{type}	0.269	0.264
$\sigma_{overlap}$	0.234	---
$\sigma_{overlap+frontal\ effects}$	---	0.267
$\Sigma_{overall}$	0.357	0.375

Table 2. Mean Standard errors.

Standard error	σ_{type}	$\sigma_{overlap}$	$\sigma_{frontal\ effects}$	$\sigma_{overlap\ +\ frontal\ effects}$	$\sigma_{overall}$
Mean	0.27	0.25	0.12	0.28	0.37

2.2 Frontal Response Estimator (FRE)

The influence of overlap on the stress/density coefficient (C_{da}) and the exponent (d_a) for segmental and triangular crush shapes was investigated. The regression exponent (d_a) was independent of overlap, but regression of $\log_e(C_{da})$ versus overlap showed high correlation. Further, C_{da} for mean overlap was found to be 58% of the full-width value and the collision recorder data showed a strong correlation between d_a and $\log C_{da}$:

$$d_a = 0.2855 \log_e C_{da\_100\%} - 1.6178 \qquad [7]$$

Given the velocity-crush characteristics of an individual car type and the car-to-car frontal effects variation or overlap plus frontal effects variation, the model yields the car type response either for specific overlap/crush shape or for all frontal collisions. In most cases, full-width barrier response data for specific car types is not available. However, if speed and maximum dynamic displacement are available, a frontal response estimator (FRE) can be computed using

$$\frac{v^2}{2} = C_{da-100\%} \cdot \left(\frac{d}{L}\right)^{1+d_a}, \qquad [8]$$

where v is the impact speed in the barrier test and d is the maximum dynamic crush (estimated from the residual crush [11]). Equations 7&8 can then be combined to calculate C_{da} and the exponent d_a for the full-width barrier case.

2.3 Application of Frontal Response Estimator

The Frontal Response Estimator was applied to 7 car types, see Figure 3 for mean and 95%ile FRE predictions compared to the real world data.

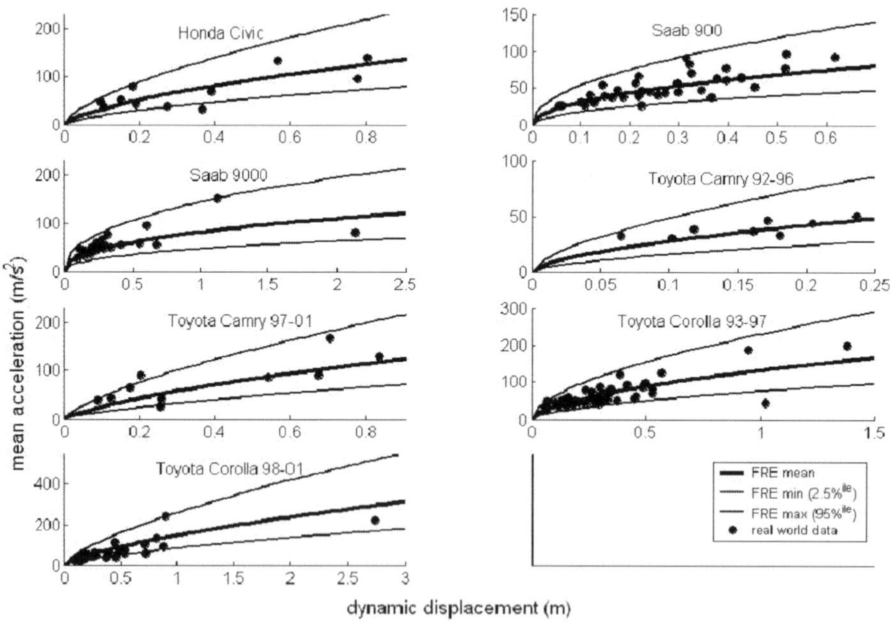

Figure 3. Frontal Response Estimator predictions.

3. DISCUSSION

Estimation of collision severity during impact is a key factor in biomechanics as it allows correlation between injuries and crash severity. Figure 1 and [6] shows the ability of the geometric crush model to predict the acceleration pulse response in offset collisions using the full-width barrier characteristic. Comparison of the model with the onboard collision recorder data (figure 2) shows the predictions encompass the crash data with good matching between the mean and the dispersion of both datasets. The mean acceleration response for individual car types and the overall population response obtained from the onboard recorder data reflect the mean frontal response. The crash data reflects the distributions of car type, overlap and collision partner interaction in frontal collisions. The influence of collision partner and frontal effects on overall dispersion is shown by comparing the standard errors due to car type, overlap and frontal effects (table 2). The influence of collision partner, etc. is less than half of that due to variation in overlap (49%) or car type (47%).

The Frontal Response Estimator, (FRE), uses the peak dynamic displacement in the full-width barrier test to derive the mean response in frontal collisions. The overall dispersion is obtained by applying the combined standard error for overlap and frontal effects to the mean response. Alternatively the response at specific degree of overlap could be obtained by applying equations 7 & 8 and the frontal effects standard error, $\sigma_{frontal\ effects}$. This procedure was applied to seven car types included in the onboard recorder data [7-9] (data for the remaining five is not in the public domain). Figure 4 shows a good visual match in all cases. A paired t test of the difference between the mean FRE prediction and real world data for six of the seven cases showed no statistical significance at a 95% confidence level. In the case of the Toyota Corolla 1998-2001, the difference was statistically significant, but a similar analysis performed on the basis of $\bar{a}$ versus Δv showed no significant difference for that car. Table 3 summarises the details.

Table 3. Results of a paired Student's t test.

Car Type	*N*	*Critical t (95%)*	*Actual t score*
Honda Civic	11	2.33	0.11
Saab 900	36	2.03	1.36
Saab 9000	23	2.07	1.63
Toyota Camry 92-96	8	2.37	1.61
Toyota Camry 97-01	10	2.26	1.51
Toyota Corolla 93-97	49	2.01	0.26
Toyota Corolla 98-01	39	2.02	2.44

In future work, the geometric model and the FRE will be used in combination with injury risk functions for frontal collisions. This will allow prediction of the effect of overlap on injury risk as well as the categorisation of the safety of individual car types.

4. CONCLUSIONS

The Geometric crush model provides an effective method to predict vehicle kinematics at varying degrees of overlap. Application of this model to the distribution of real world frontal collisions shows a very good match to the mean vehicle response. Furthermore, the model isolates the sources of variability in frontal collisions, showing that the effects of structural interactions can mask the influence of overlap. The simplicity of the FRE method, along with its success in predicting the real world responses of individual cars mean that it can now be combined with injury risk functions to find relationships between overlap and car type and injury risk.

REFERENCES

1. Kallina, I., Zeidler, F. and Scheunert, D., "Safe or unsafe in road accidents? Can this question be answered by comparing crash test results?", in *Comparative Crash Tests within the EC*, Published by TÜV Rhineland, Cologne, 1992, pp. 95-104.
2. Tsuda, Y. and Yamanoi, T., "Technical consideration on the current status of crash tests in the case of head on collisions", in *Comparative Crash Tests within the EC*, Published by TÜV Rhineland, Cologne, 1992, pp. 105-114.
3. Appel, H., Lutter, G. and Sigmund, T., "The relevance of crash tests to real road collisions", in *Comparative Crash Tests within the EC*, Published by TÜV Rhineland, Cologne, 1992, pp. 185-204.
4. Wood, D.P., Doody, M. and Mooney, S., "Application of a generalised frontal crush model of the car population to pole and narrow object impacts", SAE Paper 930894, 1993.
5. Wood D.P., Glynn C., Simms C.K., "Frontal collision behaviour: Comparison of onboard collision recorder data with car population characteristics", accepted for publication by *International Journal of Crashworthiness*, 26 November 2004, publication scheduled for Spring 2005.
6. Wood, D.P., Mooney, S. and Vallet, G., "An estimation method for the frontal impact response of cars", in *1997 IRCOBI conference*, Munich, Germany, September 24-26, 1997, pp. 321-336.
7. Ydenius, A., "Influence of crash pulse duration on injury risk in frontal impacts based in real life crashes", in *Proceedings of IRCOBI Conference*, Munich, Germany, September 18-20, 2002, pp. 155-166.
8. Wood, D.P, Ydenius, A. and Adamson, D., "Velocity changes, mean accelerations and displacements of some car types in frontal collisions", *International Journal of Crashworthiness*, 2004, Vol. 9, No. 3, pp. 311-325.
9. Wood, D.P., Adamson, D. and Ydenius, A., "Car frontal collisions: Occupant compartment forces, interface forces and stiffnesses", *International Journal of Crashworthiness*, 2003, Vol. 8, No. 6, pp. 591-603.
10. Glynn, C., "Geometric crush model of cars: Comparison with real-life responses," Final Year Thesis, Dept. of Mechanical and Manufacturing Engineering, Trinity College, Dublin, 2004, pp. 27-28.
11. Wood, D.P., Mooney, S., Doody, M. and O'Riordain S., "The influence of car crush behaviour on frontal collision safety and on the car size effect," SAE Paper 930893, 1993.

ACCIDENTS OF MOTORCYCLISTS AGAINST ROADSIDE INFRASTRUCTURE

Francisco J. López-Valdés, David García, David Pedrero and José L. Moreno
CIDAUT, Parque Tecnológico de Boecillo, P. 209, 47151 Boecillo, Valladolid, Spain; Phone: +34 983 548035, Fax: +34 983 548062, E-mail: fralop@cidaut.es

Abstract. Powered two wheelers (PTWs) riders are one of the most vulnerable groups of road users. Roadside barriers work quite effectively for passenger cars, but sometimes they are dangerous for motorcyclists, especially guard-rails. These accidents are investigated by means of an analysis of the DGT Spanish National Accident Database and reviewing different in-depth studies. Motorcyclist injury patterns have been identified. A revision of the injury criteria used for head and neck injuries has been carried out to select those suitable to be used in an impact of a motorcyclist against a roadside barrier. Head injury criterion (HIC36) for head injury has been accepted as a valid injury criterion. Regarding the neck, the Mertz criterion was considered in a first stage and some changes have been introduced in order to suit the peculiarities of motorcyclists' impacts, using adapted methods from the automobile sector related to Out-of-position (OOP) scaling methodology.

Key words: motorcyclists accidents, guard-rails, head injury criteria, neck injury criteria.

1. INTRODUCTION

In Spain, the number of motorcycles and moped has grown continuously from 2000 to 2002 due to their advantages as flexibility or enjoyment. However, these vehicles are often involved in accidents and they account for a large number of casualties. For example, in 2002, 784 motorcyclists died in road accidents in Spain (both motorcycle and moped riders). This figure implies that the 14.7% of the total number of casualties were riders, while

Michael D. Gilchrist (ed.), IUTAM Proceedings on Impact Biomechanics: From Fundamental Insights to Applications, 163–170.

the total number of powered two wheels vehicles (PTW) was the 13.3% of the total number of registered vehicles in the country.

Moreover, it also should be pointed that, although the number of cars has also increased during those years, the number of accidents of passenger cars has decreased but when focusing on motorcycles, the number of accidents in which a motorcycle is involved has even increased in the period studied.

Similar trends have been found in other different countries in Europe as The Netherlands, Italy or Germany [**1**].

2. ACCIDENTOLOGY STUDIES: INJURY PATTERNS

2.1 Spanish National Road Accident Database

Accident data from Spain have been extracted from the Spanish National Road Accident Database, intensive database performed by a public organism called DGT (General Directorate of Transport), depending on the Ministry of Interior. It contains every road accident with casualties in Spanish roads and is performed by Police forces at the scene of the accident.

In 2002, 5,979 motorcycle/moped accidents occurred in Spanish roads where 7,834 road users resulted injured. Nearly 20% of these accidents were run off the road accidents, in which the likelihood of the impact of the rider against roadside infrastructure or against a hazardous object placed close to the carriageway is extremely high. Barriers have demonstrated their effectiveness restraining passenger cars or heavier vehicles but, due to the unrestrained position of the rider and the peculiarities of motorcycle accidents, they could be dangerous as riders usually impact against the barrier with their own body. Special attention must be paid to metal guardrails and posts.

In a first stage of this study, data from the Spanish national database were analyzed in order to obtain the magnitude of the problem. However, two difficulties arise from this analysis. The first one is identifying if the motorcyclist has hit the metal barrier or not when the accident occurs using this intensive database. The second one is related to injury codification done by the Police at the moment of the accident. It can be assigned only one type of injury to each casualty, trying to code the most serious one. Sometimes this codification is not trivial and medical knowledge should be required.

Only injuries sustained by riders involved in accidents outside of urban areas have been studied, as the likelihood of hitting a barrier is higher.

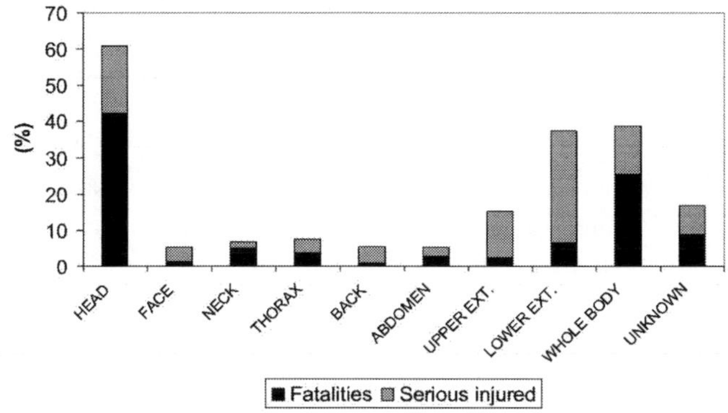

Figure 1. Main injured motorcyclist body region in fatal and serious accidents in non-urban areas. Spain, 2002.

Figure 1 shows injuries sustained by fatalities and serious injured riders. Head, extremities and whole body injuries are the most frequent injured body region.

However, it is necessary to analyze in-depth data to obtain a better insight of the sustained injuries. Available in-depth information from European and American sources has been reviewed.

2.2 In-Depth Accident Studies

Hell and Lob [2] studied 173 motorcyclist accidents in Munich from 1985 to 1990. The study considered all accident configurations. In 27 cases, riders hit a fixed object (e.g. a tree, a post, a metal barrier).

Table 1. Injured body regions (AIS2+). Motorcyclists vs. fixed objects. Munich, 1985-1990.

Body region	Frequency
Head	63%
Thorax	48%
Abdomen	19%
Pelvis	7%
Spinal cord	33%
Upper extremities	37%
Lower extremities	37%

These accidents were found to be the most sever (39 fatalities) within the studied sample. It should be pointed again that this data are not focus only in impacts against barriers but also against other objects. Again, special role is played by head and extremities, although it is also found the importance of thorax and back injuries (which are more difficult to be identified by the

Police and need a better insight into the casualties performed by medical teams). The internationally accepted *Abbreviated Injury Scale (AIS)* defined by the Association for the Advancement of Automotive Medicine (AAAM) is used to define the severity of the injuries.

Hurt et al. [**3**] found that the frequency of AIS 3+ injuries of motorcyclists depending on the body region was distributed according percentages presented in Table 2, although in this case the information is not only focused on accidents against fixed objects:

Table 2. Injured body regions (AIS3+). Motorcyclists. California.

Body region	Frequency
Head, neck and face	32.3%
Thorax	18.3%
Abdomen	10.5%
Upper extremities	15.5%
Lower extremities	23.4%

Neck injuries are not specifically mentioned in these last studies. However, Thom et al. [**4**] pointed that fatal injuries to the neck are more frequent than most of the studies shown but they need to be carefully examined to be identified. They performed the autopsy to 304 dead motorcyclists and they found 195 dislocations and 28 fractures, all of them placed at the upper region of the spine (C1-C2 level). They did not found differences in wearing helmet or not in these neck injuries. Similar findings are reported by Bishop and Wells [**5**] and Orsay et al. [**6**].

However, one of the most important motorcyclist accidents in-depth studies was the mentioned study performed by Hurt et al. [7] in California. From the accidents studied, the most severe neck injury was only a minor cervical sprain or complaint of pain. They suggested that as long as usually the load to the neck is transmitted from the head, the lower the forces on the head the lower the load on the neck.

At last, Otte [**8**] studied 876 motorcycle accidents from 1985 to 1995 in Hannover. He reported that head and lower extremities serious injuries are frequent in accidents against fix objects.

These in-depth accident studies are not focused on impacts against roadside infrastructure. Two investigations have been considered, focused specifically on these accidents. The first one has been performed by the Aerospace Engineering Department of the University of Milan [**9**]. This study focuses mainly on head, thorax, abdomen and spinal cord injuries. The second, performed by the French institute INRETS, is addressed to develop a test procedure for impacts of motorcyclists against metal barriers, based on an accidentology study at the area of Lyon during 1995 [**10**]. Nevertheless, this accident investigation was not only focused on accidents against guard rails, but also other kind of impacts. At last, it was decided to consider only head and neck in the test, as they were the most severe injuries found.

In conclusion, after the revision of the available documentation and data from different sources, it was decided to review the biomechanical criteria that could be suitable to study the impact of a motorcyclist against roadside infrastructure. This review is focused only on head and neck criteria, as they are generally mentioned as the most important injuries that could cause a fatal accident in almost all the studies.

3. HEAD INJURY CRITERION (HIC)

The brain is one of the most critical body parts due to its importance and vulnerability. Hurt et al. [7] found in their study that the regions of concern were the cranium and the enclosed brain, both protected by the helmet. Only not penetrating injuries are considered in this study, as the impacts of motorcyclist against barriers used to be of this kind. It is proposed to use HIC (Head Injury Criterion) as criterion for this head impacts. This parameter was defined based on investigations performed at the Wayne State University [**12**]. It has been revised from 1960 to 1986, when it was established that the time period considered in the measures should be 36 milliseconds.

Although this criterion has been developed for frontal impacts and only considers linear accelerations, it is commonly accepted in some lateral impact protocols as in EuroNCAP, for instance. Thus, it is proposed that, in a first stage and without trying to develop new criteria focused only in riders, HIC criterion is valid to evaluate the injuries to the head in accidents of motorcyclists against metal barriers. It is assumed that the same level as it is established in the EuroNCAP protocol can be sustained by the motorcyclist:

$$HIC_{36} \leq 1000 \Rightarrow p(AIS \quad 3+) = 20\% \tag{1}$$

However, it is necessary to consider that when testing this kind of impacts the helmet should be taken into account, so more considerations need to be done regarding the position and fixing of the helmet to the dummy. Moreover, as the aim of this paper is to give information that could be used in the development of a procedure to test impacts of motorcyclists against barriers using available tools, it is proposed that a conventional dummy should be used in the tests, so a method to attach the helmet to the head of the Hybrid III dummy has been also investigated. Further improvement of head injury criteria has been recently suggested in the project COST 327 [**13**], but there is not a clear established test procedure yet.

4. NECK INJURY CRITERIA

European regulations and associations as EuroNCAP accept injury criteria for frontal impact based on the studies carried out by Mertz with slight modifications, considering F_z (axial force, both extension and compression), F_x (shear force) and M_y (both flexion and extension torques). These limits have been accepted for the case of motorcyclists. However, there is not any established limit for the other dynamic values M_x, M_z and F_y, being the first of these three of great importance in side impact.

In USA, FMVSS 208 regulation for frontal impact uses the N_{ij} criterion. Recently, this criterion has been also developed to be applied in *Out-of-position (OOP)* situations. OOP problematic is interesting for motorcyclists because several similarities can be found (mainly, dummy receiving a lateral impact in the head). A group of biomechanical experts of the *Technical Working Group (TWG)* has established the limits for the torques M_x and M_z based on their experience for these OOP situations [**14**]. OOP tests are performed using 3-years-old, 6-years-old, Hybrid III 5th Small female and SID-IIs dummies, but not for Hybrid III 50th male. Thus, M_x values are not available for this dummy. To solve this inconvenience, it is proposed to use the same scaling method applied in frontal impact to obtain the N_{ij} criterion for the other dummies of the Hybrid III family. This scaling process is based on geometrical factors and basically, the torques Hybrid III 50th male are two times the torques in Hybrid III 5th Small female [**15**].

In summary, values for F_x, F_z and M_y are extracted from EuroNCAP (mainly based on Mertz criteria). Value for M_x is adapted from the OOP situation studied in FMVSS 208, scaling the proposed value for the Hybrid III 5th Small female to the Hybrid III 50th male. Values for F_y and M_z are not considered in this study. After this process, the proposed criterion is:

Table 3. Neck injury criterion.

F_x (N)	F_z tensile (N)	F_z compression (N)	M_y extension (Nm)	M_y flexion (Nm)	M_x (Nm)
Figure 2	Figure 3	Figure 4	57	190	134

When all these values are respected, it is considered that the neck injury criterion is fulfilled.

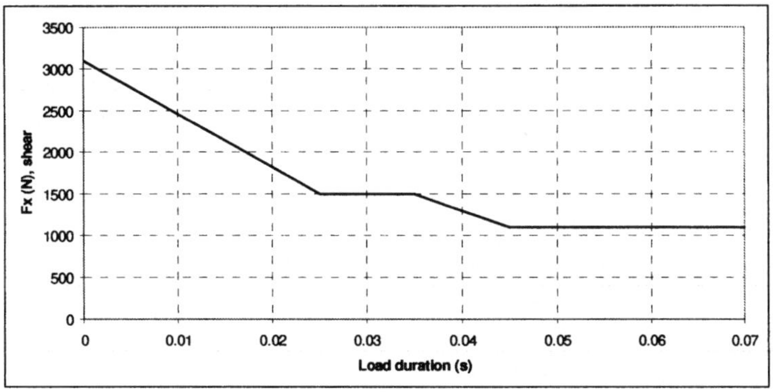

Figure 2. F_x shear force.

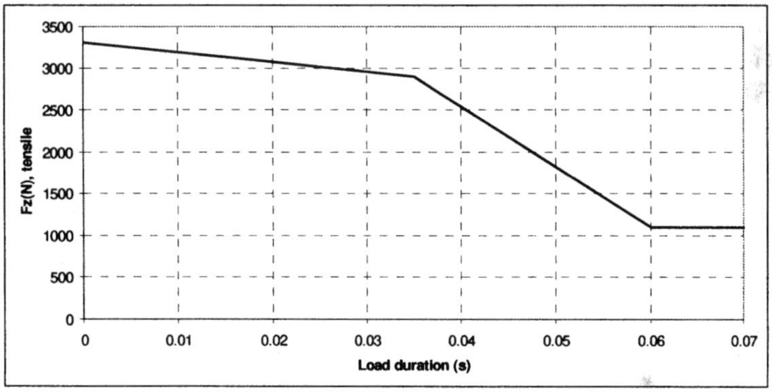

Figure 3. F_z tensile force.

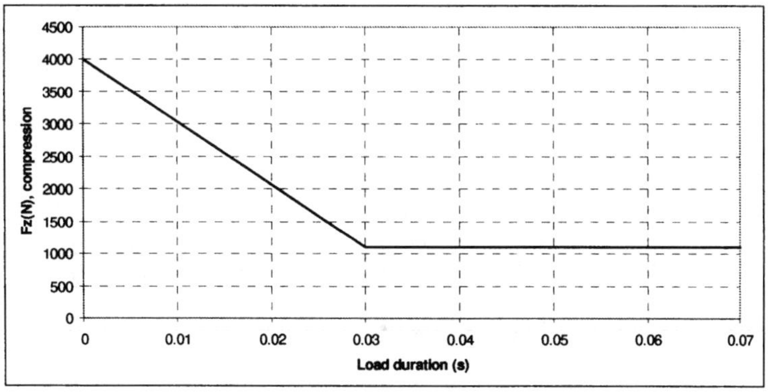

Figure 4. F_z compression force.

5. CONCLUSION

After the revision of the injury patterns of motorcyclists involved in accidents against barriers, injury criteria have been proposed for head and neck injuries.

They have been adapted from the automobile world and can be measured using a Hybrid III 50% dummy, rather than trying to develop injury criteria only focused on riders and needing specific dummies. These criteria can be used in testing of barriers to improve the safety of roadside infrastructure considering also motorcyclists and not only cars.

REFERENCES

1. AP-SP41-0001. APROSYS Sub Project 4: Motorcycle Accidents. European funded project. TIP-CT-2004-506503.
2. Hell, W. and Lob, G., Typical injury patterns of motorcyclists in different crash types – Effectiveness and improvements of countermeasures, in *37th Annual Proccedings of the AAAM,* Texas, USA, November, 1993.
3. Hurt, H., Ouelletm J.V. and Jennings G., Motorcycle crash bar effectiveness: A re-evaluation using AIS-80, in *AAAM Conference,* 1984.
4. Thom, D.R., Hurt, H.H. Jr., Smith, T.A. and Rehman, I., Atlas and axis injuries in fatal motorcycle collisions, in *AAAM Conference,* 1995.
5. Bishop, P.J. and Wells, R.P., The inappropiateness of helmet drop tests in assessing neck protection in head first impacts, *The American Journal of Sport Medicine,* **18(2)**, 1990.
6. Orsay, E.M., Muelleman R.L., Peterson T.D., Jurisisc, D.H., Kosasih, J.B. and Levy, P., Motorcycle helmets and spinal injuries: Dispelling the myth, *Annals of Emergency Medicine*, **23(4)**, 1994.
7. Hurt, H.H. Jr. and Ouellet, J.V., Effectiveness of motorcycle safety helmets and protective clothing, in *Proceedings of the 25th Conference of the AAAM,* San Francisco, California, USA, 1981.
8. Otte D., Benefit of in-depth data for analyzing injury mechanisms of accidents with bicyclists and motorcyclists, *International Journal of Crashworthiness,* **3(1)**, 1998.
9. Sala, G. and Astori, P., New concepts and materials for passive safety of motorcyclists, in *IRCOBI Conference*, 1998.
10. Quincy, R., Protocole d'essais de dispositif de retenue assurant la securite des motorcyclists, Laboratorie d'essais Inrets Equipments de la Route (LIER), 1998.
11. Chou C., Nyquist G., Analytical studies of the head injury criterion (HIC), SAE Paper No. 740082.
12. Chinn, B., Canaple B., Derler, S., Doyle, D., Otte, D., Schuller, E. and Willinger, R., COST 327: Motorcycle Safety Helmets (Final report of the action), Belgium, 2001.
13. Lund, A.K., Recommended procedures for evaluating occupant injury risk from deploying side airbags, IIHS, August 2000.
14. Frontal Offset Crashworthiness evaluation: guidelines for rating injury measures, IIHS, May 2001.

SESSION 4

BIOMECHANICAL ASPECTS OF BLUNT AND PENENTRATING HEAD INJURIES

Narayan Yoganandan, Frank. A. Pintar, Jiangyue Zhang,
Thomas A. Gennarelli, and Nathaniel Beuse*
Department of Neurosurgery, Medical College of Wisconsin, 9200 West Wisconsin Avenue, Milwaukee, WI 53226, USA
**US DOT, NHTSA, Washington, DC, USA*
Phone: 414-384-3453, fax: 414-384-3493, e-mail: yoga@mcw.edu

Abstract. The objective of this presentation is to discuss certain biomechanical aspects of head injuries due to blunt and penetrating impacts. Emphasis is given to fundamental data leading to injury criteria used in the United States (US) regulations for motor vehicle safety. Full-scale and component tests done under US Federal Motor Vehicle Safety Standards (FMVSS) are described. In addition, results providing occupant safety and vehicle crashworthiness information to the consumer from frontal and lateral impact crash tests are discussed with an emphasis on head injury assessment and mitigation. In the area of penetrating impact, newer experimental techniques are described for a better understanding of head injury secondary to penetrating impacts, with specific reference to the civilian population.

Key words: head injury criteria, Federal Motor Vehicle Safety Standards (FMVSS), brain injury, skull fracture, penetrating trauma, acceleration.

1. PURPOSE

Head injuries in the civilian population occur due to blunt or penetrating impacts. Motor vehicle crashes are a major source of blunt impact-induced head injuries. Biomechanical techniques used to establish injury criteria are helpful for assessing occupant safety and design user-friendly vehicular components. This presentation describes developments in this area along with current US standards. In the area of penetrating head trauma, a significant majority of the literature is from the military domain [1, 2]. Recently, the focus has shifted toward the civilian domain, and because of technological improvements, it is possible to conduct tests to describe biomechanical aspects of injury from this type of external insult.

Michael D. Gilchrist (ed.), IUTAM Proceedings on Impact Biomechanics: From Fundamental Insights to Applications, 173–184.

2. BLUNT IMPACT – FUNDAMENTAL DATA USED IN INJURY ASSESSMENT

Quantifications of head injuries were reported in the 1930-60 literature, although limited fundamental biomechanical studies were conducted earlier [3-7]. Linear and angular accelerations were considered governing variables to describe mechanisms of trauma and define tolerance limits [8, 9]. Translational acceleration-time histories were related to skull fracture, using tests from four isolated embalmed cadaver heads and two full-body cadavers subjected to forehead impacts on flat unyielding surfaces. The specimens were dropped from predetermined heights and resulting linear accelerations were recorded at the occiput using an accelerometer. Pulses were of short duration because of the rigid end condition at the instant of impact. Association of linear skull fracture with brain injury, i.e., concussion (80% correlation from clinical cases) was based on previous studies [3, 5]. Peak accelerations were used from three isolated and two intact whole-body cadaver tests, and mean acceleration data was used from the other isolated head to develop the original tolerance curve that had durations of up to 6 msec [8]. This curve and corrected data are shown in Figure 1. Animal experiments were used to extrapolate to longer duration acceleration-time impacts. The final response, termed Wayne State University Tolerance Curve (WSTC) using effective accelerations as the ordinate, was obtained from animal, volunteer, human cadaver, and clinical research data [10, 11]. While WSTC is applicable to the adult population, no efforts were made to develop age-dependent tolerance curves. Versace in 1971 argued that because WSTC curve was developed for average accelerations, comparisons should be made using the mean pulse of interest [12]. He proposed the head injury criterion (HIC), which was modified by NHTSA to provide a better comparison to longer duration human volunteer data [12, 13]. This criterion was adopted by FMVSS in 1972 and is still used worldwide for head injury assessment in various areas of impact biomechanics [14]. The criterion is given in equation (1).

$$HIC = \left[\frac{1}{t_2 - t_1} \int_{t_1}^{t_2} a(t)dt \right]^{2.5} (t_2 - t_1) \tag{1}$$

where t_2 and t_1, arbitrary final and initial times during the acceleration pulse, are chosen to maximize HIC and, a(t) is the resultant acceleration at the center of gravity of the head. NHTSA chose a value of 1000 as the threshold in 1972. In October 1986, the interval over which HIC was computed was limited to 36 msec (HIC_{36}) with the same threshold of 1000 for the 50th

percentile Hybrid III anthropomorphic dummy. From a theoretical perspective, Backaitis stated that "the HIC formulation contains the peak power term or the rate of change of energy as seen by the head during the impact process" [15]. Eppinger interpreted HIC "as a measure of the change of specific kinetic energy modulated by the square root of the average acceleration over the time interval," and further remarked that, "if the 2.5 power in HIC equation were instead 2, the function would represent the peak average specific power delivered to the head [16]."

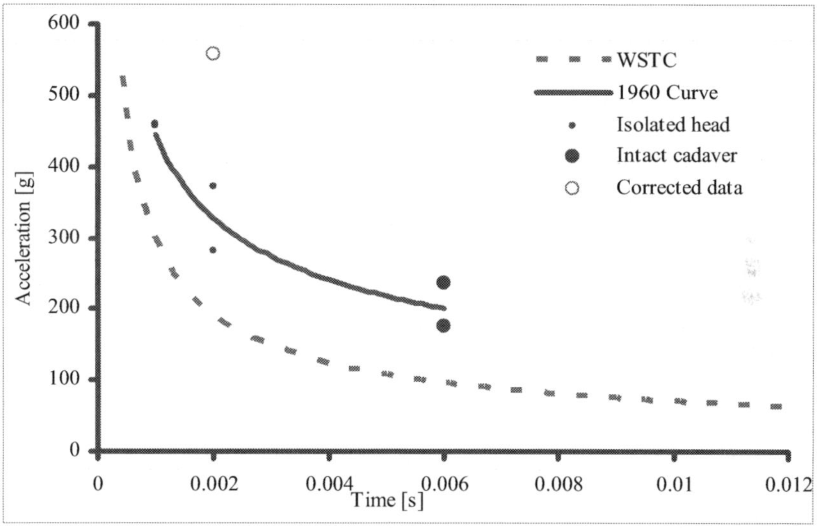

Figure 1. Acceleration responses from embalmed human cadaver tests and the original and revised tolerance curves. The 280 g data (at 0.002 sec, below the 1960 curve) was incorrectly plotted in the original publication [8]; corrected data is shown, open circle (0.001 sec, 557 g).

2.1 Adult Occupant Protection in Frontal Impact – FMVSS 208

The principal purpose of the standard is to decrease the number of injuries and fatalities by specifying crashworthiness of vehicles in terms of biomechanical variables measured in dummies tested in simulated environments. The federalized Hybrid III dummy is the anthropomorphic test device. The FMVSS 208 standard specifies the injury metric in terms of HIC for different dummies for head impact protection. The frontal impact standard calls for full-scale vehicle-to-barrier tests at a velocity of 48 km/h for belted and 40 km/h for unbelted 50th percentile male dummies. While the same fixed rigid barrier tests are specified for the 5th percentile female dummy, tests include an additional 40% left offset frontal deformable barrier test with belted driver and passenger dummies for the 5th percentile female anthropometry (Figure 2). In the case of the unbelted test with the 50th percentile adult male dummy, the vehicle is required to impact the rigid fixed barrier perpendicular to its line of travel and at any angle up to 30 degrees in either direction from the traveling line. In all tests and for both anthropometries (except offset test), the impact is always perpendicular to the path of the vehicle. For certain vehicles, an alternative unbelted test is done to evaluate airbags by sled testing at a delta V up to 48 km/h such that the sled acceleration falls within the corridors shown in Figure 3. HIC, determined using the resultant acceleration at the center of gravity of the dummy head, is computed over a 15-msec interval. The injury criterion is based on linear head acceleration data gathered for a period of 300 msec after the vehicle strikes the rigid barrier. HIC limit is chosen as 700 for both adult anthropometries, with no gender bias. For child dummies, HIC_{15} values are as follows: 12-month-old CRABI (Child Dummy AirBag Interaction) is 390, 3-year-old is 570, and 6-year-old is 700. It should be noted that the 95th percentile dummy is not specifically included in the current 208, although the HIC=700 was suggested during rulemaking. These data are in reference to advanced airbag phase-in of 208 standards.

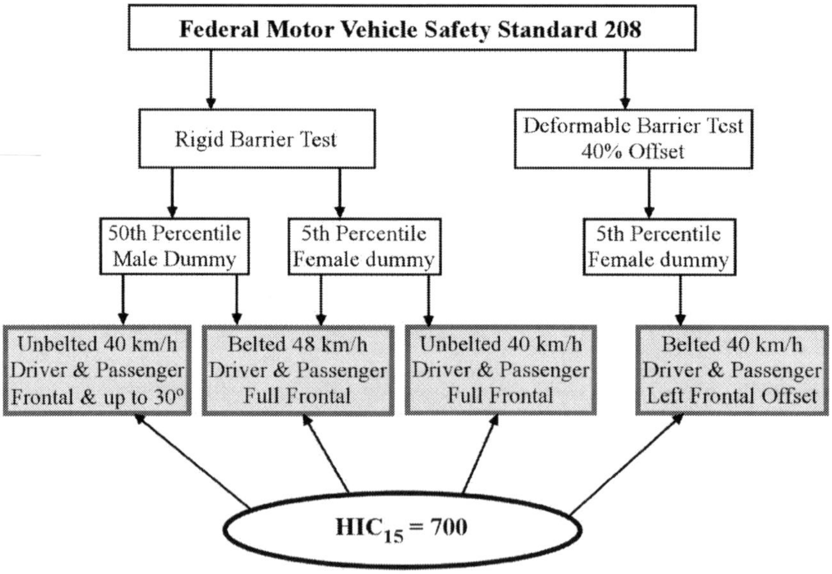

Figure 2. Test flow chart for frontal impact protection adopted by US FMVSS 208.

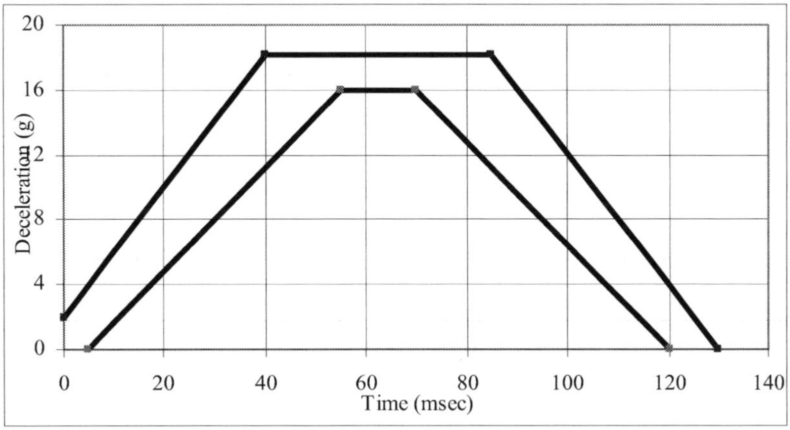

Figure 3. Deceleration pulse corridors for alternative sled tests with unbelted dummies.

The New Car Assessment Program (NCAP), referred to as consumer information tests for relative crashworthiness of vehicles, was initiated by NHTSA in 1978 [17]. The ultimate goal of NCAP is to improve safety by providing market incentives for vehicle manufacturers to voluntarily implement improved crashworthiness in vehicles, rather than through only regulations, i.e., compliance tests such as 208 and 214 standards. For frontal impacts, procedures similar to 208 are followed with the speed raised from 48 to 56 km/h. This increased speed differentiates the performance of vehicles as the energy in the NCAP test is 36% higher than the compliance 48 km/h test. Another measure of severity is the change in velocity (ΔV) of the occupant: 64 km/h (accounting for rebound) in the NCAP compared to 53 km/h in the compliance test. The NCAP currently computes HIC over a 36-msec interval compared to the 15-msec interval used in 208. However, in a recent Request for Comments (RFC) regarding the frontal program, NHTSA gave indication of its intent to begin using HIC_{15} as part of any potential upgrade. By combining injury numbers from the head and chest, a star-rating and a probability of injury are computed (Table 1). NHTSA published an RFC for establishing a high-speed regulation in 2001; it does not have a high-speed frontal offset test in the current standards. Figure 4 shows the probability of injury (AIS 4+) as a function of HIC. The head injury risk from real-world data (National Automotive Sampling System, NASS database) at the two speeds falls on the probability curve.

Table 1. Star-rating in rigid barrier frontal impact NCAP test.

Star-rating	Probability of injury
5 stars	10% or less chance of serious head/chest injury
4 stars	11-20% chance of serious injury
3 stars	21-35% chance of serious injury
2 stars	36-45% chance of serious injury
1 star	46% or greater chance of serious injury

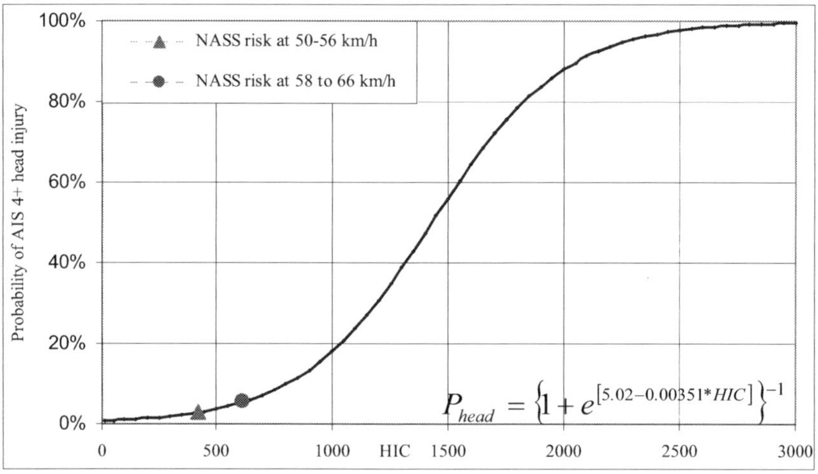

Figure 4. Probability of head injury as measured by HIC for MAIS $\geq$ 4 in frontal impacts. Solid circle and triangular symbols show the risk of head injury based on NASS analyses.

2.2 Occupant Protection in Side Impact - FMVSS 214

The FMVSS 214 standard uses the side impact dummy (SID) designed for lateral impact crashworthiness evaluations. However, a newer dummy, ES-2re, is being considered for future crashworthiness tests because of its enhanced biofidelity compared to SID; along with newer dummies, the pole test is being considered [18]. Specific limits for head injury assessment do not exist in the current version; thoracic and pelvic regions are covered by respective injury metrics. Like the frontal NCAP, the lateral impact test, LINCAP, uses a higher speed for the moving deformable barrier (62 km/h instead of 54 km/h in the compliance 214) test. These tests also use a star-rating, but the probabilities are different because the side star-rating is calculated only from injury metrics recorded in the chest/torso compared to the frontal impact that uses metrics from both the head and chest. However, since April 2002, NHTSA has noted safety concerns not reflected in the star-rating. One of those is specific to head injuries in LINCAP crash test. A safety concern remark is introduced informing the consumer about the potential for head injury in tests with HIC_{36} exceeding 1000. It should be noted that the same threshold of 1000 is chosen based on the FMVSS 201 pole test. Figure 5 shows the warning scheme that incorporates head injury measure in LINCAP tests, although probabilities are not attached with respect to specific HIC values.

Figure 5. Rating scheme used in lateral impact NCAP tests wherein a warning indicating a higher likelihood of head injury is noted if the test results in a HIC_{36} value exceeding 1000.

2.3 Child Occupant Protection in Frontal Impact - FMVSS 213

The FMVSS 213 standard focuses on child restraint systems. Unlike other standards, different dummies are used representing the growing anthropometric characteristics of the human child; table 2 shows specific dummies used for tests. Similar to 208, this standard also uses HIC although the time interval is 36 msec, and the threshold value of 1000 is independent of dummy age. The HIC time interval is currently limited but on August 1, 2005, will change to 36 msec. Specifically, depending on the type of restraint, newborn, 9-month old, 12-month, 3- and/or 6-year old dummies are subjected to a ΔV of 48 km/h in a sled environment. The newborn and 9-month old are not instrumented. The characteristic pulse shown in figure 6 has a peak acceleration of 23.5 g at 20 ms. For child seats manufactured after August 1, 2005, NHTSA adopted a trapezoidal shape corridor to define the upper and lower bounds to the pulse (Figure 7).

Table 2. FMVSS 213. Parentheses refer to the subpart specification of the part 572 dummy.

Weight (kg)	Height (mm)	Dummy to be used in child seats manufactured before August 1, 2005
≤ 5	≤ 650 mm	Newborn (part 572 K)
> 5 and ≤ 10	> 650 and ≤ 850	Newborn (K) and 9-month-old (J)
> 10 and ≤ 18	> 850 and ≤ 1100	9-month-old (J) and 3-year-old (C)
> 18	> 1100	6-year-old (I)
Weight (kg)	Height (mm)	Dummy to be used in child seats manufactured on or after August 1, 2005
≤ 5	≤ 650 mm	Newborn (part 572 K)
> 5 and ≤ 10	> 650 and ≤ 850	Newborn (K) and 12-month-old (R)
> 10 and ≤ 18	> 850 and ≤ 1100	12-month-old (R) and 3-year-old (P)
> 18	> 1100	6-year-old (N)
> 22.7	> 1100	N dummy weighted to 28.2 kg

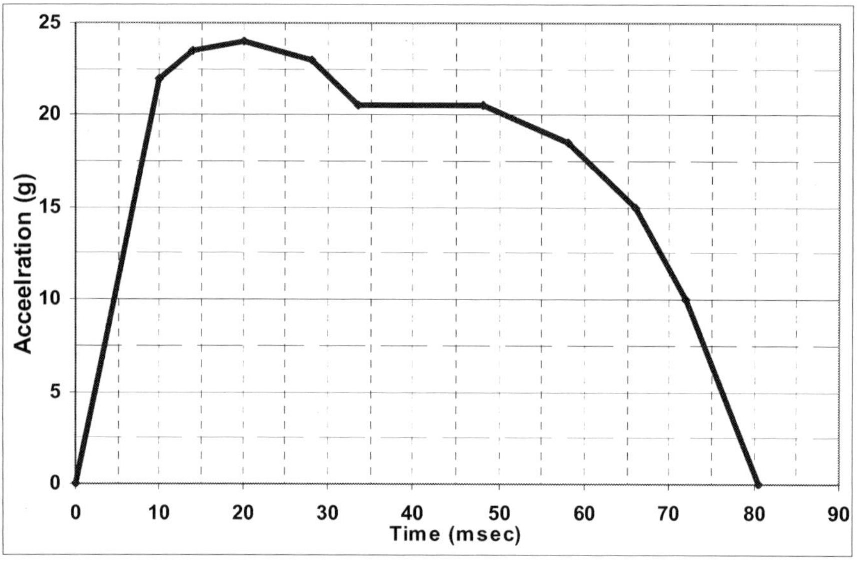

Figure 6. Acceleration-time history according to FMVSS 213 for vehicles manufactured before August 1, 2005.

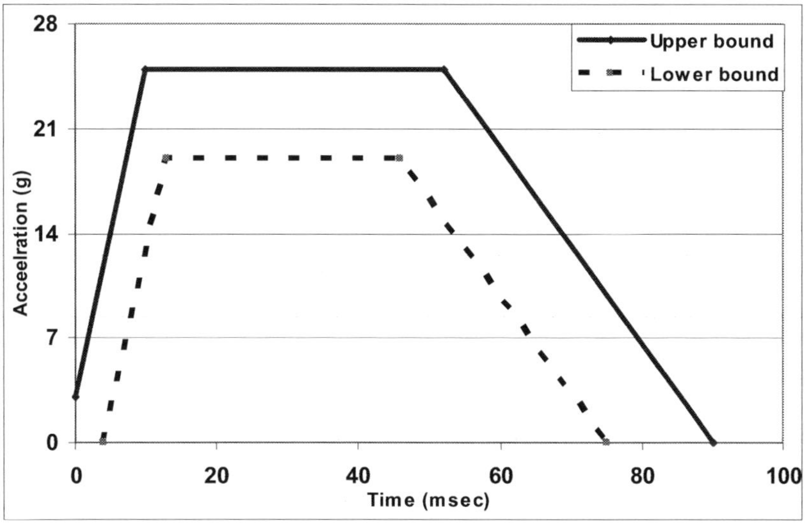

Figure 7. Acceleration-time history according to FMVSS 213 for vehicles manufactured on or after August 1, 2005.

2.4 Occupant Protection in Interior Impact - FMVSS 201

The FMVSS 201 standard has two parts. One part focuses on interior structures (A- and B-pillars and instrument panels) and uses HIC as the injury metric, with a different time limit for its computation, using a head-form for testing. The other part is an optional test for dynamically deployed head protection systems and uses a pole test. For the head-form test, a 6.8 kg, 165-mm diameter, free motion head-form impacts various points inside the vehicle with a velocity of 24 km/h. For a vehicle to pass, the deceleration of the head cannot exceed 80 g for a time period of 3 ms. For evaluating dynamically deployed head protection systems, tests are done using a SID instrumented with a Hybrid III anthropomorphic head and neck and impacting the side of a full vehicle instrumented with a 254-mm diameter stationary rigid pole at a velocity between 24 and 29 km/h. The performance criterion is HIC_{36} threshold of 1000. Injury criteria formula for the free motion head-form is given in equation (2).

$$HIC = 0.75446\ (\text{free motion head-form } HIC_{36}) + 166.4 < 1000 \qquad (2)$$

3. PENETRATING IMPACT

The majority of studies in this area is focused on the military domain and used gelatin as the simulant for injury/wound quantification, typically determined as the residual deformation following projectile penetration [1, 2]. In addition, the shape of the model is primarily confined to rectangular cross sections. Therefore, their applicability to human head trauma is limited. Recent tests from our laboratories have focused on civilian projectiles, used a more realistic brain simulant, and employed high-frequency pressure transducers coupled with very high-speed digital videography to capture the sequence of temporary cavities, and adopted a model better approximating the shape and boundary conditions of the head.

Briefly, two agents of a silicone dielectric gel, Sylgard 527 A and B were mixed and poured into a diameter globe. A hole in the center of the globe approximated the foramen magnum. A layer of neutral density reference lines was embedded in the "mid-sagittal plane" of the globe to monitor temporal movements of the projectile and gel material. Four pressure sensors were inserted into the globe through predrilled symmetrical holes. Two transducers were close to the entry, and two were close to the exit of the projectile. These pairs were referred to as entry- and exit-transducers for data interpretation. All transducers were approximately at the mid-height of the globe. They were inserted 3.5 cm from the outer surface of the globe so that all sensors were mutually orthogonal to each other. Nine-mm and 25-caliber projectiles were discharged to ensure penetration at the mid-diameter of the globe. The test was photographed using a digital video camera at 20,000 frames/sec. A digital data acquisition system was used to capture the transducer signals at 308 kHz. A fresh globe was used for each test. Parallel tests were conducted by replacing the Sylgard gel with gelatin. Pressure data are summarized in Table 3.

Table 3. Peak average pressures (kPa) comparing responses from two brain simulants for two projectiles. Averages were computed from the two entry- and two exit-transducer sets.

Projectile	25 caliber		9 mm	
Simulant	Entry-transducers	Exit-transducers	Entry-transducers	Exit-transducers
Gelatin	245	151	569	484
Sylgard gel	179	242	645	630

In general, results from both simulants indicated significantly higher energy and wounding power for the 9-mm than the 25-caliber projectile; this was determined by the formation and collapse of the temporary cavities and pressure distributions. The 9-mm and 25-caliber projectiles had entry velocities of 378 and 238 m/s. The Sylgard gel responded with higher changes in pressures than the gelatin (e.g., entry: 466 versus 324 kPa) when the projectile was changed from 25-caliber to 9-mm, demonstrating greater sensitivity. Because material properties of the gel are closer to the human brain, i.e., increasing dynamic modulus with increasing loading frequency, and because this simulant responded with more differentiable responses compared to the gelatin, the gel may be the most appropriate simulant for brain injury penetrating trauma studies. Temporal pressure distributions at various locations can be used to validate computer models aimed to delineate stress analysis-related variables for brain injury quantifications. Numerical models using the finite element technique are being pursued in our laboratories.

4. CONCLUSIONS

Certain biomechanical aspects of head injuries due to blunt and penetrating impacts are discussed. Head injury criterion based on linear acceleration still remains as the metric for injury assessment and vehicular crashworthiness evaluations in the US, although HIC has been modified in recent years. With specific reference to the civilian population, to understand the biomechanics of penetrating impact, newer experimental techniques are beginning to emerge.

ACKNOWLEDGEMENTS

This study was supported in part by DTNH22-H-03-04117, PHC CDC R49 CCR 519614, and VA Medical Research.

REFERENCES

1. Yoganandan, N. and Pintar, F., Biomechanics of penetrating trauma, CRC Review Biomed Eng, 1997, 25:485-501.
2. Yoganandan, N., Pintar, F.A., Kumaresan S., et al., Dynamic analysis of penetrating trauma, J Trauma, 1997, 42:266-272.
3. Gurdjian, E.S., Webster, J.E. and Lissner, H.R., Observations on the mechanism of brain concussion, contusion, and laceration, Surg Gynecol Obstet, 1955, 101:680-690.
4. Gurdjian, E.S., Lissner, H.R., Evans, F.G., et al.. Intracranial pressure and acceleration accompanying head impacts in human cadavers. Surg Gynecol Obstet., 1961, 113:185-190.
5. Gurdjian, E.S. and Schwan, H., Management of skull fracture involving frontal sinus, Am J Surg 1932, 95, 22-32.
6. Messerer, O., Uber Elasticitat und Festigkeit der Menschlichen Knochen, Stuttgart, Germany, 1880:79.

7. Evans, F.G., Lissner, H.R. and Lebow, M., The relation of energy, velocity, and acceleration to skull deformation and fracture, Surg Gynecol Obstet, 1958, 107:593-601.
8. Lissner, H.R., Lebow, M. and Evans, F.G., Experimental studies on the relation between acceleration and intracranial pressure changes in man, Surg Gynecol Obstet, 1960, 111:329-338.
9. Holbourn, A.H., Mechanics of head injuries, Lancet, 1943, 2:438-441.
10. Gurdjian, E.S., Lissner, H.R. and Patrick, L.M., Concussion-mechanism and pathology, 7th Stapp Car Crash Conference, 1963, 58-60.
11. Society of Automotive Engineers, Human tolerance to impact conditions as related to motor vehcile design-J885, Warrendale, PA: SAE, 1980.
12. Versace, J., A review of the severity index, Stapp Car Crash Conf. Coronado, CA, 1971, 771-796.
13. Kleinberger, M., Sun, E., Eppinger, R., et al., Development of improved injury criteria for the assessment of advanced automotive restraint systems, Washington, DC: NHTSA, 1998, 115 pp.
14. NHTSA. www.nhtsa.dot.gov [Department of Transportation], October 2003.
15. Backaitis, S.H., The head injury criterion, Head and Neck Injury Criteria: A Consensus Workshop, Washington, DC: US Department of Transportation, 1981, 175-177.
16. Eppinger, R.H., Discussion of injury criteria, Head and Neck Injury Criteria: A Consensus Workshop, Washington, DC: US Department of Transportation, 1981, 204-249.
17. Hackney, J.R., Kahane, C.J. and Quarles, V.R., The new car assessment program -- historical review and effect, 1994, 75-89.
18. Federal Register, Docket No. NHTSA-2004-18864. 49 CFR Part 572, Anthropomorphic Test Devices; ES-2re side impact crash test dummy, September 15, Washington, DC: US Government Printing Office, 2004:55550-55571.

BIOMECHANICS OF FRONTAL SKULL FRACTURE

H. Delye[1], P. Verschueren[2], B. Depreitere[1], C. Van Lierde[2], I. Verpoest[3], D. Berckmans[4], J. Vander Sloten[2], G. Van der Perre[2] and J. Goffin[1]
[1]*Division of Experimental Neurosurgery and Neuroanatomy, K.U. Leuven, Leuven, Belgium*
[2]*Division of Biomechanics and Engineering Design, K.U. Leuven, Leuven, Belgium*
[3]*Divison of Metallurgy and Materials Engineering, K.U. Leuven, Leuven Belgium*
[4]*Laboratory for Agricultural Buildings Research, K.U. Leuven, Leuven, Belgium*

Abstract. The main purpose of the present study was to investigate whether an energy failure level would apply to the skull fracture mechanics in unembalmed post mortem human heads under dynamic frontal loading conditions. A double-pendulum set-up was used to conduct frontal impact tests on specimens from eight unembalmed post mortem human subjects. The specimens were isolated at the occipital condyle level and pre-test computed tomography images were obtained. The specimens were rigidly attached to an aluminum pendulum in an upside down position and obtained a single degree of freedom, allowing motion in the plane of impact. A steel pendulum delivered the impact and was fitted with a flat-surfaced, cylindrical aluminum impactor, which distributed the load to a force sensor. The relative displacement between the two pendulums was measured using a laser sensor and used as a measure for the deformation of the specimen in the plane of impact. Two impact velocity conditions were created: low (3.60±0.24 m/s) or high (5.18±0.04m/s) velocity. Computed tomography and dissection techniques were used to detect pathology. If no fracture was detected, repeated tests on the same specimen were performed with higher impact energy until fracture occurred. Eventually all specimens were fractured. Peak force, displacement and energy variables were used to describe the biomechanics. These preliminary data suggest a positive correlation between impact velocity and energy to fracture. Further experiments are necessary to elucidate the possibility of an energy criterion for skull fracture in head impacts.

Key words: skull fracture, frontal impact loading, energy criterion.

Michael D. Gilchrist (ed.), IUTAM Proceedings on Impact Biomechanics: From Fundamental Insights to Applications, 185–193.

1. INTRODUCTION

In western countries, head injury is a major cause of morbidity and mortality. While traumatic injuries are the leading cause of death under the age of 45 years [**1**], up to half of the trauma deaths and the majority of cases of permanent disability are directly related to a sustained head injury [**2**]. Road traffic accidents are the most frequent cause of head injury in western countries, accounting for up to 60% of cases, followed by falls, assaults and sport and recreation associated injuries [**1, 3**]. To reduce the burden of head injury, prevention, besides optimal neurosurgical care for the head-injured patients, is mandatory. The importance of prevention is also underlined by the relatively high number of head-injured victims that die before reaching the hospital [**4**]. Therefore, biomechanical and crashworthiness communities have conducted many studies to understand the mechanisms of injury, develop tolerance criteria and provide fundamental data to mathematical analogues such as the finite element model, in order to improve preventive measures. Among others, the biomechanical characteristics of skull fracture have been studied intensively, as the skull bone is the main protector of the soft brain tissue against deformations secondary to external forces. Yoganandan and Pintar recently published a very thorough review of literature [**5**]. Despite all efforts, the exact mechanical pathogenesis of skull fracture has not been completely clarified. It was suggested by Yoganandan et al. [**6**] in a series of 12 head impacts that the energy to failure was a possible tolerance criterion. This was however not confirmed in a series of 61 skull impacts at our institution [**7**]. The present study was conducted to study the energy absorbability of intact unembalmed human heads during fracture under dynamic frontal loading.

2. MATERIALS AND METHODS

2.1 Specimen Preparation

Frontal impact tests were conducted on specimens from eight unembalmed post mortem human subjects, within 5 days after decease. The specimens were isolated at the occipital condyle level, keeping the intracranial contents intact. Pre-test computed tomography (CT) images were obtained at 1.0-mm intervals to rule out pre-existing anomalies and to acquire geometrical data. During scanning, a spine phantom with known bone mineral densities was scanned together with each specimen. The CT-scan data were processed using Mimics™ and Magics™ software (Materialise, Belgium). Routines were developed in Matlab™ to calculate scalp and skull thicknesses at the impact site. The average number of Hounsfield Units of the cranial vault was calculated as a measure for the bone density of the cranial vault and a Matlab™ routine was developed to calculate the local bone density of the impact site. The largest coronal plane diameter (LR diameter), the largest axial plane diameter (AP diameter), circumference and the distance between the nasion and occiput (Nasion-Occiput) were measured as shown in Table 1.

Table 1. Specimen characteristics.

ID	**1**	**2**	**3**	**4**	**5**	**6**	**7**	**8**
Age (years)	73	91	82	87	72	73	81	79
AP diameter (mm)	183	180	177	188	176	181	192	181
LR diameter (mm)	150	142	134	151	134	141	153	146
Nasion-occiput (mm)	278	342	279	350	297	276	375	320
Circumference (mm)	562	601	518	547	495	525	553	521
Specimen weight (kg)	4.6	3.3	3.2	3.7	3.0	3.6	4.0	3.8
Bone density impact site (mg/cm^3)	1125	937	848	905	905	848	888	968
Bone density cranial vault (mg/cm^3)	1297	1100	1166	1248	1168	1149	1187	1204
Bone thickness impact site (mg/cm^3)	7.2	10.6	6.6	6.8	13.0	11.1	7.3	6.2
Scalp thickness impact site (mg/cm^3)	4.5	3.3	3.1	4.8	3.3	2.5	2.3	3.4

The medulla oblongata, cerebellar tonsils and a small caudal part of the cerebellar hemispheres were removed, creating a small intracranial cavity just above the foramen magnum. Four screws were inserted around the foramen magnum, pointing towards the centre of this cavity. A steel profile (length 70 mm, surface 20x20 mm) was rigidly secured in the foramen magnum using a polyester resine (V-11™, Voss Chemie), which totally filled the cavity, encompassing the screws. The steel profile fitted on the interface at the lower end of the aluminum pendulum of the test set-up (see below).

2.2 Experimental Set-up

The test set-up consisted of two pendulums, mounted over a steel bar that was fixed to the wall at one end and supported by a steel structure at the other. One pendulum was made of aluminum and had a mass of 5.6 kg and a length of 128 cm. This pendulum supported the specimens by means of a steel profile which fitted on the interface at the lower end of the pendulum. Being rigidly attached to the pendulum in an upside down position, the specimens obtained a single degree of freedom, allowing motion in the plane of impact. The second pendulum was made of steel and had a mass of 14.3 kg, and a length of 128 cm. This pendulum was used to strike the blow. It was fitted with a flat-surfaced, cylindrical aluminum impactor with a diameter of 70 mm, which distributed the load to a force sensor (impact ICP® force sensor, model 200C20, PCB Piezotronics Inc., range 0 – 88.960 N, sensitivity 56.2 mV/kN). The impact energy and velocity could be changed by altering the angle at which the pendulum was released or by adding/removing mass by means of steel weights. A laser sensor (displacement sensor M5L/20, Gepa, range 20 mm) was fitted on the steel pendulum on the same height as a small reflective surface on the aluminum pendulum. This allowed measuring for the relative displacement between the two pendulums as a measure for the deformation of the specimen in the plane of impact (after adjustment for the difference between the height of the sensor and the exact height of the skull impact). A picture of the set-up is shown in Figure 1.

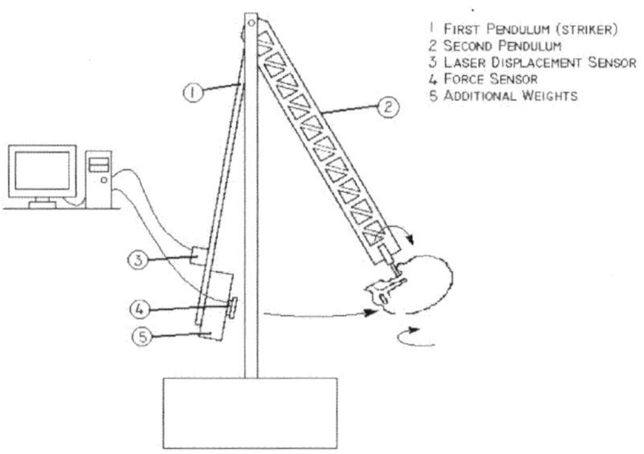

Figure 1. Experimental set-up.

The test matrix consisted of a single test on each specimen, either at low (3.60±0.24 m/s) or high (5.18±0.04 m/s) velocity. Computed tomography and dissection techniques were used to detect pathology. If no fracture was detected, repeated tests on the same specimen were performed with higher impact energy until fracture occurred.

2.3 Data Acquisition and Analysis

Data sampling was done at a frequency of 65,000 Hz. For data acquisition and management Testpoint™ software was used. Biomechanical data processing included the transformation of the force-time and deflection-time histories to a force-deflection response. This response was used to compute the energy absorbing characteristics of the structure. The force and the displacement (specimen deformation) at the time the fracture was initiated (F# and d# respectively) was identified by a discontinuity in the force-displacement curve. The area under the curve to that discontinuity was calculated as a measure for the energy absorbed by the specimen until the initiation of the fracture (Eabs#). Prior to the specimen impact tests, dummy tests were performed to establish the displacement error due to the deformation of the pendulums, which is inherent to the measurement set-up. This displacement error was linearly correlated with the force load and also depended on the exact dummy position. It rose to a maximum of 0.2 mm at 10 kN. A linear multiple regression analysis was carried out to investigate the influence of the skull properties (geometric parameters, weight and bone density) and kinetic energy input on F#, d# and Eabs#. Prior to the regression analysis the correlation between all candidate predictor variables was investigated by calculating Pearson correlation coefficients. The predictor variables used in the models were selected in order to keep interdependencies minimal. The following set of predictor variables was chosen: kinetic energy input, impact site bone density, impact site bone thickness, impact site scalp thickness and head circumference. The analysis was performed separately for F#, d# and Eabs# as dependent variables. P-values < 0.05 were considered significant.

3. RESULTS

Six specimens were initially tested at low velocity, resulting in only one fractured specimen, no.1. Specimen no.2 and no.7 were initially tested at high velocity, resulting in fracture in both specimens. All specimens were eventually fractured and fracture patterns were typically linear. Table 2 includes a summary of biomechanical data for each specimen.

Table 2. Force, displacement and energy data for each specimen.

ID	1	2	3	4	5	6	7	8
Impactor release angle (°)	55	90	90	90	90	90	90	90
Impactor velocity (m/s)	3.39	5.21	5.19	5.19	5.17	5.18	5.20	5.19
Impactor weight (kg)	37.3	37.3	37.3	48.3	40.3	40.3	40.3	40.3
Kinetic energy input (J)	214.3	506.2	502.4	655.5	645.5	648.0	653.0	650.5
F# (N)	6079	12320	7651	6981	12586	14425	13813	12167
d# (mm)	4.34	5.96	3.55	5.35	4.35	3.37	3.85	4.93
Eabs# (J)	9.3	41.2	11.8	21.9	24.4	18.0	30.3	27.0

Mean values of 11420±2927 N, 4.48±0.97 mm and 24.93±9.37 J were found for the peak force, maximum displacement and absorbed energy until initiation of fracture for high velocity impacts, respectively. For low velocity impact, values of 6079 N, 4.34 mm and 9.27 J were noted. Force-displacement curves exhibited nonlinear characteristics typical of biological materials. A representative force-deflection response is shown in Figure 2.

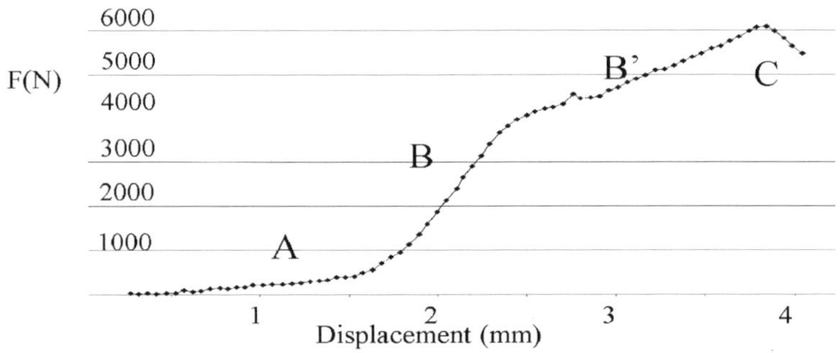

Figure 2. Typical force-displacement response, illustrating the different regions and peak impact force.

After an initial toe region (A), we observed a nearly linear increase in force (B,B') until the specimen reached its peak impact force (C) and initiation of fracture occurred. The linear region seems to consist of two parts, differing in slope. The linear multiple regression analysis could not reach statistical significance for any of the chosen predictors, due to the small sample size. However, we found for F# a positive correlation with impact site bone thickness (Pearson r =0.53) and a negative correlation with impact site bone density (Pearson r =-0.48) and impact site scalp thickness (Pearson r= -0.75). For d#, we found a positive correlation with impact site scalp thickness (Pearson r=0.63) and head circumference (Pearson r=0.62). For Eabs#, we found a positive correlation with head circumference (Pearson r =0.74).

4. DISCUSSION AND CONCLUSION

4.1 Experimental Set-up

When using a piston testing device impacting a fixed specimen, it is possible to accurately determine the area of contact at the impact site and to measure the specimen deformation in a direct way. However, the stress distribution, and thus the way the skull is deformed, is not the same in a restrained and unrestrained condition. When using free fall-techniques, the specimen is completely unconstrained. This methodology precludes coupling effects of the cervical column and inferior structures on the biomechanical response of the head. However, determining the exact area of contact at the impact site requires additional sensors and the deformation of the specimen has to be calculated from acceleration data. When using our double pendulum set-up, the specimen is allowed to move with one degree of freedom and at the same time, this set-up allows exact measurement of specimen deformation and area of contact at the impact site. It combines the advantages of both other techniques.

4.2 Biomechanical Data

Compared to the earlier performed skull experiment [**7**], we found higher values for all parameters. In this earlier experiment, 61 defleshed, embalmed skulls were impacted at different velocities and different impact locations, using the double-pendulum set-up. Fourteen skulls were impacted on the frontal skull bone. For these tests, we found mean values for F#, d# and Eabs# of respectively 3242 ± 699 N, 1.69 ± 0.39 mm and 3.49 ± 0.39 J for low velocity loading. For high velocity loading we found 6114 ± 2436 N, 1.44 ± 0.53 mm and 4.52 ± 1.91 J respectively. We think that the presence of the scalp and intracranial contents is responsible for the higher values of all parameters in this study. The exact influence of the scalp and intracranial contents however is not clear. When analyzing the individual force-displacement histories, we typically found a toe region, followed by a nearly linear region consisting of two parts differing in slope (Figure 2). The toe region reflects the influence of the scalp. Upon impact, the scalp will absorb energy while becoming compressed. Being compressed to its maximum, its energy absorbing capacity is exceeded, and further energy absorption will depend on deformation of the underlying skull. This is reflected in the nearly linear region of the curve. Why this linear region consists of two parts differing in slope, and thus stiffness, is not well understood and currently under investigation. Yoganandan et al. [**6**] found in the quasi-static loading tests a similar yielding phenomenon, which they contributed to the sequential fracturing of the outer cortex and inner cortex. Such a phenomenon was not observed in the dynamic loading conditions, and they suggested this was due to the high rate of onset that may have resulted in simultaneous fracture propagation. However, they sampled dynamic test data at a frequency of 8000 Hz only. It is possible that such a yielding phenomenon also exists under dynamic loading conditions, but can only be found when data sampling is performed at a much higher frequency, i.e. 65 kHz. We think that this biphasic region reflects characteristics of the skull, rather than the scalp or intracranial contents, because this biphasic linear region was also present in some of the force-displacement curves from the earlier performed skull experiment. Our data seems to correlate well with other studies: Hodgson et al. [**8**] noted for frontal loading with a flat plate at 2.81 m/s a mean peak force of 7.6 kN, Stalnaker et al. [**9**] reported for frontal loading with a rigid impactor at 6.26 m/s a peak force of 14.6 kN and Yoganandan et al. [**6**] noted for dynamic frontal loading values of 13600 N, 4.01 mm and 23.51 J for F#, d# and Eabs#, respectively. Yoganandan et al. [**6**], who were the first to study the influence of the load velocity on the fracture force, found that the critical force increased when increasing the impact velocity. A dynamic scale factor of approximately two was reported. The energy to failure, however, remained within the same range, suggesting that the energy to failure was a possible tolerance criterion for skull fracture. In our earlier skull experiment [**7**], the fracture force increased with increasing impact velocity. However, the energy to failure increased with increasing impact velocity too. This means that, in a situation where the skull is allowed to move, no fixed energy threshold predicting the occurrence of fractures can be defined. Because of the use of embalmed, defleshed skulls in the earlier experiment, it was the main purpose of the present study to investigate whether an energy failure level would apply to the skull fracture mechanics in unembalmed post mortem human heads under dynamic frontal loading conditions. However, as only one specimen fractured at low velocity conditions, we cannot draw a statistically significant conclusion. Still, unlike the findings of Yoganandan et al. [**6**], the presented preliminary data suggest that the positive correlation between impact velocity and energy to fracture, as seen in the earlier skull experiment, remains present in the context of dynamic impacts on intact mobile cadaver head specimens. Further experiments are necessary to elucidate the possibility of an energy criterion for skull fracture in head impacts.

REFERENCES

1. Jennett, B., Epidemiology of head injury, *J Neurol Neurosurg Psychiatry,* **60**, 1996, 362-369.
2. Kraus, J., Epidemiology of head injury, in. *Head injury,* Cooper P (ed), William Wilkins; Baltimore, 1993, pp. 1-25.
3. Kraus, J., McArthur, D.M. et al., Epidemiology of brain injury, in *Neurotrauma,* Narayan, R., Wilberger, J. and Povlishock, J. (eds.), McGraw Hill, New York, 1996, pp.13-30.
4. Klauber, M.R., Barrett-Connor, E. et al., The epidemiology of head injury: a prospective study of an entire community - San Diego County, California, 1978, *Am J Epidemiol,.* **113(5),** 1981, 500-509.

5. Yoganandan, N. and Pintar, F.A., Biomechanics of temporo-parietal skull fracture, *Clin Biomech*, **19,** 2004, 225-239.
6. Yoganandan, N., Pintar, F.A. et al., Biomechanics of skull fracture, *J Neurotrauma,* **12(4)**, 1995, 659-667.
7. Depreitere, B., A rational approach to pedal cyclist head protection, *Doctoral Thesis,* Faculty of Medicine, Division of Experimental Neurosurgery and Neuroanatomy, Katholieke Universiteit Leuven, 2004, pp. 53-67.
8. Hodgson, V. and Thomas, L.M., Breaking strength of the human skull vs. impact surface curvature, *HS-800-583*, US Department of transportation, Springfield, VA.
9. Stalnaker, R., Melvin, J. et al., Head impact response, in *Proceedings of the 21st Stapp Car Crash Conference*, New Orleans, Louisiana, USA, October 19-21, 1977, pp. 305-335.

NUMERICAL MODELING OF THE HUMAN HEAD UNDER IMPACT: NEW INJURY MECHANISMS AND TOLERANCE LIMITS

Daniel Baumgartner and Rémy Willinger
University of Strasbourg, IMFS, 2 rue Boussingault, F-67000 Strasbourg, France;
E-mail: baumgart@imfs.u-strasbg.fr

Abstract. The undergoing research establishes human head tolerance limits against impact. The methodology consists in a finite element modeling of 64 accidents of helmeted motorcyclists and American footballers, and pedestrians. The finite element model of the head of the University of Strasbourg is controlled with the kinematics, derived from a HYBRID III dummy head and an analytical model of the pedestrian, sustained by each victim. A statistical correlation study between the calculated head loadings and the observed injuries leads to the identification of injury criteria thanks to risk curves. A global strain energy of the subarachnoidal space of 5.4 J leads to subdural or subarachnoidal haematoma. A brain Von Mises stress of 18 kPa generates moderate neurological lesions which become severe from 38 kPa. A global strain energy of the skull of 2.2 J leads to skull fractures which are besides well predicted by the model.

Key words: human head, injury mechanisms, tolerance limits, biomechanics of impact, Finite Element Method (FEM), accident reconstruction.

1. INTRODUCTION

The undergoing research belongs to the human head protection against impact. In almost all kinds of traffic accidents, the head injuries represent very frequent and severe injuries. Head injuries reduction is therefore a high priority in the field of traffic safety improvement. Over the past 40 years, the biomechanical research provided criteria for tissue tolerance against impact which led to safety standards. Thus, head protection in road traffic accidents

Michael D. Gilchrist (ed.), IUTAM Proceedings on Impact Biomechanics: From Fundamental Insights to Applications, 195–203.

is usually designed against an experimental derived criterion: the Head Injury Criteria (HIC). The prediction ability of that criterion is criticized because of its empirical derivation and its limited capacity to predict brain injuries for example. The present research work endeavors to establish head tolerance limits for specific injury mechanisms (moderate and severe brain neurological lesions, subdural and subarachnoidal haematoma (SDH and SAH), skull fractures) thanks to numerical real world accidents reconstructions. The numerical tools used rely on the finite elements method (FEM). Such tools can be considered as the best ones in order to investigate the dynamical response of the head in a great variety of impact conditions, and thus establish the expected criteria through numerical accident reconstruction. In the framework of accident reconstruction, Newman et al. [1] propose a detailed methodology in order to assess mild traumatic brain injuries based on accidents reconstructions of American footballers head collisions during games. Newman et al. [2] use Zhou et al. [3] FEM of the head to suggest that mild traumatic brain injuries occur from a brain Von Mises stress of 20 kPa and a brain pressure of 300 kPa. In the same way, Willinger et al. [4] develop a FEM of the head called the ULP FEM which is used to reconstruct 13 helmeted motorcyclists' accidents in order to undertake a statistical approach of head injury mechanisms and afferent tolerance limits thanks to histograms. That recent study establishes a brain neurological lesions occurrence criterion through by the ULP FEM of the head predicted brain Von Mises stress. The tolerance limit of the brain is determined as about 20 kPa. Being based on this first study, the undergoing research attempts to work out new head tolerance limits relatively to specific injury mechanisms based on a great number of varied accident reconstructions which allow a statistical approach.

2. METHODS

The method used is based on 64 experimental and analytical real world accidents replications followed by the numerical reconstructions of the observed head injuries. Two kinds of accidents are considered:

- Road traffic accidents involving helmeted motorcyclists (13 cases – males aged between 18 and 40 years) and pedestrians (29 cases – males and females aged between 14 and 70 years).
- Head collisions during games between helmeted American footballers (22 cases – males aged between 18 and 30 years).

In each accident, the head is struck and its main impact is clearly identified by both the medical report and the accident scenario investigation report. For each accident, the ULP FEM of the head is submitted to the recorded kinematics of the head.

For the motorcyclists and American football players, these kinematics are obtained thanks to experimental accidents replications using an instrumented HYBRID III dummy head. The experimentally recorded accelerations are then applied to the skull of the head FEM. It is assumed that the accelerations measured during the accident replication and applied to the skull of the head FEM are the same as the accelerations sustained by each victim during the accident.

For the pedestrians, these kinematics are obtained thanks to an analytical full rigid body model of the pedestrian. That analytical model allows it to evaluate the relative position and velocity between the head and the windscreen at the beginning of the impact. Both relative position and velocity between the head and the windscreen are then applied to the head.

The determination of the head injury risk curves for specific injury mechanisms is based on a correlation study between the calculated head loadings on the one hand and the observed injuries on the other hand. The method used to infer such risk curves relies on logistic regression as detailed further.

3. MATERIALS: THE ULP FEM OF THE HUMAN HEAD

The ULP FEM of the head used in the present study is three dimensional with a continuous mesh. It contains 13208 elements divided in 10395 brick elements and 2813 shell elements and it weights 4.8 kg. The ULP FEM of the head includes the main anatomical components of the head which main characteristics and material properties are the following:

- A linear elastic skin (ρ=1000kg/m^3–E=16.7MPa–ν=0.42) in 2296 brick elements.
- A linear elastic face (e=10mm–ρ=2500kg/m^3–E=5000MPa–ν=0.23) in 529 shell elements.
- An elastic plastic brittle skull (Cortical (external layers): e=2mm–ρ=1900kg/m^3–E=15000MPa–ν= 0.21–K=6200MPa–UTS=90MPa–UCS=145MPa / Trabecular (internal layer): e=3mm–ρ=1500kg/m^3–E=4600MPa–ν=0.05–K=2300MPa–UTS=35MPa–UCS=28MPa) in 1813 three layered composite shell elements.
- A viscous elastic brain and cerebellum (ρ=1040kg/m^3–K=1125MPa–G_0=0.049MPa–G_{inf}=0.0167MPa–β=145s^{-1}) in 5508 brick elements.

- A linear elastic subarachnoidal space (ρ=1040kg/m^3–E=0.012MPa–ν=0.49) in 2591 brick elements.
- A falx of the brain and a tentorium of the cerebellum (e=1mm–ρ=1040kg/m^3–E=31.5MPa–ν=0.45) in 471 shell elements.

The ULP FEM of the head illustrated in Figure 1 is validated in a large time domain going from 2 ms to 30 ms against experimental data from Nahum et al. [5] and Trosseille et al. [6] in terms of intra cranial accelerations and pressures, and against experimental data from Yoganandan et al. [7] regarding skull fractures.

In the whole accident reconstruction procedure, the head is assumed to be free in its six degrees of freedom shall it be in rotation and in translation. Indeed, the neck influences the head's dynamical response 30 ms after the beginning of the impact as shown by a complementary study not yet published. The here considered accidents never last more than 20 ms. It is therefore allowed to assume that the neck does not influence the head's dynamical response to an impact. Moreover, the acceleration of gravity (1 g) is neglected towards the high levels of accelerations sustained by the head (over hundreds of g) during each impact. The initial position of the head is therefore of no importance.

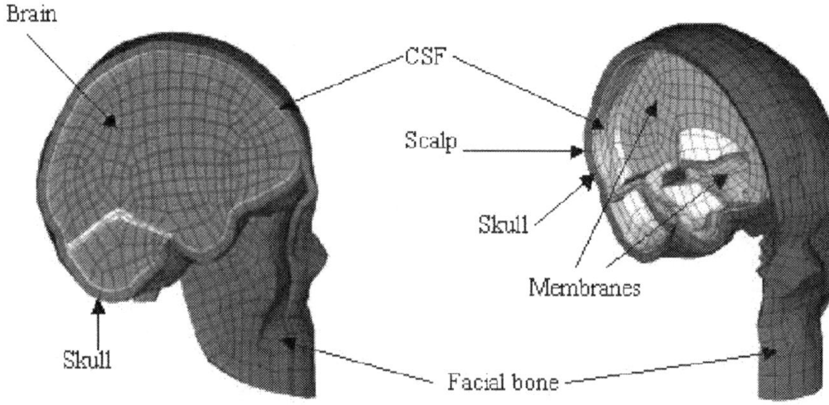

Figure 1. ULP finite element model of the human head.

4. RESULTS

The main interesting calculated loadings are the following (into brackets the extreme values reached by each parameter in the 64 considered accidents):

- The brain Von Mises stress: VM (2 kPa – 85 kPa).
- The internal energy of the subarachnoidal space: IESS (350 mJ – 26500 mJ).
- The internal energy of the skull: IESK (330 mJ – 29200 mJ).
- The number of deleted elements on the skull: DE (0 – 504).

The established risk curves rely on a logistic regression called "Modified Maximum Likelihood Method" that was elaborated by Nakahira et al. [8].

4.1 Brain Neurological Lesions

24 victims reveal moderate and 11 victims reveal severe brain neurological lesions whereas 29 victims do not present that kind of injury. It appears that the victims with low values (< 21 kPa ≅) of brain Von Mises stress do not reveal brain neurological lesions. Moreover, victims with middle values (< 39 kPa and > 16 kPa ≅) of brain Von Mises stress reveal moderate brain neurological lesions. Eventually, victims with high values (> 30 kPa ≅) of brain Von Mises stress reveal severe brain neurological lesions. To conclude, brain Von Mises stress can be considered as a good injury criteria regarding brain neurological lesions shall they be moderate with a tolerance limit of 18 kPa or severe with a tolerance limit of 38 kPa as shown by the risk curve illustrated in Figure 2.

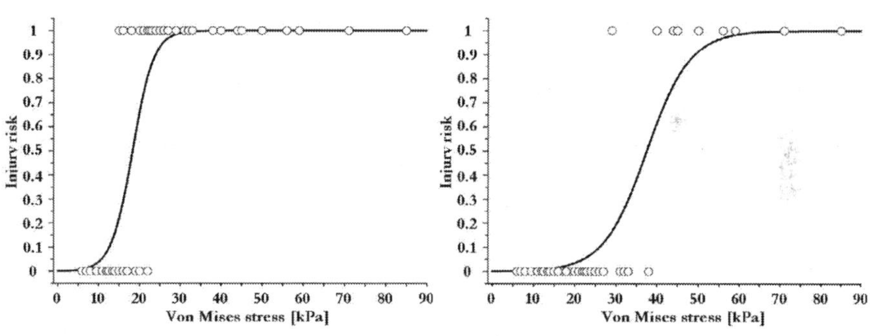

Figure 2. Brain moderate and severe neurological lesions risk curves.

4.2 Subdural and Subarachnoidal Haematoma

Seven victims reveal subdural or a subarachnoidal haematoma whereas 57 victims do not present that kind of injury. It appears that the victims with low values (< 5 J ≅) of internal energy of the subarachnoidal space do not reveal that kind of injury whereas victims with high values (> 4 J ≅) of internal energy of the subarachnoidal space present that kind of injury. To conclude, internal energy of the subarachnoidal space can be considered as a good injury criteria regarding subdural or subarachnoidal haematoma with a tolerance limit of 5.4 J as shown by the risk curve illustrated in Figure 3.

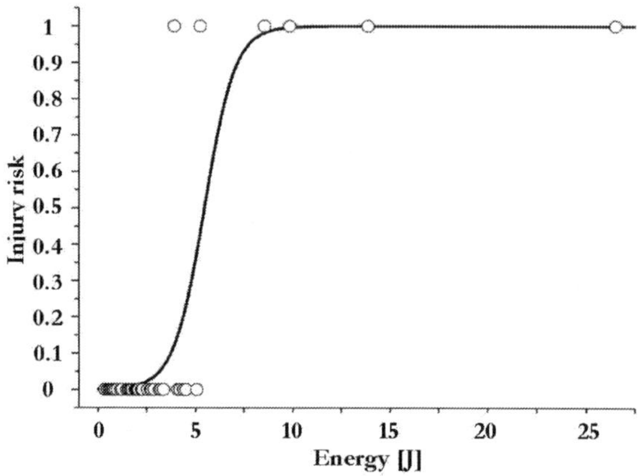

Figure 3. Subdural and subarachnoidal haematoma risk curve (one injured case is not represented since energy (> 40 J) out of scale).

4.3 Skull Fractures

Only the pedestrians are considered in this part of the study since they represent the accidents for which skull fractures are observed. 19 victims reveal skull fractures whereas 10 victims do not present that kind of injury. It appears that the victims for whom DE has a 0 value do not reveal the considered injury whereas victims for whom DE has a non 0 value present that kind of injury. Moreover, the on the ULP FEM of the head deleted elements, are precisely located where skull fractures are observed. Eventually, the extent of the considered injury is equally comparable to the extent of the observed injury. To conclude, the ULP FEM of the head is able to predict skull fractures if they are not located on the base of the skull. It appears that the victims with low values (< 3 J ≅) of internal energy of the skull do not reveal the considered injury whereas victims with high values (> 1 J ≅) of internal energy of the skull present that kind of injury. To conclude, internal energy of the skull can be considered as a good injury criteria regarding skull fractures with a tolerance limit of 2.2 J as shown by the risk curve illustrated in Figure 4.

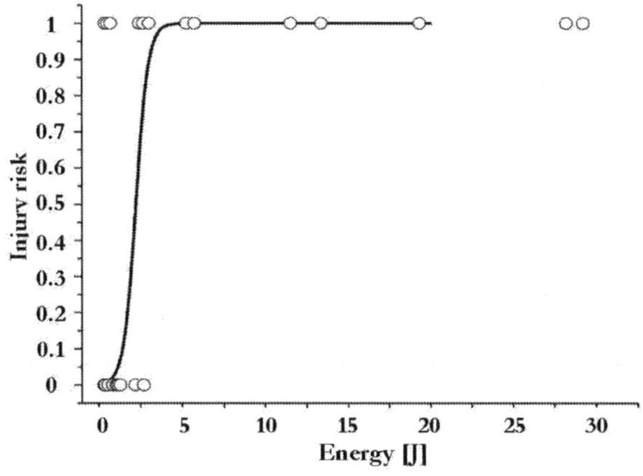

Figure 4. Skull fractures risk curve.

5. DISCUSSION

The tolerance limits that are proposed are clearly specific to the ULP FEM of the head since they rely on mechanical parameters (Von Mises stress, global strain energy, etc.) that depend on the mechanical behavior of the modeled tissues. Nevertheless an attempt of comparison to other tolerance limits determined in other studies but by using finite element models can be undertaken.

Concerning the brain neurological lesions, the tolerance limit established in this study rises with 18 kPa for moderate injuries and 38 kPa for severe injuries in terms of brain Von Mises stress. Past studies based on the same numerical method as the one used in the present work propose brain tolerance limits in terms of brain Von Mises stress regarding brain neurological lesions as follows:

- 11 kPa by Zhou et al. [9] who reconstruct a car accident using the head FEM from Ruan et al. [10].
- 15 kPa by Kang et al. [11] who reconstruct a motorcycle accident using an initial version of the ULP FEM of the head.
- 20 kPa by Newman et al. [2] who reconstruct an American football player accident using the FEM of the head from Zhou et al. [3].
- 27 kPa by Anderson [12] who reconstructs 16 head impacts on living sheep.

These values are in the same range as the ones obtained in the present study by using the same numerical method.

Concerning subdural and subarachnoidal haematoma, the tolerance limit established in this study rises with 5.4 J in terms of internal energy of the subarachnoidal space. In past studies on subdural and subarachnoidal haematoma or more generally on vascular head injuries, no authors express tolerance limits in terms of internal energy of the subarachnoidal space. Nevertheless, since always, the relative motion between the skull and the brain is considered as a generator of such kind of injuries as explained by Al-Bsharat et al. [13] and Hardy et al. [14] who measure the maximal displacement between the brain and the skull by using high speed X rays.

Concerning skull fractures, the tolerance limit established in this study rises with 2.2 J in terms of internal energy of the skull. In past studies on that kind of injuries, only Gurdjian et al. [15] propose an experimental minimal strain energy to deform the skull through impacting cadaver heads. The limit values of internal energy of the skull obtained by Gurdjian et al. [15] are the following: from 1.59 J to 2 J for the frontal bone and 0.91 J for the occipital bone. These experimental derived values are in the same range as the ones obtained in the present study through a numerical method.

6. CONCLUSION AND PERSPECTIVES

This research work belongs to the head protection against impact and proposes a new approach in the determination of head tolerance limits to specific injury mechanisms. The method used is based on an experimental accident replication and a numerical reconstruction of real world head injuries. 64 real world accidents involving helmeted motorcyclists and American footballers, and pedestrians are therefore considered. For each accident, the ULP FEM of the head is submitted to the recorded kinematics of the head during its impact obtained thanks to an experimental or analytical accident replication using a HYBRID III dummy head. The determination of the head tolerance limits relatively to specific injury mechanisms is based on a correlation study between the calculated head mechanical loadings and the observed injuries thanks to risk curves. Consequently, the established head tolerance limits relative to specific injury mechanisms are the following:

- For moderated brain neurological lesions, a brain Von Mises stress reaching 18 kPa.
- For severe brain neurological lesions, a brain Von Mises stress reaching 38 kPa.
- For subdural and subarachnoidal haematoma, an internal energy of the subarachnoidal space reaching 5.4 J.
- For the skull fractures, an internal energy of the skull reaching 2.2 J.

Besides, the ULP FEM reasonably predicts the linear skull fractures if they are not located at the base of the skull. These first results demonstrate the interest of the proposed approach as well as the need to analyse the by the head sustained injuries by specific injury mechanisms and not simply by Abbreviate Injury Scale (AIS) or HIC values.

REFERENCES

1. Newman, J., Beusenberg, M., Fournier, E., Shewchenko, N., Withnall, C., King, A., Yang, K., Zhang, L., McElhaney, J., Thibault, L. and McGinnis, G., A new biomechanical assessment of mild traumatic brain injury – Part 1 – Methodology, *IRCOBI*, 1999, pp. 17-36.
2. Newman, J., Barr, C., Beusenberg, M., Fournier, E., Shewchenko, N., Welbourne, E. and Withnall C., A new biomechanical assessment of mild traumatic brain injury – Part 2 – Results and conclusions, *IRCOBI*, 2000.
3. Zhou, C., Khalil, T. and King, A., A 3D human finite element head for impact injury analyses, *Symposium of Prevention through Biomechanics*, 1995.
4. Willinger, R. and Baumgartner, D., Numerical and physical modelling of the human head under impact – toward new injury criterion, *International Journal of Vehicle Design*, Vol. 32(1), 2001, pp. 94-115.
5. Nahum, A., Smith, R. and Ward, C., Intracranial pressure dynamics during head impact, *STAPP*, 1977, pp. 339-366.
6. Trosseille, X., Tarrière, C., Lavaste, F., Guillon, F. and Domont, A., Development of a FEM of the human head to a specific test protocol, SAE 922527, 1992.
7. Yogonandan, N, Pintar, F., Sances, A., Walsh, P., Ewing, C., Snyder, T. and Snyder, R., Biomechanics of skull fracture, *Head Injury Symposium*, 1994, pp. 227-236.
8 Nakahira, Y., Furukawa, K., Niimi, H., Ishihara, T., Miki, K., and Matsuoka, F., A combined evaluation method and modified maximum likelihood method for injury risk curves, *IRCOBI*, 2000, pp. 147-156.
9. Zhou, C., Kahlil, T. and Dragovic, L., Head injury assessment of a real world crash by finite element modelling, *AGARD*, 1996.
10. Ruan, J., Kahlil, T. and King, A., Human head dynamic response to side impact by finite element modelling, *Journal of Biomechanical Engineering*, 1991.
11. Kang, H., Willinger, R., Diaw, B. and Chinn, B., Validation of a 3D anatomic human head model and replication of head impact in motorcycle accident by finite element modelling, *STAPP*, 1997.
12. Anderson, R., A study of the biomechanics of axonal injury, PhD Dissertation, University of Adelaide, 2000.
13. Al-Bsharat, A., Hardy, W., Yang, K., Khalil, T., Tashman, S. and King, A., Brain-skull relative displacement magnitude due to blunt head impact: New experimental data and model, *STAPP*, 1999, pp. 321-332.
14. Hardy, W., Foster, C., Mason, M., Yang, K., King, A. and Tashman, S., Investigation of head mechanisms using neutral density technology and high-speed biplanar x-ray, SAE 2001-22-0016, 2001.
15. Gurdjian, E., Webster, *Head Injury*, Little Brown Company, 1958.

MECHANISMS AND TOLERANCE CURVES OF TRAUMATIC DIFFUSE AXONAL INJURY (DAI)

Erich Schuller and Iris Niemeyer
Institute for Legal Medicine, Munich University, Munich, Germany

Abstract. Diffuse Axonal Injury (DAI) is a common brain lesion in all kinds of fatal traffic accidents, such as car occupants, pedestrians, motorcycle and bicycle riders. The objective of this real accident study is to complete knowledge of significant injury mechanisms causing DAI, and, in particular, to identify tolerance limits. 16 fatalities were examined, all selected from autopsy cases of Munich University Institute for Legal Medicine. Based on extensive microscopic neuropathological examination, "brain maps" were developed to illustrate DAI location and frequency and to identify significant injury mechanisms related to causative head loading parameters determined by accident reconstruction. Findings suggest that DAI in brain stem and medulla oblongata occurs in cases exposed to axial linear acceleration of the head (±gz). DAI in the corpus callosum region is caused when rotational head acceleration around vertical z-axis is predominant. Shear forces due to extensive skull deformation and tensile forces, in particular parietal, are causing DAI in the hemispheres. DAI tolerance limits are identified as follows: (1) Mean linear acceleration 130 g for short duration (10 ms) and 110 g for longer duration (20 ms); (2) Rotational velocity 38 rad/s at 7.500 rad/s^2, and (3) Rotational acceleration 5,500 rad/s^2 for short duration (10 ms) and 3,500 rad/s^2 for longer duration (20 ms).

Key words: Diffuse Axonal Injury (DAI), real accident analysis, tolerance curves.

Michael D. Gilchrist (ed.), IUTAM Proceedings on Impact Biomechanics: From Fundamental Insights to Applications, 205–212.

1. INTRODUCTION

Diffuse Axonal Injury (DAI) occurs when the brain is subjected to strain, e.g. caused by rapid change of head motion (direct impact, inertial loading), and represents widespread microscopic damage of the axons due to the resulting brain tissue shear strain [1, 2]. Previous studies have examined this particular intracranial lesion in clinical surveys, animal experiments, real accidents, and computer (FE)-simulations [3–6]. However, for a better understanding of DAI there is still the need of extended basic data, both to complete knowledge of significant injury mechanisms, and, in particular, to develop tolerance criteria. To contribute to that task is the objective of this real accident study.

2. MATERIAL AND METHODS

Sixteen fatalities were included in this DAI study, all selected from autopsy cases investigated at the Munich University Institute for Legal Medicine. Selection criteria were defined as follows: (1) Determination of relevant physical head loading parameters must be possible; (2) Acute DAI is included only; (3) Cases showing primary vascular lesions with relevant haemorrhage are excluded.

For each case selected a forensic autopsy was performed to survey the overall injury pattern and, in particular, the brain was subject to an extensive microscopic neuropathological examination to detect DAI location and frequency. "Brain maps" were developed to illustrate DAI location and frequency in order to assign lesion structure to significant injury mechanisms.

Common methods of accident reconstruction (including numerical replication tools, such as PC-Crash®, MADYMO®) were applied case by case to determine the following physical head loading parameters: (1) linear acceleration, (2) rotational velocity, (3) rotational acceleration, and (4) impact duration.

3. RESULTS

Table 1 presents a summary of the 16 casualties investigated. 11 of them were male, 5 female, years of age vary from 14 to 78 years.

Fifty per cent of cases were traffic accidents of different types, i.e. pedestrian, cyclist, motorcyclist, and car driver. Seven more cases represent

Table 1. Summary of casualties investigated.

Case #	Accident Type	Gender / Age	Cause of Death
01	Pedestrian - Tram	Male / 56	Head injury
02	Parachutist 4000 m	Male / 28	Polytrauma
03	Fall 4,4 m	Male / 14	Head Injury
04	Fall Stairs 2,7 m	Male / 67	Head Injury
05	Ski Racer	Female / 26	Medullar Rupture
06	Fall Standing height	Male / 54	Heart Failure
07	Motorcyclist - Car	Male / 28	Head Injury
08	Cyclist - Car	Male / 41	Polytrauma
09	Car Driver - Front Crash	Male / 27	Heart Rupture
10	Motorcyclist - Car	Male / 28	Spinal Cord Rupture
11	Pedestrian - Car	Female / 73	Head injury
12	Fall Stairs 2,8 m	Female / 78	Heart Failure
13	Cyclist - Motorcycle	Female / 35	Head Injury
14	Pedestrian - Train	Male / 17	Polytrauma
15	Fall 15. Floor	Female / 16	Polytrauma
16	Parachutist 2700 m	Male / 27	Polytrauma

falls from different heights, starting at standing height up to extremes experienced by two parachutists falling free from 2700 m and 4000 m. Case No. 5 was a fatal ski racing accident.

Cause of death was diagnosed head injury in 8 cases and disconnection of the medulla oblongata and the cervical cord in 2 cases. The 4 "Polytrauma" cases had extensive lesions of the whole body. 3 persons died from heart failure and heart rupture.

Physical head loading parameters derived from accident reconstruction are listed case by case in Table 2. Indicating the principal direction of linear acceleration (Lin) the corresponding mean value of linear head acceleration a [g] is presented and, in addition, principal axes of head rotation (Rot) linked with mean rotational acceleration $\acute{\omega}$ [rad/s^2], mean rotational velocity ω [rad/s], and impact duration t [ms].

For better understanding Figure 1 illustrates the directions of acceleration and axes of rotation defined in the head coordinate system.

As an example, the findings for Case No. 6 shall be demonstrated: A 54 year old man falls backwards from an upright position and hits his back of the head (more on the left side) on the pavement. Mean head accelerations are as follows: Linear 140 g in +x and –z direction, rotational 12 krad/s^2 around +y and +z axes at 46 rad/s. Impact duration is 4 ms.

Table 2. Physical head loading parameters.

Case #	Lin	a [g]	Rot	$\dot{\omega}$ [rad/s²]	ω [rad/s]	t [ms]
01	-y	245	+x(+z)	21,005	61	3
02	+z	289	+y	2,965	74	27
03	-z	440	+y/-z	37,638	81	2
04	+y	266	-x(+z)	22,797	63	3
05	+z	80	-x	84	8	40
06	+x(-z)	140	+y(+z)	12,020	46	4
07	-x	330	-y(-z)	33,912	188	6
08	-z	133	+x/-z	5,679	55	10
09	-x	510	-y	43,611	87	2
10	-z	265	+y/±z	15,143	73	5
11	-y	157	+x	13,433	48	4
12	+z	70	-y	23,881	64	3
13	-y	279	-x	7,565	36	5
14	+x	207	+y(+z)	17,751	56	3
15	+z	112	+y	2,885	78	27
16	+z	379	+y/+z	3,111	31	26

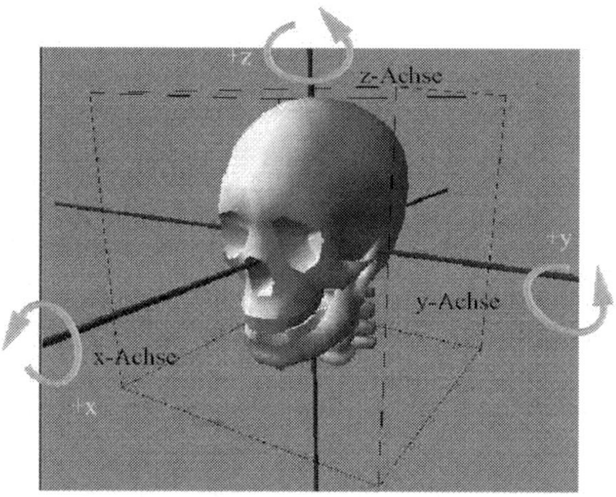

Figure 1. Head coordinate system.

The findings of the microscopic neuropathological examination are summarized in Table 3.

DAI detected positively in certain brain regions is indicated by +, distinguishing only left side (+-), only right side (-+), and bilateral (++) location. Otherwise it is marked negative (--). Rupture of brain tissue is indicated by < (on left side), > (on right side) and < > (on both sides). Brain regions included are the corpus callosum (CC), brain stem (BS), spinal cord (SC), occipital (OC), parietal (PA), and hemisphere (HE).

Table 3. DAI location.

Case #	CC	BS	SC	SG	OC	PA	HE
01	++	++	--	++	++	--	++
02	++	++	--	++	+-	--	++
03	++	++	--	++	--	--	++
04	++	++	--	++	--	--	++
05	--	<++>	--	--	--	--	--
06	++	++	--	++	++	--	++
07	++	--	--	++	++	--	++
08	++	++	--	++	++	--	++
09	--	--	--	++	--	++	++
10	++	++	<++>	++	--	++	++
11	--	--		--	--	--	--
12	--	++	<++>	++	--	++	++
13	--	--	--	--	--	--	--
14	--	-+	--	--	--	--	--
15	--	<++>	<++>	--	--	--	--
16	++	<-->	--	++	++	--	++

CC= Corpus Callosum, BS = Brain Stem,
SC = Spinal Cord, SG = Stem Ganglia,
OC = Occipital, PA = Parietal, HE = Hemisphere

One example of a "Brain Maps" established is shown in Figure 2 (Case No. 6). The upper graph sketches a frontal section through longitudinal tracts including the brain stem, the lower graph a frontal brain section with the rostral part of the corpus callosum. Indicated by dots DAI is observed bilaterally in the corpus callosum (CC), stem ganglia (SG), and brain stem (BS). DAI in the occipital hemisphere is not visualized in this particular graph.

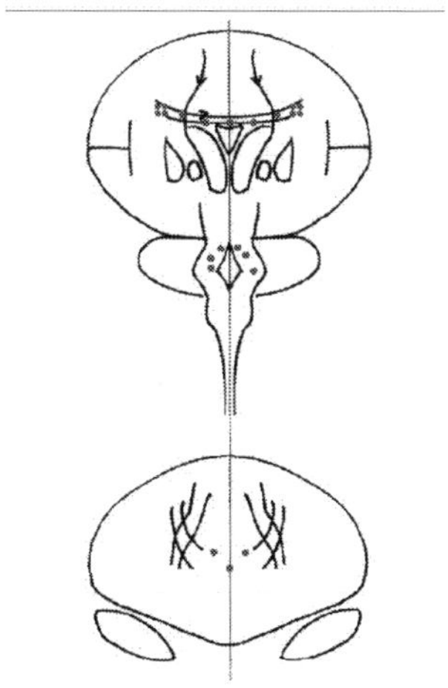

Figure 2. "Brain Map" Case No. 6.

4. CONCLUSIONS

Analysing the incidence of DAI and head loading characteristics the following conclusions can be drawn:

(1) In the brain stem and the medulla oblongata DAI occurs in cases exposed to axial linear acceleration of the head (±z), while this location of DAI is uncommon in cases subjected to head impact loading in the transversal plane (±x frontal and occipital, ±x lateral).

(2) In the corpus callosum region DAI occurs when rotational head acceleration around vertical z-axis is predominant. Shear forces due to extensive skull deformation and tensile forces, in particular parietal, are causing DAI in the hemispheres.

Physical head loading parameters listed in Table 2 are illustrated in Figures 3 to 5 in order to identify DAI tolerance limits relating to linear acceleration, rotational velocity, rotational acceleration, and impact time.

(1) Linear (mean) head acceleration versus impact time is presented in Figure 3. The DAI tolerance curve indicates the following tolerance limits: 130 g for short duration (10 ms) and 110 g for longer duration (20 ms).

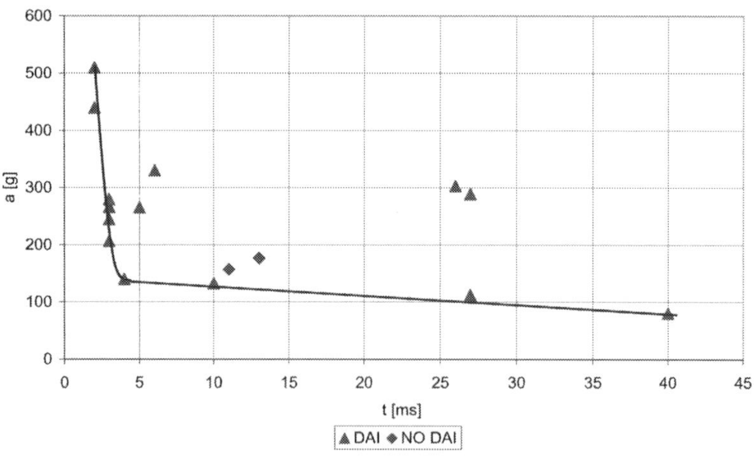

Figure 3. Linear head acceleration vs. time.

(2) Rotational (mean) head acceleration versus rotational velocity is shown in Figure 4. The tolerance limit for DAI is the following: Rotational velocity 38 rad/s at 7,500 rad/s².

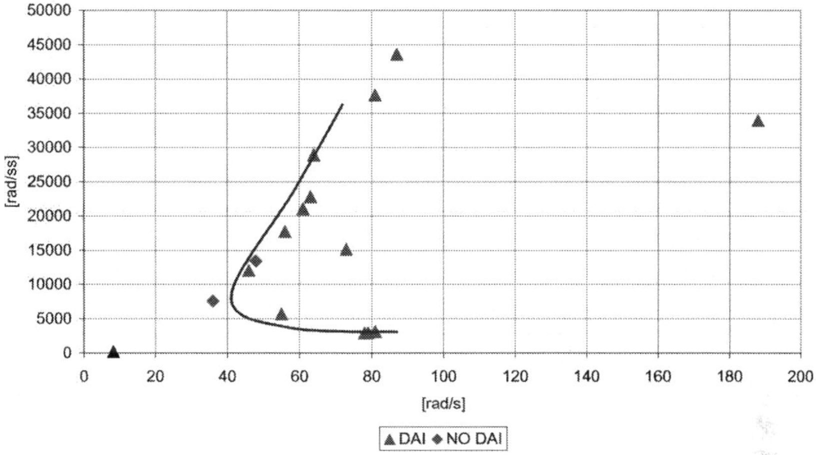

Figure 4. Rotational velocity and rotational acceleration.

(3) Rotational (mean) acceleration versus impact time is demonstrated in Figure 5. The tolerance curve indicates tolerance limits as follows: Rotational acceleration 5,500 rad/s² for short duration (10 ms) and 3,500 rad/s² for longer duration (20 ms).

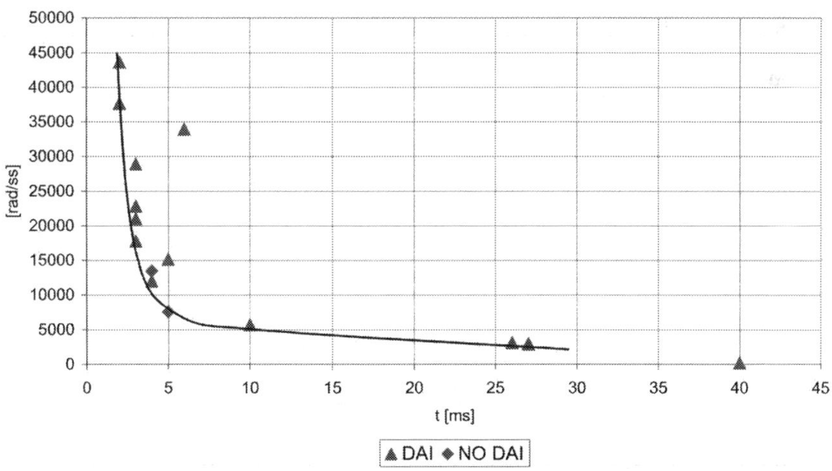

Figure 5. Rotational acceleration vs. time.

REFERENCES

1. Adams, J.H., Doyle, D., Ford, I., Gennarelli, T.A., Graham, D.I. and McLellan, D.R. (1989) Diffuse axonal injury in head injury: Definition, diagnosis and grading, *Histopathology* 15: 49-59.
2. Povlishock, J.T. (1992) Traumatically induced axonal injury; pathogenesis and pathobiological implications, *Brain Pathology* 2:1-12.
3. Adams, J.H., Graham, D.I., Gennarelli, T.A. and Maxwell, W.L. (1991) Diffuse axonal injury in non-missile head injury, *J Neurol Neurosurg Psychiatry* 54: 481-483.
4. Thibault, L. and Gennarelli, T.A. (1993) Biomechanics of diffuse brain injuries, in *Biomechanics of Impact Injury and Tolerances of the Head-Neck Complex,* Backaitis, S.H. (ed.), SAE PT-43, pp. 555-561.
5. Gennarelli, T.A., Thibault, L.E., Tomei, G., Wiser, R., Graham, D. and Adams, J.H. (1993) Directional dependence of axonal brain injury due to centroidal and non-centroidal acceleration, in *Biomechanics of Impact Injury and Tolerances of the Head-Neck Complex,* S.H. Backaitis (ed.), SAE PT-43, pp. 595-599.
6. Ryan, G.A. and Vilenius, T.S. (1995) Field and analytic observations of impact brain injury in fatally injured pedestrian, *J Neurotraum.* 12(4): 627-634.

SENSORY DISORDERS OF THE AUDITORY AND VESTIBULAR SYSTEMS FOLLOWING BLUNT HEAD TRAUMA

Arne Ernst

Dept. Otolaryngology at UKB, Hospital of the University of Berlin (Charité Medical School), Warener Str. 7, 12683 Berlin, Germany; Phone +49-30-5681 2901; Fax +49-30-5681 2903; Email: ArneborgE@ukb.de

Abstract. It is the aim of the present review to correlate head and neck trauma and audiovestibular disorders. Acceleration/deceleration forces (during rear-end collisions of the "whiplash" type), blunt trauma of the head (without fracturing, e.g. falling from height, being hit by a solid bar) can impair the audiovestibular system at different sites. The (peripheral) auditory and vestibular receptor cells within the fluid-filled labyrinth (within the petrous bone) can transiently or permanently be impaired by mechanical dysarray or fluid concussion. Moreover, permanent lesions can arise from axonal injury of the central (integrative) parts of the auditory pathway. The major auditory complaints of the patients include an (acute) transient hearing loss and tinnitus. In a minority of patients, tinnitus, hearing loss and hyperacusis can become chronic symptoms. Most of the vertiginous complaints occur within 24 hours after the trauma (acute disorders) or after 3 weeks – 3 months (chronic disorders).

The acute disorders include labyrinthine concussion, rupture of the round window membrane, cervicogenic, postural instability. The chronic disorders include otolith disorders, delayed endolymphatic hydrops, canalolithiasis. Extensive medical diagnostics is required to differentiate the type of underlying disorder. In general, there is no significant correlation between trauma mechanisms and type of audiovestibular disorder. However, acute audiovestibular disorders have a good prognosis in general, while chronic disorders have a poorer one.

Key words: hearing loss, vertigo, trauma of the head and neck, whiplash, neurotology.

Michael D. Gilchrist (ed.), IUTAM Proceedings on Impact Biomechanics: From Fundamental Insights to Applications, 213–220.

1. INTRODUCTION

Blunt trauma of the head, neck and craniocervical junction can result from falls, whiplash-type injuries, contact injuries after impact with a solid object etc. The trauma mechanisms can vary as well as the impact forces (1). The physician is usually unable to thoroughly reconstruct the impact mechanisms after an interview with the patient. In addition to those descriptions, the extent of damage of the cars involved, some patients have an amnesia related to the accident.

It is therefore important to diagnose the auditory and vestibular systems by objective electrophysiological and other techniques, particularly in those patients with relevant clinical complaints (2). Neuroimaging (MRI, CT) usually reveals no evident intracranial pathology, skull base fractures etc. in blunt trauma of the head and neck (3). However, the auditory and vestibular systems can be impaired at different sites. The impact of the head or the "whiplash"-triggered impact leads to a pressure transfer into the petrous bones (2). The petrous bone surrounds the so-called "labyrinth" with the inner ear – the auditory receptor cells (organ of Corti , inner and outer hair cells) and the vestibular receptors (semicircular canals, maculae sacculi and utriculi) (Figure 1). The labyrinth is fluid-filled (perilymph/endolymph) and the petrous bone contains large, pneumatized spaces which are ideal prerequisites for such a pressure transfer in blunt trauma of the head and neck (1) (Figure 2). Acceleration/deceleration forces (coup/contre-coup) or a direct blow over the temporal squama cause a concussion of the fluid-filled labyrinth and a receptor dysarray (hair cell damage, basilar membrane rupture or break-up, otolith displacement, shearing of the tectorial membrane, small bleedings into vascular stria or into the semicircular canals etc.) (4). Late sequaelae of such a concussion involve the audiovestibular nerve (shearing/ tearing forces) or the central auditory pathway (central vestibular structures are known, but there is no "classical pathway" up to the cortex) (5).

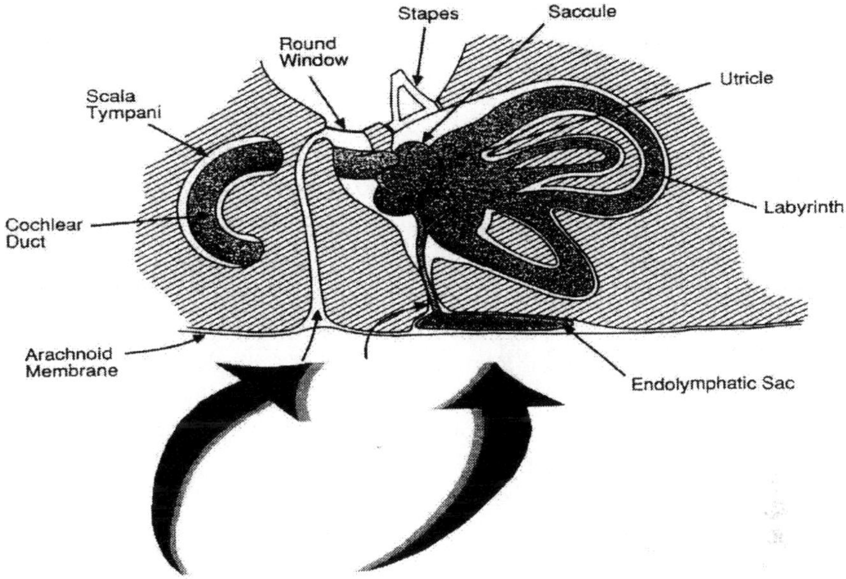

Figure 1. Neuroanatomical pressure transfer routes upon blunt head trauma from the CSF spaces to the membraneous labyrinth filled with perilymph and endolymph (i.e., inner ear and vestibule).

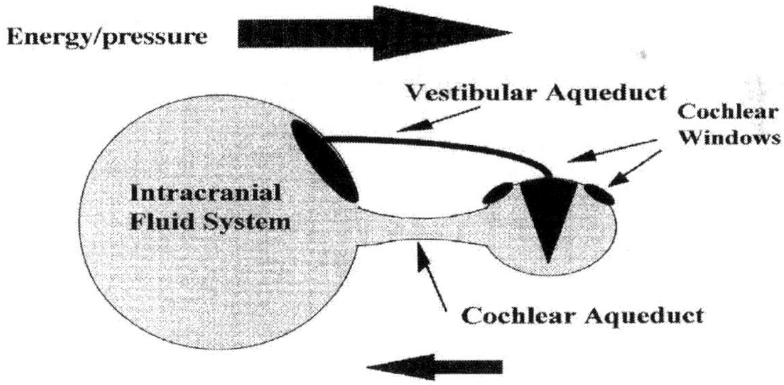

Figure 2. Energy transfer routes upon blunt head trauma from the CSF spaces to the membraneous labyrinth filled with perilymph and endolymph (i.e., inner ear and vestibule).

2. NEUROTOLOGICAL DISORDERS OF THE VESTIBULAR SYSTEM: DIAGNOSTICS AND MANAGEMENT

2.1 Diagnostics of Neurotological Disorders of the Vestibular System

After questioning the patients about the vertiginous complaints ("dizziness", "unsteadiness", "slipping away", "rotational or positional vertigo"), the following neurotological tests should be applied (6):

a. vestibulospinal tests (Romberg, Unterberger test)
b. head thrust test
c. VOR testing (caloric irrigation, rotational chair testing)
d. dynamic posturography testing
e. otolith testing (VEMP recordings and otolith-ocular responses (OOR)
f. electrocochleography (optional)
g. trunk sway measures of postural stability by Sway Star (optional)

The different tests cover all aspects of peripheral and central vestibular pathology, i.e. disorders of the semicircular canals (b, c), the endolymphatic sac (f) otolith organs (e), vestibulospinal and cervicogenic disorders (a, d, g).

2.2 Management of Acute and Chronic Disorders

The *acute disorders* which affect the patients within the first 24 hours after trauma include:

- *BPPV* (benign positional vertigo) is characterized by freely floating otolith crystals in the semicircular canals which are broken off their macular surface. Primary treatment should consist of repositioning manoeuvers (7). After reoccurrence, patients can also be operated on (8). A positional, torsional nystagmus upon Dix-Hallpike's manoeuver should be taken as confirmation of the diagnosis.
- *Labyrinthine concussion* is a state of functional loss of receptor function due to the impact forces (compatible to cerebral concussion) without permanent cellular damage. It will normalize over the first 5 days with drug treatment (sedatives, antemetics) on average. The occurrence of spontaneous nystagmus, initial unilateral hyporesponsiveness upon caloric irrigation, pathological rotational chair test result (peripheral lesion) and deviation upon

vestibulospinal testing can be taken as confirmation of the diagnosis (9, 10).

- *Perilymphatic fistulae* is characterized by a loss of perilymph due to the partial rupture of the oval and /or round window bordering the fluid filled-labyrinth (Figure 2). A sudden onset of dizziness in combination with sensorineural hearing loss (SNHL) of varying degree are typical clinical signs. These patients should be treated surgically by patching the round and oval window niches with connective tissue and fibrin glue. The occurrence of spontaneous nystagmus, initial unilateral hyporesponsiveness upon caloric irrigation, pathological rotational chair test result (peripheral lesion), deviation upon vestibulospinal testing, a sensorineural hearing loss of different extent (in 50% more than 40 dB over all frequencies) with/without tinnitus can be taken as confirmation of the diagnosis (2, 4).
- *Central vestibular disorder* can be characterized by a mixed pathology with the leading symptom of "tumbling". The patients should be instructed to rest, treated by drugs (sedatives, antemetics). These patients recover within a few days, at least one week. They can be given an additional individualized home training. Different types of nystagmus (vertical nystagmus, skew deviation, ocular torsion), horizontal body rotations upon vestibulospinal tests, pathological SOT over all conditions (I-VI) can be taken as confirmation of the diagnosis (10).

The following *chronic disorders* can be diagnsoed after about 3 weeks – 3 months (Table 1):

- *Delayed endolymphatic hydrops* is a diseases of the endolymphatic duct and sac which is accompanied by a functional loss of pressure transfer and ionic dysequilibrium. Morphologically, a bulged Reissner's membrane can be found as hydropic characteristics. The patients report a classical rotational vertigo with attacks, but without hearing loss. EcoG should be used to confirm the presence of a hydrops. Most of the patients are free of vertigo under drug treatment, a few patients have to be operated on by endolymphatic sac surgery (11, 12).
- *Cervicogenic, postural instability* occurs widely, but not only after a "whiplash-type" of injury. The diagnosis can be confirmed by SwayStar testing (increase in body sway), by exclusion of other disorders and by pathological dynamic posturography testing. These patients usually give mixed symptoms which are usually reinforced when the neck pain and the limitation of cervical spine mobility is at a maximum. In general, the extent and expression of vertigo varies in preparation of the physical condition of the patient. The patients can be primarily treated by physiotherapy (e.g. osteopathy,

exercises), analgesics, antirheumatics and/or vestibular habituation training (6, 13, 14).

- *Otolith disorders* are characterized by a functional loss of the human gravity sensors (saccular and utricular maculae). This can be assessed by VEMP testing (saccular) or eccentric rotation (utricular). A combination of saccular and and utricular dysfunction is usually found. The patients with otolith disorders generally present after a longer period-of-time because most of them do not consider their "unsteadiness" as a specific symptom, but rather as common sequaelae of the trauma. These patients can be provided with individualized training routines on a tilting platform 6, 15, 16).

Nearly every second of those patients after blunt trauma of the CCJ present with a combination of different types of neurotological disorders (e.g. initially BPPV, followed by a delayed endolymphatic hydrops) (6).

3. AUDIOLOGICAL DISORDERS AFTER BLUNT TRAUMA OF THE HEAD: DIAGNOSTICS AND MANAGEMENT

3.1 Audiological Diagnostics

Audiological diagnostics should be limited to pure-tone audiometry (PTA) (a) in the acute phase. PTA reliably reveals a (transient) hearing loss as it frequently occurs after labyrinthine concussion or perilymphatic fistulae. In the chronic phase, objective measures should be additionally included to assess central auditory structures, incl. auditory brainstem response audiometry (ABR) (b), transiently evoked otoacoustic emissions (TEOAE) (c), stapedial reflex recordings (d). Psychoacoustic measures – loudness discomfort thresholds (LDL) (e) and tinnitus suppression/characteristics (f) – can be applied if required (2, 5).

3.2 Management of Posttraumatic Hearing Disorders

Acute hearing disorders with a hearing loss (a) are primarily the result of a labyrinthine concussion (disordered cochlear micromechanics) or a perilymphatic fistula (see above). The treatment is either medical or surgical (see above). The prognosis is excellent (6).

Chronic hearing disorders are characterized by (permanent) hearing loss, tinnitus or hyperacausis (loudness discomfort). While a permanent hearing loss is very rare, tinnitus and hyperacusis occur in about 10% of all patients with permanent complaints after blunt trauma of the head and neck. Diagnostic criteria are defined by (b-f). The underlying pathomechanism is an axonal injury with partial functional and cellular losses of the central auditory pathway. The prognosis is poor, habituation training and psychosomatic therapies can be applied (5).

4. CONCLUSION

The present review is intended to evaluate the impact of acceleration or forces applied to the head, neck region and to the craniocervical junction (blunt trauma). Most interestingly is the frequently lacking correlation between an individual trauma mechanism/the extent of applied forces and a particular vestibular/audiological disorder. The variety of disorders as outlined above can be diagnosed concomitantly or successively in some of the patients. There is a long-standing controversy in the literature as to whether blunt trauma to the head and neck, "whiplash" type of injury or other types of injury can affect the vestibular system "specifically" (e.g. 2, 9, 17 and the comments on this article). In our series, the patients with cervicogenic, postural instability underwent various trauma mechanisms (including brain concussion with cervical compression, having the head and craniocervical junction hit by another solid item, apart from "whiplash") and were frequently affected by another or even two other types of neurotological disorders (6). When considering the medium-term prognosis (one year) of each type of vestibular disorder, the acute disorders have the best outcome (6, 18). The poorest outcome is found in the subpopulation of patients with cervicogenic, postural instability and in otolith disorder. The first is presumed to be the result of the interindividual variation in the musculo-skeletal system (and the ability to develop coping strategies by habituation/training) (14, 17) and the highly variable conditions of the cervical spine and the soft tissue (muscles, ligaments, joints). Otolithic vertigo can at present only be treated with very limited success in the medium-term perspective. This could be based on the specific nature (gravity sensor) of this type of receptor (16) and its ancient phylogenetic origin which makes it and the central vestibular correlates quite resistant to external influences, including training and habituation.

The acute hearing disorders have the same (good) prognosis as vestibular disorders, but the chronic tinnitus, hyperacusis can lead to a substantial loss of the quality of life and to secondary psychiatric illness (depression, post-traumatic stress syndrome etc.). All therapeutic efforts have a very limited success (5).

ACKNOWLEDGEMENTS

Supported by HVGB grant, St. Augustin (Germany) and the Sonnenfeld Foundation, Berlin.

REFERENCES

1. Shaw, N.A., The neurophysiology of concussion. Progress Neurobiol 2002; 67:281-344
2. Fitzgerald, D.C., Head trauma: hearing loss and dizziness. J Trauma 1996; 40:488-496.
3. Gaetz, M., The neurophysiology of brain injury. Clin Neurophysiol 2004; 115:4-18.
4. Ernst, A., Marchbanks, R. and Samii, M., Intracranial and intralabyrinthine fluids - basic aspects and clinical applications. Springer Berlin Heidelberg New York, 1996.
5. Nolle, C., Todt, I., Seidl, R.O. and Ernst, A., Pathophysiological changes of the central auditory pathway after blunt trauma of the head. J Neurotrauma 2004; 21:251-258.
6. Ernst, A., Basta, D., Seidl, R.O., Todt, I., Scherer, H. and Clarke, A., Management of posttraumatic vertigo. Otolaryngol Head Neck Surg 2004; 131:225-235.
7. Parnes, L.S. and Price-Jones, R.G., Particle repositioning manoeuver for BPPV. Ann Otol Rhinol Otol 1993; 102:325-331.
8. Parnes, L.S. and McClure, J.A., Posterior semicircular canal occlusion in the normal hearing ear. Otolaryngol Head Neck Surg 1991; 104:52-57.
9. Guyot, J.P., Liard, P., Thielen, K. et al., Isolated vestibular areflexia after blunt head trauma. Ann Otol Rhinol Laryngol 2001; 110:562-565.
10. Vibert, D. and Hausler, R., Acute vestibular deficits after whiplash injuries. Ann Otol Rhinol Laryngol 2003; 112:246-251.
11. Shea, J.J., Xianxi, G. and Orchik, D.J., Traumatic endolymphatic hydrops. Am J Otol 1995; 16:235-240.
12. Ge, X. and Shea, J.J., Transtympanic electrocochleography: a 10-year experience. Otol Neurotol 2001; 22:465-470.
13. Allum, J.H.J., Aitkin, A.L. and Ernst, A., Differences between trunk sway characteristics on a foam support surface and on the Equitest ankle-sway-referenced support surface. Gait & Posture 2002; 16: 264-270.
14. Sjostrom, H., Allum, J.H.J., Carpenter, M.G. et al., Trunk sway measures of postural stability during clincal balance tests in patients with chronic whiplash injury symptoms. Spine 2003; 28:1725-1734.
15. Colebatch, J.G., Halmagyi, G.M. and Skuse, N.F., Myogenic potentials generated by a click-evoked vestibulocollic reflex. J Neurol Neurosurg Psychiatry 1994; 57:190-197.
16. Clarke, A.H. and Engelhorn, A., Unilateral testing of utricular function. Exp Brain Res 1998; 121:457-464.
17. Rubin, A.M., Wooley, S.M., Dailey, D.V. et al., Postural stability following mild head or whiplash injury. Am J Otol 1995; 16: 216-221.
18. Basford, J.R., Chou, L.S., Kaufmann, K.R. et al., An assessment of gait and balance deficits after traumatic brain injury. Arch Phys Med Rehabil 2003; 84:343-349.

BIOMECHANICAL, RADIOGRAPHIC AND OSTEOLOGIC OBSERVATIONS OF LOWER CERVICAL SPINE INJURIES

Leon Kazarian and Randy Shively
P.O. Box 306, 674 McBee Rd., Bellbrook, OH 45305, USA;
E-mail: leon@kazarian.com, *Phone: 937-426-3004, Fax: 937-426-4281*

Abstract. This paper discusses the nature and gamut of vertical midsagittal fracture with and without bilateral laminae fracture in terms of the "teardrop fracture". The occurrence of midsagittal fracture patterns of adjacent vertical levels were identified. The characteristics and mechanics of the midsaggital fracture are hypothesized relative to vertebral body architecture.

Key words: cervical biomechanics, teardrop fracture, laminae, lamina fracture.

1. INTRODUCTION

Catastrophic lower cervical spine injuries with irreversible neurologic sequelae are often the consequences of direct head impact, and have been reported in both contact and non-contact sports. Contact sports include football, hockey, rugby, and wrestling, while examples of non-contact sports are skiing, track and field, diving and surfing. The injuries under study are also associated with falls from heights and crashes of motor vehicles when the head of the victim strikes the roof or the windshield.

Lower cervical spine injuries result when the top of the head strikes a solid object and the flexed neck reacts in an attempt to stop the inertial energy of the moving trunk and of the extremities involved. The forces and moments generated by the deceleration are complex, particularly when added to the response of the cervical spine to impact loading. When exposed to high velocity loading the cervical spine shows a primary level of injury

Michael D. Gilchrist (ed.), IUTAM Proceedings on Impact Biomechanics: From Fundamental Insights to Applications, 221–228.

along with bony injuries of lesser magnitude at contiguous or secondary levels. One commonly reported pattern of injury is the teardrop fracture. As the name suggests, this is a triangular fragment of bone at the anterior inferior corner of the lower cervical body [1, 2].

A retrospective study was conducted with the purpose being to review imaging configurations, develop an inventory of fractured elements, and detail the incidence and association of midsagittal centrum and posterior neural arch fractures as integral components of the teardrop fracture. The spinal levels, sagittal and coronal vertebral fractures and fracture of the posterior arch were identified. It was assumed that the evaluation of the characteristics and mechanisms of injury is an important component in any biomechanical, computational or forensic analysis.

2. MATERIALS AND METHODS

This study is based on 33 cases, a total of 9 female and 24 male individuals, ranging in age from 18 to 28. Witness statements, paramedics' reports, emergency department records, initial medial records (up to and including surgery) and discharge summaries were reviewed and coded. All of the subjects sustained teardrop fractures resulting in irremediable spinal cord damage. Initial plain x-rays, CT scans, and in some cases MRIs and/or tomograms were reviewed and correlated to the available imaging. The types and frequency of accidents were as follows: motor vehicle, 8; shallow water diving, 16; football, 3; wrestling, 3; falling, 2; surfing, 1.

3. RESULTS

The location and distribution of injuries at the primary fracture site (i.e. site of compromise of bony failure) were as follows. There were 4 instances of C4 teardrop fractures, 26 instances of C5 teardrop fractures, and 3 instances of C6 teardrop fractures. Lesser magnitude contiguous fractures were noted at spinal levels immediately adjacent to the primary fracture site. Radiographs of the cervical spine showed the mechanism of injury in all cases to be the result of compression forces with varying degrees of flexion.

3.1 The Teardrop Fracture

Forms of teardrop fractures are shown in Figure 1.

Anterior Fragment: The size of the anterior fragment was usually one third or less of the vertebral centrum at the primary fracture site. The triangular anterior inferior vertebral fragment fracture line began at the lower end plate of the centrum and its orientation extended obliquely upward toward the anterior margin of the affected vertebral body. The triangular anterior inferior vertebral fragment remained somewhat in alignment with the anterior margin of the subjacent vertebrae and disks.

Posterior Fragment: The larger posterior fragment was displaced and / or rotated posteriorly reducing the space available for the spinal cord.

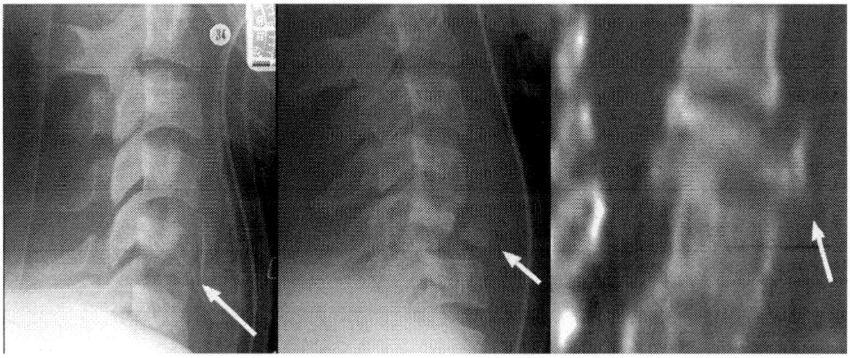

Figure 1. Various forms of C5 teardrop fractures. The triangular anterior inferior corner fractured fragments are identified by the arrows. The larger posterior fragment has been forced rearward into the spinal canal.

3.2 Midsagittal Centrum and Bilateral Arch Fractures

Of the 33 cases, 27 were instances of teardrop anterior inferior corner fracture with midsagittal vertebral centrum fracture and bilateral arch fractures. The anterior inferior corner fracture with midsagittal vertebral centrum fracture and bilateral arch fractures were all noted to be at the primary site of fracture. The frontal view (Figure 2) may or may not reveal a vertical radiolucent line extending from the superior end plate to the inferior end plate at the primary fracture site. The neural arch was often bilaterally fractured through its posterior arch.

In 18 of the 27 cases, subjects sustained secondary midsagittal vertebral centrum fractures. Some of these also exhibited bilateral posterior arch fractures of lesser magnitude at contiguous levels.

In 6 of the 33 cases, the subjects sustained teardrop fractures unaccompanied by midsagittal vertebral centrum fractures or posterior arch fractures at the primary level of trauma.

In 27 cases there was complete separation of the vertebral centrum in the coronal and sagittal planes. The teardrop fracture separated the centrum into anterior and posterior components. The vertical fracture of the vertebral centrum in the sagittal plane split the vertebral body into right and left halves. The neural arch was bilaterally fractured frequently at the lamina. The transverse diameter (interpedicular distance) of the vertebral body increased; the spinous process was free floating.

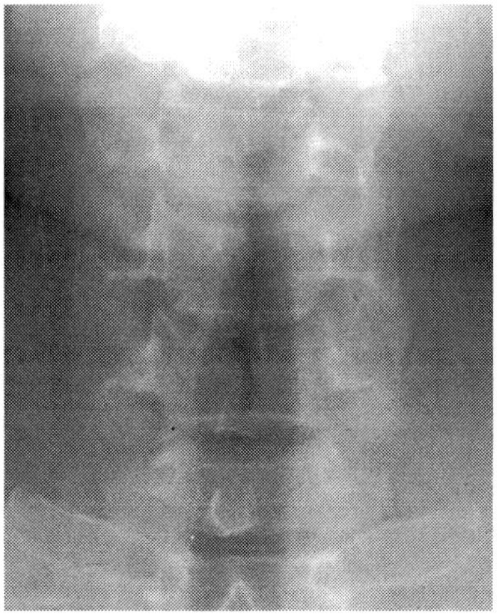

Figure 2. Midsagittal split at two levels.

4. DISCUSSION

The structure of the lower cervical spine is anatomically distinct, kinematically unique and biomechanically complicated. The teardrop fracture occurs in the lower cervical spine and is the result of indirect forces. It is most prevalent at the C4-C6 region of the spine. The C5 vertebral body is the most commonly involved vertebral level. The teardrop constitutes a highly unstable condition and in the cases presented it is the result of higher energy impact. As already indicated, the term "teardrop fracture" usually connotes a general pattern of spinal injury. It has no agreed upon definition biomechanically or clinically [3]. Its etiologic its radiographic characteristics and mechanisms of injury have been reported by Harris [4], Torg [5], Babcock [6], Lee [7, 8], Abitbol [9]. Its fracture characteristics are complex and more often than not, the teardrop fracture has secondary fracture components at neighboring levels. The teardrop fracture is a recurring injury, a major cause of morbidity and merits separate biomechanical description and classification.

When a triangular shaped teardrop fracture of the vertebral body in the lower cervical spine is identified, there is a high probability of coexisting lesions in the centrum and/or in the neural arches over neighboring spinal levels.

The primary level of injury consists of an anterior inferior corner fracture fragment, the displacement of the inferior margin of the fractured vertebral body backwards into the spinal canal, the midsagittal vertebral centrum fracture and the posterior arch fractures. Multiple fractures of lesser magnitude are commonly present at contiguous levels neighboring the primary level of injury, as shown in Figure 4. The vertical sagittal fracture and the posterior arch fractures can occur independently of each other.

Triangular Anterior Inferior Corner Fracture Fragment: The bony perimeter of the displaced triangular anterior inferior vertebral fragment was left intact. There was no deformation or distortion in its shape. The bony trabeculation within the displaced triangular anterior inferior vertebral fragment was not compressed or crushed. There was no increase in density, but rather the fragment appears to have been sheared off of the anterior inferior beak of the centrum in compression.

Larger Posterior Fragment: The larger posterior portion of the vertebral body was displaced backwards into the spinal canal intruding upon the space occupied by the spinal cord. Its posterior inferior corner may or may not be fractured.

Midsagittal Vertebral Body Fracture: The midsagittal vertical fracture and bilateral neural arch fractures occur commonly as a component of the teardrop fracture. The midsagittal cleavage fracture crack was initiated at the posterior wall and between the pedicular roots of the centrum. The fracture line propagated in more or less a straight line from the posterior centrum wall toward the anterior centrum wall. The vertebral body was separated into two halves. The radiographic imaging (axial CT) demonstrated the pattern of midsagittal cleavage at the site of the teardrop fracture to be common. The midsagittal vertebral body fracture accompanied with an anterior inferior corner fracture (in addition to the larger posterior fragment) appears to be a hallmark associated with severe spinal cord damage.

Neural Arch:Fractures Laminae fractures are components of the teardrop / sagittal fracture. The laminae are thin flat plate-like struts of bone that converge and join in the midline at an angle to form the spinous process. The location and appearance of these fractures vary. The laminae may fracture at their base near the transverse process. Fractures may be symmetrical, asymmetrical or may be horizontally or vertically oriented. Laminae may be split at a single or multiple locations. If the laminae are fractured the spinous process is free floating. Should this occur, the neural canal may be encroached upon from any direction and at any angle.

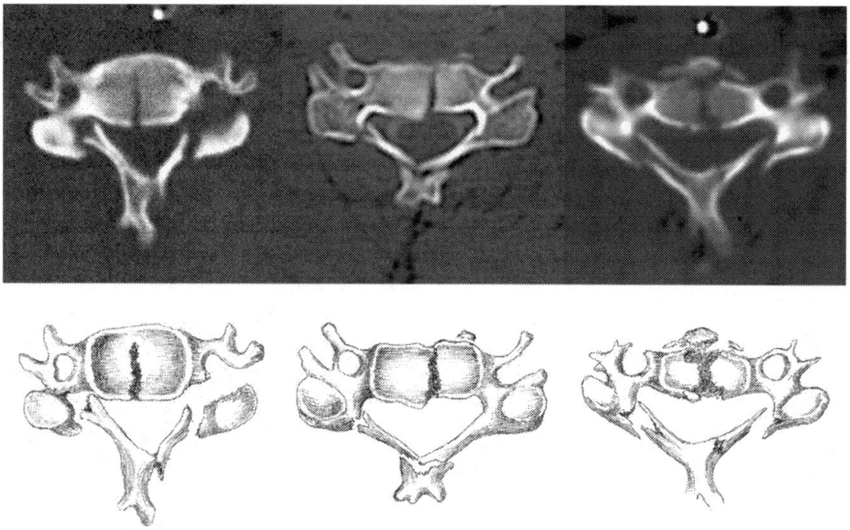

Figure 3. Typical variants of the locations of the planes of fracture separations together with their direction of longitudinal propagation are shown above and illustrated below. (primary level fractures).

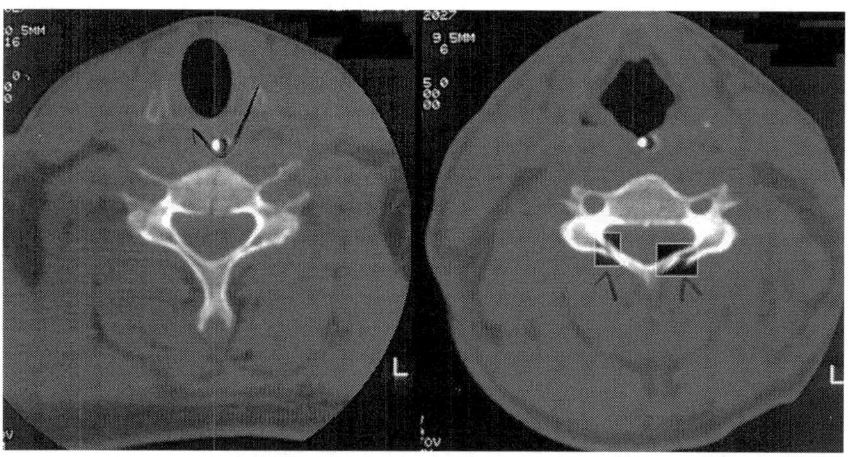

Figure 4. Hairline midsagittal fracture without posterior arch fractures. Bilateral lamina fractures without midsagittal fracture. (secondary level fractures).

The vertebral bodies immediately adjacent to the site of the primary fracture site revealed the presence of similar midsagittal hairline fractures only of lesser magnitude. The hairline fractures may or may not extend from the posterior towards the anterior vertebral centrum wall. The site of fracture initiation begins at the posterior wall between the pedicular roots before it propagates or branches.

The midsagittal vertebral centrum fracture may occur in isolation at adjacent cervical levels without bilateral neural arch fractures. Bilateral neural arch fractures may occur in isolation without a midsagittal vertebral centrum fracture at neighboring levels. The width of the separation observed on plain and axial CT imaging varied and appeared to be dependent upon the presence of posterior neural arch fractures.

4.1 The Etiology of Fracture

Anterior Inferior Fragment: In the lateral view, the inferior surfaces of the C3-C6 vertebral body are concave, skewed anteriorly, and form a sickle shaped bony lip or beak in the anterior median plane. In flexion, the sickle shaped bony lip of the upper vertebral body tilts and slides anteriorly; it appears that the fragment was "squeezed" or pinched off by the superior and inferior vertebral bodies as hypothesized by Harris [10,4], Kim [7] and others.

Larger Posterior Fragment: Once the triangular fragment of bone is fractured the remaining posterior larger fragment may become wedged between the adjacent superior and inferior vertebral bodies. This results in a decrease of the space available for the spinal cord.

Midsagittal Fracture Fragments: The midsagittal fracture was a hallmark associated with the teardrop fracture. The vertical fractures have been previously reported by Richman [11], Rogers [12], Lee [8], Torg [3] and others.

When the vertebral centrum is fractured in its coronal and sagittal planes, the fracture results in at least four fragments; its transverse diameter is increased. The position, origin and configuration of these fragments make it possible to infer the mechanism of injury. Its coronal and sagittal plane fractures are the result of compression forces in the region of the anterior elements of the vertebral body.

The etiology of the midsagittal fracture is unclear. Fracture initiation and propagation appear to coincide with the cylindrical truncated channels for the basivertebral veins. This suggests a vertical compressive load results in an outward splaying of the centrum. One conjecture is that the centrum fractures in a plane approximately perpendicular to the direction of loading. The fracture appears to be primarily created by a symmetrical compressive force applied in the sagittal plane. Another conjecture is that the mechanism of vertebral splitting is thought to be due to acute herniation of the intervening disk into the vertebral body. A third conjecture is thought to be this: a downward compressive force on the vertebral unit decreases the height of the disk, increasing its hoop tension and as a consequence, forcing the uncinate processes outward; which in turn split the vertebral centrum as well as the posterior arch.

REFERENCES

1. Scher, A.T., “Tear-Drop’ fractures of the cervical spine - radiological features, SA Medical Journal, 1982, 6:355-356.
2. Schneider, R. and Kahn, E., Chronic neurological sequelae of acute trauma to the spine and spinal cord. 1. The significance of the acute flexion or "tear drop" fracture dislocation of the cervical spine, J. Bone Joint Surg. (Am.), 1956, 38:985-997.
3. Torg, J., Pavlov, H. and Glasgow, S., Radiograhic evaluation of athletic injuries to the cervical spine, in Disorders of the Cervical Spine, O'Leary P. Williams & Wilkins (eds), 1992.
4. Harris Jr., J.H., Edeiken-Monroe, and Kopaniky, D.R., A practical classification of acute cervical spine injuries, Orthopedic Clinics of North America, 1986, 17(1):15-30.
5. Torg, J.S., Quedenfeld, T.C., Moyer, R.A., et al., Severe and catastrophic neck injuries resulting from tackle football, J.A.C.H.A., 1977, 25:224-226.
6. Babcock, J.L., Introduction to symposium on cervical spine injuries, Arch. Surg., 1976, 111: 637.
7. Lee, C., Kim, K.S. and Rogers, L.F., Sagittal fracture of the cervical vertebral body, AJR, 1982, 139:55-60.
8. Lee, C., Kwang, K.S. and Rogers, L.F., Triangular cervical vertebral body fractures: Diagnostic significance, AJR, 1982, 138:1123-1132.
9. Abitbol, J.-J. and Kostuik, J.P., Flexion injuries to the lower cervical spine, in The Cervical Spine, Third Edition, Committee TCSRSE (ed.), Lippincott-Raven Publishers: Philadelphia, 1998, pp. 457-464.
10. Harris Jr., J.H., Acute injuries of the spine, Seminars in Roentgenology, 1978, 13(1):53-68.
11. Richman, S. and Friedman, R.L., Vertical fractures of cervical vertebral bodies, 1954, 62:536-543.
12. Rogers, W.A., Fractures and dislocations of the cervical spine, The Journal of Bone and Joint Surgery, 1957, 39-A(2):341-376.

SESSION 5

IMPACT INJURY IN SPORT

Andrew S. McIntosh
*The University of New South Wales, School of Safety Science, UNSW, Sydney, Australia, 2052. phone: *612 9385 5348, fax: *612 9385 6190, email: a.mcintosh@unsw.edu.au*

Abstract. Impacts in sport can cause severe and fatal head, spinal and thoracic injuries. As sports have developed during the last half century, methods have been developed to control injury risks, such as rule modifications, helmets, padded clothing and training. The biomechanics of severe injury is well understood through investigations of motor vehicle accident trauma. Research into sports injury can assist in reducing injury risks and also identify mechanisms and tolerance limits for lower severity injury. The paper provides an overview of impact injury risks in a range of sports, including concussion, commotio cordis and spinal cord injury. The main focus of the paper is injury control methods, their biomechanical bases and their success in reducing injury. Successful methods have been developed that include breakaway bases and helmets. Biomechanics plays an important role in quantifying physical hazards, developing interventions, methods for testing their performance, and assisting in field research.

Key words: sports, impact injury, biomechanics, safety, helmets.

1. INTRODUCTION

While the greatest focus of impact injury related research and technology development has been in understanding and preventing motor vehicle related injury, there has been a parallel interest in sports related impact injury. Many esteemed researchers and research groups who have established their careers in motor vehicle safety have also made substantial contributions to studying sports injuries and their prevention. Alongside the intrinsic value of reducing sports injury, the study of the mechanisms of these injuries can both inform the development of methods to prevent lower severity injury to motor

Michael D. Gilchrist (ed.), IUTAM Proceedings on Impact Biomechanics: From Fundamental Insights to Applications, 231–245.

vehicle occupants, in recreation and occupational settings, and improve our overall understanding of human tolerance to impacts.

The fundamentally different aspects of sport in comparison to motor vehicle related injury are that the health benefits of participation in sport are considered valuable for the individual and society, and that in many cases the impact energies are related to gravity and velocities reached under human power. A public health dilemma exists between the benefits of participation in sport and the immediate and long term costs in terms of injury. Injury in professional sport influences team and individual performance and the fortunes of players and others in a wide radius from the sports field; it was reported in Der Spiegel that the share price of Borussia Dortmund football club was affected when it was announced that one of their star players had been diagnosed with a brain tumor [1]. Therefore, there is a broad interest in understanding and preventing sports injury.

Sport gives rise to numerous events in which high energy impacts may occur, e.g. projectile ball sports (cricket, baseball, squash and hockey), high velocity sports (skiing, equestrian and cycling), and contact sports (ice hockey, rugby and American football). There are many common injuries across all sports as well as injuries that are unique to specific sports, e.g. commotio cordis in baseball. In very broad terms there are two types of injury mechanisms in sport, acute overload and overuse. However in the causation of injury, both mechanisms may be present. Impact events explain many acute overload injuries. From a biomechanical perspective the cause of the injury must explain how the event produced loads that exceeded the 'strength' of the tissue, whether the 'strength' was normal or impaired, and the specific mechanism.

This paper will provide an overview of injury risk management in sport, including epidemiology of injury in sport and impact injury, followed by sections on the range of loading in sport and an analysis of injury risk control methods.

2. EPIDEMIOLOGY OF SPORTS INJURY

2.1 Research Methods

Studies of sports injury are hampered by the definition of injury and it is often difficult to compare injury rates and patterns between and within sports. Common definitions are: injuries resulting in a player missing a training session or game; and, an injury requiring on-field attention. The latter gives rise to much higher rates per period of exposure and the profile of injuries with the former definition tend to be more serious and include muscle, bone and joint injuries. However, return to play guidelines, team management, individual recovery and motivation, and family input also influence when a player returns to participation following injury. Data have been collected through a range of methods: prospective studies with on–field recorders; team staff; insurance data; and, telephone interviews, for example. Biomechanical data are even more difficult to collect in the field, but the availability of high quality match video has provided researchers with at least qualitative data and in some cases quantitative input data for numerical and physical models. Many laboratory projects have assessed the relative loads, between well executed skills and techniques, and against known load tolerance limits, but in the absence of injury risks. The multifactorial nature of injury causation in sport requires the collection of a substantial quantity of data on individuals, teams, the environment and their exposure, in order to identify the role of biomechanical factors in injury causation [2]. The end effect is that research is lengthy, laborious and expensive.

Table 1. Profile of head injury risks in selected sports.

Sport	Head injury as % of all injuries [references]	Special injury issues
Rugby	14-25% [3]	Concussion 5-15% [3]
Ice Hockey	4-18% in professional level, 1995-2001 [4]	Intracranial injury and blinding
American Football	6.1% concussion [5]	Severe and fatal injury
Baseball	28% face, 11% head, 5% mouth in 5 to 14 year olds and emergency visits; [6]	Leading cause of sports related eye injury in the USA [6]
Soccer/Football	4-20% [7]	Fatal injuries in children due to falling goal posts. Debate over concussion and heading
Boxing	16% of injuries in professional boxing concussion [8]	Risk of severe acute and long term brain injury
Cricket	9-25% [9]	Oro-facial and eye injuries
Professional Jockeys	19% head, neck and face [10]	Severe spectrum of injuries from falls and collisions
Skiing & Snowboarding	30% traumatic brain injuries [11]	Collision with rigid objects, e.g. trees, may be fatal

2.1.1 Injury profiles

The following is a brief summary of injury profiles in a selection of sports. Table 1 presents a profile of head injury in sports for which helmets are available. In general the head/face/neck, followed by the lower limb and upper limb represent the hierarchy of injury by region.

Injury rates vary greatly, due to the sport, level of participation and injury definition: Youth soccer - 0.6 to 19.1 per 1,000 player hours [12] 29 injuries per 1,000 player hours [13] in Norwegian under 21 team matches; Rugby - 19.4 lost time injuries per 1,000 player hours (under 15 to international combined) [3], 100 injuries per 1,000 player hours at 2003 Rugby World Cup (unpublished work of author); American Football - 40 per 1,000 athletic exposures [14]; Ice Hockey - 96 per 1,000 player hours (junior) [15], 53 per 1,000 player game hours [16]; Professional Horse Racing- 606 injuries per 1000 jockey years (1993-1996, USA) [10]; and, Skiing and Snowboarding , 2-6 injuries per 1,000 skier days [11,17]

2.1.2 Severe impact injury

Severe impact injury in sports include: spinal cord injury (SCI), mild to severe head injury, commotio cordis, fractures and dislocations. Fatal injuries do occur, although in some cases an event exposes an underlying anatomical or physiological abnormality. The history of fatal injuries in American football for period after 1945 is complicated [18]. In total 497 persons died playing American football, of these 69% died from fatal brain injuries and 16% SCI. The authors reported that the introduction of laws to control the use of the head in blocking and tackling and the implementation of NOCSAE helmet standards resulted in a significant decrease in fatalities since 1976 [18]. A study of deaths in football in Victoria, Australia, identified nine cases of intracranial injury resulting from head impacts in the period 1968 to 1999 [19].

Even though the overall rate of SCI in rugby is low, there is a distinct SCI risk in rugby associated primarily with the tackle and scrum [20]. In a recent review of SCI in Australia between 1986 and 1996, six of the 31 SCI cases occurred in schoolboys, with an annual incidence for schoolboys of 1.7 compared to 4.8 per million adult players for this period [21]. The risk of SCI was 10 to 12 fold greater with adult players than schoolboys [22]. In the USA during the period 1970-1996, 36 of the 62 cervical spine injuries in rugby occurred in the scrum, including 14 junior players [23, 24]. Scrum-related SCI affected the front row with the hooker, one out of 15 players, suffering 30% of SCI [25]. There is a higher risk of SCI on scrum engagement than scrum collapse. There is also a risk of SCI in American football which has been associated with spear tackling, in which the tackler dives head first into the opponent [26]. Body checking, especially from behind, in ice hockey was similarly blamed for SCI in ice hockey [4].

Commotio Cordis has been observed in projectile sports: baseball, ice hockey and lacrosse [27]. The potentially fatal physiological injury arises due to a chest impact that occurs just prior to the peak of the T wave causing ventricular fibrillation. On average there were four deaths in youth baseball in the USA per annum from 1973 to 1995 [28]. Slightly less than two in every four deaths were due to commotio cordis and one in four resulted from ball to head impacts.

Horse racing is another sport with a high risk of head, neck and thoracic injury due to falls from the horse, impacts with the rail and with the horse. Between 1975 and 2000 there were 9 fatalities in horse racing. The causes of death were in general brain, cerebrovascular and thoracic injury [29].

Table 2.. Biomechanical loads in a selection of sports and skills.

Sport / Skill	Type	Magnitude (as Force, body weight &/or Energy)	References
Running	vGRF (feet)	1.41 - 4.16 BW	[30]
Jump/take off	vGRF (feet)	2.68 BW	[33]
Jump/landing	Drop landing from 60cm	3.4 - 4 BW	[34, 35]
volleyball	Landing	3.3 - 6 BW	[36, 37]
Long jump	Take off	10 BW	[36]
	Drop landing from 2.55m bar	8.6-11.6BW	[38]
Gymnastics	vGRF (feet)	1.2 – 14.4 BW	[31, 32]
	vGRF (hands)	0.6-2.2	[32]
Netball	vGRF (feet) step jump task	3.12-3.41	[39, 40]
Basketball	Run and stop	2.35 BW	[41]
Baseball	Ball impact	0.142 kg, up to over 45 m/s & 160 J	[42]
Cricket	Ball impact	0.156 kg up to over 45 m/s & 160 J	[42]
Ice Hockey	Head impact	35g measured within helmet, Speed of skater can reach 50 km/h	[42, 43]
	Puck Impact	0.163 kg, up to over 45 m/s & 160 J	[42]
Rugby	Scrum	Total force across front row 4.4 kN to 8 kN	[44]
American Football	Head collision	29.2g measured within helmet in high school	[43]
Soccer	Heading	16 to 55 g	[43, 45]
Boxing punch		3-5 kN skill dependent	[46, 47]

3. BIOMECHANICAL LOADS IN SPORT

Athletes impact the ground repeatedly while running, jumping and landing. Table 2 presents a selection of external biomechanical loads from a range of sports. Vertical ground reaction forces in distance running, for example, have been measured in the range of 1.4 to 4.2 times body weight [30]. So an 80 kg runner could easily experience vertical ground reaction forces of greater than 2 kN. In gymnastics ground reaction forces in take off and landing applied to the feet have been measured up to 14 times body weight and to the hands at twice body weight [31, 32]. Poor skill execution can result in injury. The energy and impact forces measured in boxing, scrum engagement, and from projectiles are high and known to cause injury to the head and chest.

4. CONTROLLING IMPACT INJURY RISKS IN SPORT

4.1 Elimination, Substitution & Engineering

In the context of formal sports it is possible to eliminate some injury risks by elimination or substitution of hazards, or engineering methods. A number of examples will be given: substituting the boundary fence in cricket, substitution of balls in baseball, breakaway bases, and the role of ground hardness. Banning a sport is an additional approach. Removing rigid structures from the playing field or replacing them are obvious methods for preventing impact injuries.

4.1.1 Baseball balls

In baseball, softer balls have been shown to reduce the risk of head injury in comparison to standard balls [48, 49]. Marshall et al observed a 28% reduction in the risk of injury in baseball for games using the reduced impact ball compared to the standard ball (rate ratio = 0.72) [49]. The softest impact ball was observed to be associated with the lowest risk of injury (adjusted rate ratio = 0.52) and the authors reported on a study noting that adult and child players found it difficult to identify the differences between standard and safety balls in pitching, throwing and batting. Unfortunately, the benefits of safety balls in preventing commotio cordis may be minimal based on impact testing using the Viscous Criterion as an assessment tool for chest injury risk [50].

4.1.2 Breakaway bases in baseball and softball

The design and evaluation of breakaway bases for baseball and softball is a highlight of the sensible application of engineering controls to address a common injury risk. The breakaway base consists of a mother (base) plate fixed into the ground by a post and a detachable base top. If a sufficiently large shearing force is applied to the base top grommets release and it slides away. Therefore, the maximum leg force that is experienced by the player is equal to the release force of the base top. The breakaway force can be designed to limit the force applied to the leg to a level below which injury will not occur. Field studies of breakaway bases have shown that they are very successful in reducing injury in softball and baseball Janda et al found that breakaway bases could almost eliminate injury related to sliding in softball, when a prospective trial of breakaway bases was conducted [51-53]. The rate ratio was 22.8 injuries on a field with stationary/fixed bases to one on a field with breakaway bases. In a follow up study in college and professional baseball an 80% reduction in the risk of sliding related injuries was observed on breakaway base compared to stationary base equipped fields. In cricket similar problems have been addressed by introducing a rope inside the fence to represent the boundary of the cricket field.

4.1.3 Environment – ground hardness and surface

Australian Rules Football is a fast, kicking and running game player at a professional and amateur level across Australia. The game is associated with a high rate of lower limb musculoskeletal injuries and collision injuries, e.g. concussion. Recent studies have debated whether ground hardness is a direct factor in the etiology of injury or whether it indirectly contributes to higher injury rates as it enables higher running speeds and higher energy impacts. Norton et al concluded that higher game speeds resulted in more collisions and a higher risk of injury [54]. Ground hardness and turf will also affect lower limb dynamics during running, stepping and cutting.

4.2 Training

Training has two main components in sport, fitness and skills. Fitness training in the area of strength and power is important in the preparation of participants, and may lead to increases in the strength of bone, muscle and ligaments, as well as improved cardio-respiratory capacity. However, there are potentially a number of detrimental effects for the participant and opponent/s due to the increased mass and velocity of the 'fit' participant in contact sports. Neuromuscular programs have been shown to be effective in reducing joint injuries in some sports. For example, a player stopping suddenly and turning, in the absence of body contact, can reduce the risk of knee ligament injury by improving their balance and control through neuromuscular programs [55, 56]. The aim of the programs is to reduce the loads applied to sensitive joint structures through improving the alignment of body segments and, thus, reduce muscle and joint forces. Myklebust et al observed a reduction in the rate of anterior cruciate ligament injuries in elite female team handball players after exposure to a neuromuscular training program [56].

4.3 Administrative Controls (Laws and Rules)

Rules are one of the most common methods used to prevent injury. Rules and administrative procedures, e.g. playing field preparation, penalizing hazardous play, PPE use, coaching and referee accreditation, and team selection, have a role in preventing injury.

4.3.1 The rugby scrum

The rugby scrum has received substantial attention over the years with regards to SCI. A rugby scrum consists of eight players divided into the three front row and two back row and three ‘loose’ forwards. The front row of each team’s scrum engages through their heads and shoulders in a forceful driving motion. This can result in high axial compressive neck forces combined with a bending moment and/or shear forces; during the 2003 Rugby World Cup there was one case of cervical dislocation during a scrum engagement that was a career ending injury. Milburn measured the forces applied to an instrumented scrum machine, and found that the total horizontal forward force on engagement ranged between 4.4 kN for high school players to 8 kN for the Australian national team [44]. After the initial engagement, the sustained force reduced by approximately 20%. The forces on engagement have the potential to exceed axial neck load and bending moment tolerance limits. Scrum laws for under 19 year olds eliminate an impact on engagement and permit each front row to orient itself well, thus reducing neck loads.

4.3.2 The tackle and body checking

The tackle in rugby, Australian and American football, and body checking in ice hockey are associated with risks of minor to severe impact injury, including skull fractures, brain injury and SCI. In rugby, approximately half of all injuries arise in the tackle. Factors that may give rise to injury risks, including SCI and concussion, in the tackle include: high tackles; high velocity tackles; tackles in which the tackler may have been in the peripheral vision of the ball carrier; 'big hits' in which the ball carrier is tackled by more than one player; and, a general lack of skill for the tackler [25, 57, 58]. Apart from high tackles and spear tackles, where the ball carrier is speared head first into the ground, the other types of tackle are legal. Body checking in ice hockey, especially involving the head, has led to a concern that this form of contact should be eliminated through rule changes due to its association with head injury [59]. As with spear tackling in American football there is a view that the use of helmets has increased hazardous play due to players' perceptions that they are at a lower risk of injury, often referred to as risk compensation.

4.4 Personal Protective Equipment (Helmets)

Helmets and padded headgear are used in many projectile and contact sports. Recently, there has been much debate over the use of padded headgear in soccer. Helmets are designed to attenuate the impact energy and distribute the impact force applied to the head. If a helmet can reduce the head impact force and head's acceleration to below relevant tolerance levels and under sport specific impact scenarios, then a helmet can function to reduce the risk of injury. The widespread use of helmets indicates that training and laws are not sufficient to prevent head injury, as even the most talented athletes suffer from head injuries. Factors in helmet design and performance, effectiveness and efficaciousness, are described below

Beyond helmets, protective eyewear and some chest protectors the safety performance of the majority of sporting products is not covered by recognized product standards. Standards, such as the Australian/New Zealand, British, ASTM, NOCSAE, ISO and Canadian, cover the performance of a variety of helmets for sports including cricket, ice hockey, baseball, equestrian, American football and cycling. Some sports administer their own standards for padded clothing, e.g. the International Rugby Board. However, consumer laws do not always prevent the sale of helmets or products that do not comply with recognized standards. In addition, differences in standards and marketing contribute to a situation in which the consumer may not be sufficiently well informed to purchase the most effective safety products. It is fair to say that most helmet standards are designed to produce helmets that prevent severe head injury, e.g. involving skull fractures, and that their role in preventing mild injury such as concussion, is unclear. Further, it is difficult to design safety devices that are equally effective for all levels of impact severity. The impact performance requirements of a selection of helmet standards are presented in Table 3.

Table 3. Summary of selection of helmet standards.

Sport	Standard	Impact Test	Energy	Anvil	Criteria
Rugby	IRB [60]	Drop test	13.8 J	Flat rigid anvil	Between 200 and 550 g
Ice Hockey	CAN/CSA Z262.1-M90 [61]	Drop test (& penetration test)	50 J	Flat rigid anvil	< 275 g, (GSI < 1500)
American Football	ASTM F429 & F717 [62]	Drop test (24 impacts)	75 J (5 kg, 5.47 m/s)	MEP	< 275 g for average of 3 tests at 6 locations. < 300g for final 6 tests
Snow Sports	ASTM F2040-02 [63]	Drop tests	6.2 m/s (flat) (96 J with 5 kg)	Rigid flat, hemispherical & edge	< 300 g
Baseball	NOCSAE [64]	Ball fired at helmet	52 J (27 m/s)	NOCSAE headform	SI < 1200
Cricket	AS 4499 [65]	Striker with ball (1.5 kg)	30 J	Striker dropped onto helmet	25% reduction in acceleration compared to bare headform test

NB: GSI is the Gadd Severity Index, GSI and SI are functions of headform acceleration and time.

There has been a gradual evolution of helmets in American football and ice hockey from padded headgear to helmets comprising a hard shell, a liner and a visor. Canadian and NOCSAE standards for ice hockey and football helmets, respectively, were established in the early 1970s. Both types of helmets are assessed using a standard free-fall drop test onto an anvil utilizing a headform acceleration of 275 g. as the pass/fail criterion, however impact energies and anvils differ (Table 3). The Modular Elastomeric Programmer (MEP) anvil is generally used in NOCSAE methods with a higher impact energy in contrast to rigid anvils and lower energies in other standards. Tests on two models of standards compliant ice hockey helmets in 2001 revealed that the rigid anvil impacts compared with MEP tests showed greater differences in the impact energy attenuation properties of the helmets as the energy increased [42]. The denser and thicker expanded polypropylene (EPP) liner in one helmet maintained the headform acceleration below 200g for all impacts up to 1.2 m drop height, while with rigid anvil impacts the softer and thinner nitrile acetate liner helmet exceeded 250 g from 1 m drops. With MEP tests headform accelerations for both helmets remained below 150 g. In the context of injury prevention this raises the important issue of setting standards to address specific injury risks, either by severity or incidence rates. The tests indicated that the EPP helmet would offer greater protection under severe impacts and prevent serious injury more effectively than the other helmet.

Compared with the Australian standard for cricket helmets, the NOCSAE standard for baseball helmets provides a realistic impact and assesses helmet performance against an absolute injury tolerance criterion. The ASTM standard for snow sport helmets includes a variety of anvils that may reflect hazards in ski fields.

Levy et al reported that there was a 74% reduction in fatalities and an 84% reduction in serious head injury in American football following the implementation of NOCSAE helmet standards [18]. A study of the biomechanics of concussion in helmeted professional American footballers using video analysis of head impacts and laboratory reconstructions found that the change in velocity of the head was 7.2 m/s compared to 5 m/s with resultant headform accelerations of 98 g. compared to 60 g, averaged for concussion and uninjured cases, respectively [66]. The authors concluded that helmets prevented the more serious spectrum of brain injury and skull fracture, however the helmets were not preventing concussion in the higher speed impacts. In comparison, and also using video analysis, the change in velocity of the head was observed to be 4.5 m/ for a larger sample of concussed unhelmeted elite Australian footballers [67]. Preliminary results from a new phase of analysis of the Australian football cases using MADYMO modeling of the head impacts, reveals head accelerations of 70 g. for Cantu scale grade 1 concussions and 82 g for grade 3 concussions [68]. Considering the differing methods and their limitations, there is a strong similarity in head accelerations in concussion cases. This suggests that while helmets prevent severe head injuries, helmeted players may simply adapt to the new level of protection and expose themselves to higher energy impacts than unhelmeted footballers.

Studies into the effectiveness of helmets in sports where the use is not mandatory are emerging. Ski helmets have been found to be effective for adult and child skiers and snowboarders reducing the risk of severe head injury by up to 56% and 29% for all head injury [69, 70]. Research to date indicates that padded headgear does not reduce the incidence of concussion in Rugby Union football [71-73] Survey responses of under 15 male rugby players and, as yet unreported, studies of the author on a wider age group suggested that players believe that they can tackle harder and play more confidently while wearing headgear [72]. As the action of tackling is responsible for half of all rugby injuries, this combination of perception and biomechanical performance is of concern. Laboratory testing of a thicker and denser version of padded headgear demonstrated that impact energy attenuation could be improved, but possibly not to a level consistent with the impacts observed to produce concussion in unhelmeted footballers [74].

5. CONCLUSIONS

A wide range of injury risks are present in sport. The performance of PPE and other interventions requires improvement through consideration for impact hazards, injury risks, and test methods, combined with the promotion of 'safe' devices. The biomechanical quantification of hazards, in terms of impact interfaces and energies, for specific injuries and injury severities combined with further analysis of injury mechanisms and tolerance is required to develop systems and devices to reduce sports injury. Epidemiological, medical and human factors research methods are required in combination with biomechanical and technological approaches to reduce further injury risks in sport.

REFERENCES

1. Geyer M, Der oeffentliche Krebs, Der Spiegel, 2001, September 38/15.
2. R Bahr and Holme I, Risk factors for sports injuries—a methodological approach, *Br J Sports Med*, 37, 2003, 384–392.
3. McIntosh AS, Best JP, Orchard J and Savage T, Rugby Union Injury Surveillance Study: 2000–2002, The University of New South Wales, Sydney, 2004.
4. Biasca N, Wirth S and Tegner Y, The avoidability of head and neck injuries in ice hockey: an historical review, *Br J Sports Med*, 36, 2002, 410–427.
5. Meeuwisse WH, Hagel BE, Mohtadi NG, et al, The distribution of injuries in men's Canada west university football, *Am J Sports Med*, 28, 2000, 516–523.
6. Yen KL and Metzl JD, Sports-specific concerns in the young athlete: Baseball, *Pediatric Emergency Care*, 16, 2000, 215–220.
7. Kirkendall DT, Jordan SE and Garrett WE, Heading and head injuries in soccer, *Sport Med*, 31, 2001, 369–386.
8. Zazryn TR, Finch CF and McCrory P, A 16 year study of injuries to professional boxers in the state of Victoria, Australia, *Br. J. Sports Med*, 37, 2003, 321–324.
9. Stretch RA, Bartlett R and Davids K, A review of batting in men's cricket. *J Sp Sci*, 18, 2000, 931–949.
10. Waller AE, Daniels JL, Weaver NL and Robinson P, Jockey Injuries in the United States, *JAMA*, 283, 2000, 1326–1328.
11. Levy AS, Hawkes AP, Hemminger LM and Knight S, An analysis of head injuries among skiers and snowboarders, *J Trauma*, 53, 2002, 695–704.
12. American Academy of Pediatrics, Injuries in Youth Soccer: A subject review, *Pediatrics*, 105, 2000, 659–661.

13. Andersen TE, Larsen Ø, Tenga A, Engebretsen L and Bahr R, Football incident analysis: a new video based method to describe injury mechanisms in professional football, *Br J Sports Med,* 37, 2003, 226–232.
14. Marshall S, Waller AE, Dick, RW, Pugh, CB, Loomis, DP and Chalmers, DJ, An Ecologic Study of protective equipment and injury in two contact sports. *Int. J of Epidemiology*, 31, 2002, 587–592.
15. Stuart MJ and Smith A, Injuries in junior A ice hockey. A three year rospective study, *AmJ Sports Med,* 1995; 23: 458-461
16. Tegner Y and Lorentzon R, Ice hockey injuries: incidence, nature and causes, *Br J Sports Med,* 25, 1991, 87–89.
17. Hunter RE, Skiing injuries, *American Journal of Sports Injury*, 27, 1999, 381–389.
18. Levy ML, Ozgur BM, Berry C, et al, Analysis and evolution of head injury in football, *Neurosurgery*, 34, 2004, 649–655.
19. McCrory PR, Berkovic SF and Cordner SM, Deaths due to brain injury among footballers in Victoria, 1968-1999, *MJA*, 172, 2000, 217–219.
20. McIntosh AS, Rugby injuries, in Epidemiology of Pediatric Sports Injuries: Team Sports, Maffulli N, Caine DJ (eds) *Med. Sport Sc*, Karger, Basel, 49, 2005, pp. 120–139.
21. Spinecare Foundation and the Australian Spinal Cord Injury Units, Spinal cord injuries in Australian footballers, ANZ *J. Surg*, 73, 2003, 493–499.
22. Scher AT, Rugby injuries to the cervical spine and spinal cord: a 10-year review, *Clin Sports Med,* 17, 1998 195–206.
23. Wetzler MJ, Akpata T, Laughlin W and Levy AS, Occurrence of cervical spine injuries during the rugby scrum, *Am J Sports Med,* 26, 1998, 177–180.
24. Wetzler MJ, Akpata T, Albert T, Foster TE and Levy AS, A retrospective study of cervical spine injuries in American rugby, 1970 to 1994, *Am J Sports Med,* 24, 1996, 454–458.
25. Quarrie KL, Cantu RC and Chalmers DJ, Rugby union injuries to the cervical spine and spinal cord, *Sports Med,* 32, 2002, 633–653.
26. McElhaney JH and Myers BS, Biomechanical aspects of cervical trauma, in *Accidental Injury biomechanics and prevention*, Nahum AM and Melvin JW, Springer Verlag, New York, 1993. pp. 311–361.
27. McCrory P, Commotio Cordis, *Br J Sports Med,* 36, 2002, 236–237.
28. American Academy of Pediatrics, Committee on Sports Medicine and Fitness, Risk of Injury From Baseball and Softball in Children, *Pediatrics*, 107, 2000, 782–784.
29. Turner M, McCrory P, Halley W, Injuries in professional horse racing in Great Britain and the Republic of Ireland during 1992–2000, *Br J Sports Med,* 36, 2002, 403–409.
30. Williams KR, The dynamics of running. In *Biomechanics in sport: performance enhancement and injury prevention.* V Zatisorsky (ed). Blackwell Science, London, 2000, pp 161–183.
31. Panzer VP, Dynamic assessment of lower extremity load characteristics during landing. College of Human Development and Performance, University of Oregon, Eugene, 1989.
32. Beatty and McIntosh, *Proceedings of the Fifth Australasian Biomechanics Conference*, Sydney, NSW, Australia, December 9-10, 2004, pp 72–73.
33. Young, W, McLean B, et al, Relationship between strength qualities and sprinting performance, *J Sports Med. Phys Fit*, 35, 1995, 13–19.
34. Decker MJ, Torry MR, et al, Landing adaptations after ACL reconstruction, *Med Sci Sp Ex*, 18, 2002, 1408–1413.
35. Decker MJ, Torry MR, et al, Gender differences in lower extremity kinematics, kinetics and energy absorption during landing, *Clin Biomech*, 18, 2003, 662–667.

36. McNitt-Gray JL, Musculoskeletal loading during landing. In *Biomechanics in sport: performance enhancement and injury prevention.* V Zatisorsky (ed) Blackwell Science, London, 2000, pp 523–549.
37. Salci, Y, Kentel, BB, et al, Comparison of landing manoeuvers between male and female college volleyball players, *Clin Biomech*, 19, 2004, 622–628.
38. Ozgüven N and Berme N, An experimental and analytical study of impact forces during human jumping, *J Biomech*, 12, 1988, 1061–1066.
39. Hopper DM, McNair P, et al, Landing in netball: effects of taping and bracing the ankle, *Br J Sports Med*, 33, 1999, 409–413.
40. Cowling EJ, Steele JR, et al, Effect of verbal instructions on muscle activity and risk of injury to the anterior cruciate ligament during landing * Commentary, *Br J Sports Med*, 37, 2003, 126–130.
41. Nyland JA, Shapiro R, et al, Relationship of fatigued run and rapid stop to ground reaction force, lower extremity kinematics, and muscle activation, *J Orthop Sp Phys Ther,* 20, 1994, 132–137.
42. McIntosh AS and Janda D, Cricket helmet performance evaluation and comparison with baseball and ice hockey helmets, *Br J Sports Med*, 37, 2003, 325–330.
43. Naunheim RS, Staneven J, Richter C and Lawrence ML, Comparison of impact data in hockey, football and soccer, *J Trauma*, 48, 2000, 938–941.
44. Milburn PD, The biomechanics of rugby scrummaging, University of Wollongong, Wollongong, 1990.
45. Naunheim RS, Bayly PV, Standeven J, et al, Linear and angular head accelerations during heading of a soccer ball, *Med Sci Sports Ex*. 35, 2003, 1406–1412.
46. Smith MS, Dyson RJ, Hale T and Janaway L, Development of a boxing dynamometer and its punch force discrimination efficacy, *J Sp Sc.* 18, 2000, 445–450.
47. Joch W, Fritche P, Krause I, Biomechanical analysis of boxing, In *Biomechanics VII-A*, Morecki K, Fidelius K, Kdzior K, Wit A (eds), University Press, Baltimore MD 1981, pp 343–349.
48. Janda D, The Prevention of Baseball and Softball Injuries, *Clin Orthop Rel Res*, 409, 2003 20–28.
49. Marshall SW, Mueller FO, Kirby DP and Yang J, Evaluation of safety balls and faceguards for prevention of injuries in baseball, *JAMA*, 289, 2003, 568–574.
50. Janda DH, Bir CA, Viano DC and Cassatta SJ, Blunt chest impacts: assessing the relative risk of fatal cardiac injury from various baseballs, *J Trauma*, 44, 1998, 298–303.
51. Janda DH, Wojtys EM, Hankin FM and Benedict ME, Softball sliding injuries – a prospective study comparing standard and modified bases, *JAMA*, 259, 1988, 1848–1850.
52. Janda DH, Wojtys EM, Hankin FM, Benedict ME and Hensinger RN, A three-phase analysis of the prevention of recreational softball injuries, *Am J Sports Med,* 18, 1990 632–635.
53. Janda DH, Maguire R, Mackesy D, et al, Sliding injuries in college and professional baseball – a professional study comparing standard and break-away bases, *Clin J Sports Med,* 3, 1993, 78–81.
54. Norton K, Schwerdt S and Lange K, Evidence for the aetiology of injuries in Australian football *Br J Sports Med,* 35, 2001, 418–423.
55. Griffin LYE, Neuromuscular training and injury prevention in sports, *Clin Orthop Rel Res*, 409, 2003, 53–60.
56. Myklebust G, Engebretsen L, Brækken IH, et al, *Clin J Sports Med,* 13, 2003, 71–78.

57. Noakes TD, Jakoet I and Baalbergen E, An apparent reduction in the incidence and severity of spinal cord injuries in schoolboy rugby players in the Western Cape since 1990, *SAMJ*, 89, 1999, 540–545.
58. Garraway WM, Lee AJ, Macleod DAD, et al, Factors influencing tackle injuries in rugby union football, *Br J Sports Med,* 33, 1999 37–41.
59. Eliminate Head-Checking in Ice Hockey, *Clin J Sports Med,* 11, 2001, 211–213.
60. IRB Law 4M-98g - Standard Performance Specification for Specific Items of Players' Clothing, Dublin, 1998.
61. Canadian Standards Association, Ice hockey helmets, CAN/CSA, Z262.1-M90, Toronto, 2002.
62. American Society for Testing and Materials, Standard Test Method for Shock-Attenuation Characteristics of Protective Headgear for Football (F429-01) and Standard Specification for Football Helmets (F717-89), West Conshohocken.
63. American Society for Testing and Materials, Standard Specification for Helmets Used for Recreational Snow Sports (F2040-02), West Conshohocken.
64. Standard performance specification for newly manufactured baseball/softball batter's helmets, NOCSAE DOC (ND) 022-03m04, 2004.
65. Australian/New Zealand Standard, Protective headgear for cricket, Part 1: Helmets, AS/NZS 4499.1:1997, Standards Australia, Sydney, 1997.
66. Pellman EJ, Viano DC, Tucker AM, et al, Concussion in professional football: reconstruction of game impacts and injuries, *Neurosurgery*, 53, 2003, 799–814.
67. McIntosh AS, McCrory P and Comerford J, The dynamics of concussive head impacts in rugby and Australian rules football, *Med Sci Sp Ex*, 32, 2000, 1980–1984.
68. Frechede B and McIntosh AS, *Proceedings of the Fifth Australasian Biomechanics Conference*, Sydney, NSW, Australia, December 9-10, 2004, pp 76–77.
69. Hagel BE, Pless IB, Goulet C, et al, Effectiveness of helmets in skiers and snowboarders: case-control and case crossover study, *BMJ*, Jan 2005; 10.1136/bmj.38314.480035.7C.
70. Macnab AJ, Smith T, Gagnon FA and Macnab M, Effect of helmet wear on the incidence of head/face and cervical spine injuries in young skiers and snowboarders, *Injury Prevention,* 8, 2002, 324–327.
71. McIntosh AS and McCrory P, Effectiveness of headgear in under 15 rugby union football, *Br J Sports Med,* 35, 2001, 167–169.
72. Finch C, McIntosh AS and McCrory P, What do under 15 year old schoolboy rugby union players thick about protective headgear ? *Br J Sports Med.* 35, 2001, 89–94.
73. McIntosh AS and McCrory P, Impact energy attenuation performance of football headgear, *Br J Sports Med,* 34, 2000, 337–341.
74. McIntosh AS, McCrory P and Finch C, Performance enhanced headgear – a scientific approach to the development of protective headgear, *Br J Sports Med,* 38, 2004, 46–49.

CLINICAL AND BIOMECHANICAL RESEARCH FOR BICYCLE HELMET OPTIMISATION

C. Van Lierde[1], B. Depreitere[2], P. Verschueren[1], H. Delye[2], D. Berckmans[3], I. Verpoest[4], J Goffin[2], J. Vander Sloten[1] and G. Van der Perre[1]

[1]*Division of Biomechanics and Engineering Design, K.U. Leuven, Belgium*
[2]*Division of Experimental Neurosurgery and Neuroanatomy, K.U. Leuven, Belgium*
[3]*Laboratory for Agricultural Buildings Research, K.U. Leuven, Belgium*
[4]*Department of Metallurgy and Materials Engineering, K.U. Leuven, Belgium*

Abstract. Epidemiological studies on bicycle accidents show that a substantial fraction of the cyclists that call for medical aid, are suffering from skull and brain damage. The aim of the research performed at the K.U.Leuven is to reduce the risk of serious head injuries by the creation of a new type of bicycle helmet. To achieve this goal, a clinical review of pedal cyclist head lesions has been performed, a 3D accident simulation program has been developed, skull behaviour during impact testing has been studied and the current protective effect of a typical series of bicycle helmet has been evaluated. The absence of a temple cover allows a potentially dangerous contact between the temple of the head and the impact surface **[1]**.

Key words: bicycle helmet, impact modelling, skull fracture.

1. INTRODUCTION

Pedal cyclists are a vulnerable group among road users. The most common region injured in bicycle accidents is the head. 21 to 61% of the victims of bicycle accidents who seek medical care have a head injury **[2]**. Moreover, cranio-cerebral trauma is a direct cause of death for the majority of fatal accidents **[3]**. To reduce the risk of serious head injuries in bicycle accidents, a clinical review of pedal cyclist head lesions has been performed and a 3D accident simulation program has been developed. Also, skull behaviour during impact testing has been studied and the current protective effect of a typical series of bicycle helmet has been evaluated.

Michael D. Gilchrist (ed.), IUTAM Proceedings on Impact Biomechanics: From Fundamental Insights to Applications, 247–254.

2. MATERIALS AND METHODS

A data bank containing records related to real bicycle accidents was constructed. Each record was associated with one of 86 pedal cyclists that had undergone a neurosurgical intervention at the University Hospital of Leuven in the period between January 1990 and June 2000 as a result of a head injury. The collected data pertained to the accident circumstances, the sustained injuries and the outcome.

In parallel with the statistical analysis of the pedal cycling accidents from a clinical point of view a mathematical accident simulation model was developed with the Dynamic Analysis and Design System™ from LMS. It consists of a 3D multi-body model representing the cyclist and two solitary 3D entities representing the bicycle and the vehicle respectively.

To study skull behaviour, a double pendulum set-up was used, which would offer one rotational degree of freedom to the skull while ensuring the feasibility of performing controlled measurements (Figure 1). The impacting

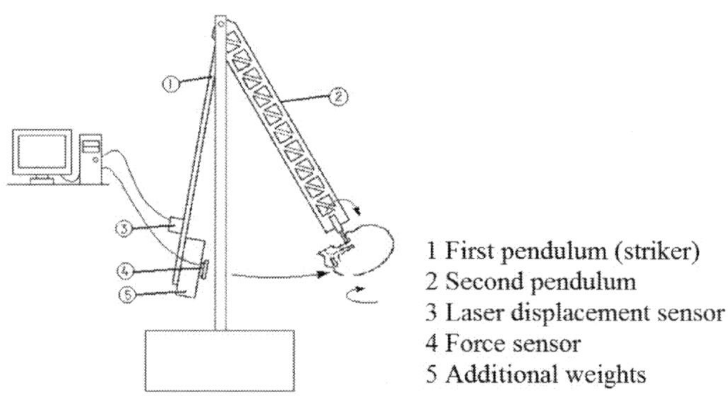

Figure 1. Experimental set-up.

pendulum had a length of 1.48 m, was made out of steel and had a mass of 14.3 kg, which could be increased up to either 26.4 or 34.7 kg by adding additional weights. The second pendulum was made of aluminum, having a mass of 5.6 kg and a length of 1.28 m. 112 skulls were collected from the department of anatomy of the K.U.Leuven. After decease the bodies of the donors all had been embalmed with a solution of formaldehyde. Upon receipt of the specimen, the scalp and intracranial contents were removed by immersing the heads in water at 80° C during approximately 6 hours. To prevent decay, they were stored in a solution of 1% formaldehyde in water. To enable characterization of certain geometrical features and material properties, computed tomography scans were taken of all 112 skulls. Impacts were given with a low (1.6 to 3.0 m/s) or high (3.9 to 4.6 m/s) velocity.

The test setup to study bicycle helmet effectiveness consisted of the impactor as described above, which was used to deliver lateral impacts onto the head of a helmeted cadaver (Figure 2). Its mass was increased to 36.5 kg.

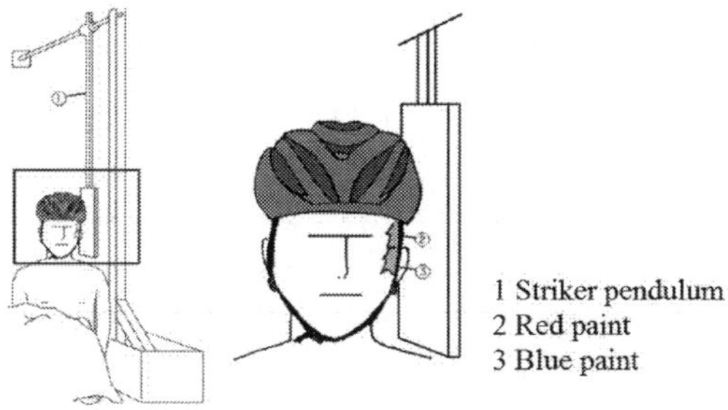

Figure 2. Schematic representation of the experimental set-up.

The cadavers were positioned in front of the pendulum in a sitting position with the midsaggital plane of the head parallel to the plane of the impactor. To determine whether direct contact was possible between the impactor and the temporal or zygomatic area of the head, 6 impact tests were performed at an impact velocity of 6 m/s. The skin overlying the zygomatic arch was painted blue and the skin overlying the temporal squama and ala major red. After the impact the presence of paint on the impactor was used to assess if indeed direct contact had occurred and if so, with which part of the head. Six bicycle helmet types were evaluated (Figure 3). Five had a more or less similar, but not identical design (model A), being representative of the common bicycle helmets intended for adults. One helmet had a different design and provided a larger coverage of the temporal area (model B). This latter helmet type was intended to be worn by children. All helmets used in the study were conform to the European EN 1078 standard.

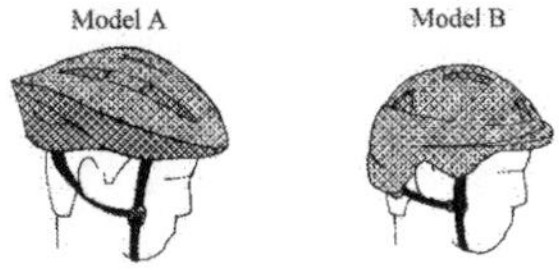

Figure 3. Drawing of the two types of bicycle helmet used in the study.

3. RESULTS

The frequency of the different types of injuries is summarised in Table 1. Fisher Exact-tests were performed to evaluate the coexistence of injuries. Cerebral contusions, acute subdural haematomas, subarachnoid haemorrhages and brain swelling proved to be associated with each other in a statistically significant manner.

Table 1. Overview of head injury types in accidents with and without motor vehicle involved.

Injury	Total group (N=86)	No motor vehicle (N=42)	Motor vehicle involved (N=44)
Skull fracture	74 (86.0%)	39(92.9%)	35(79.6%)
Extradural haematoma	34(39.5%)	17(40.5%)	17(38.6%)
Acute subdural haematoma	29(33.7%)	15(35.7%)	14(31.8%)
Cerebral contusion	63(73.2%)	32(76.2%)	31(70.5%)
Intracerebral haematoma	15 (17.4%)	11(26.2%)	4(9.1%)
Diffuse axonal injury	11(12.8%)	3(7.1%)	8(18.2%)
Subarachnoid haemorrhage	45(52.3%)	21(50.0%)	24(54.5%)
Intraventricular haemorrhage	11(12.8%)	3(7.1%)	8(18.2%)

The simulation model was used in the reconstruction of 12 real-life cycling accidents involving motorized vehicles, selected from the 44 motor vehicle involved cases in the databank mentioned above. The selection was based on the completeness of the available documentation and the level of detail with which the circumstances of the accident were described. The results of the simulations -described by the peak values for head acceleration and loading- and the corresponding injuries were juxtaposed in an attempt to correlate the head impact response in terms of its mechanical output to the occurrence of specific lesions. Irrespective the impact site and the impact pulse duration, a correlation was seen between the occurrence of fractures and the peak values for linear acceleration (Figure 4).

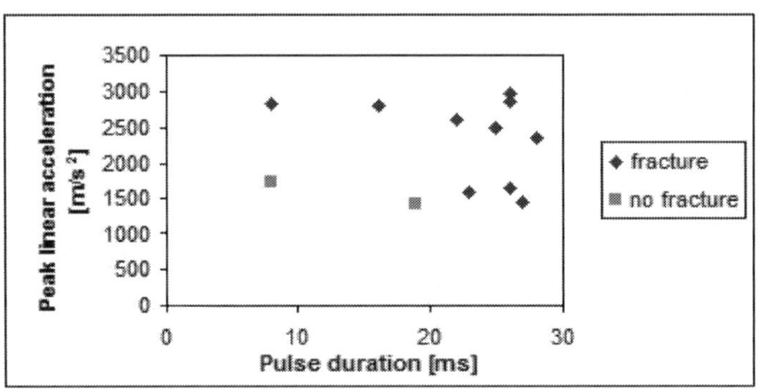

Figure 4. Relation between the peak linear acceleration and the occurrence of skull fracture.

Despite the limited number of data points a threshold value of approximately 1500 m/s^2 can be distinguished. With respect to occurrence of frontal or temporal cortical contusions, several observations can be made (Figure 5). Firstly, it appears that for parietal impacts both a high peak rotational acceleration (approx. 16000 rad/s^2) and a long pulse duration (>20 ms) are required to induce these types of lesions. Secondly, contusions are seen at lower levels of rotational acceleration in frontal and temporal impacts as compared to parietal impacts. For other lesions no obvious relations were found with any of the calculated mechanical parameters.

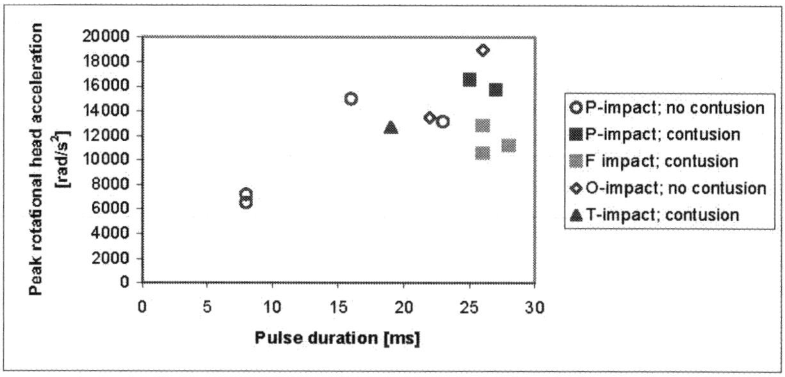

Figure 5. Relation between the peak rotational acceleration of the head on the occurrence of contusions. (F=frontal, O= occipital, P= parietal, T=temporal).

Sixty-one skull impact tests resulted in a clear fracture of the cranial vault and produced adequate force-displacement curves (12 temporal, 16 frontal, 23 parietal and 10 occipital impacts). A typical curve is shown in Figure 6.

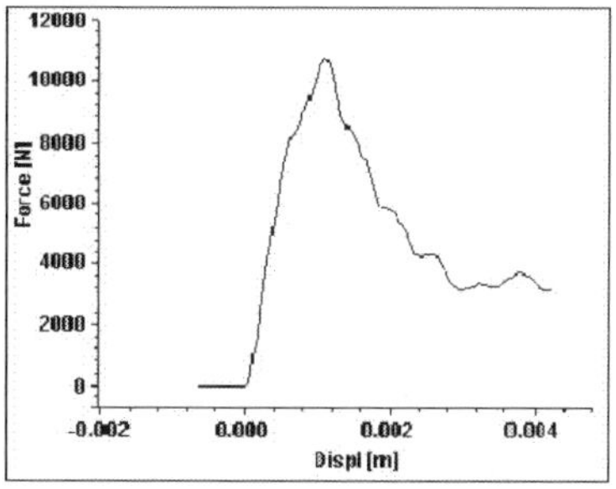

Figure 6. Typical force-displacement curve for cranial vault fracture test.

Fracture of the skull was believed to occur at the point of maximum force. For each impact location the mean force at fracture ($F_{\#}$), skull deformation at fracture ($d_{\#}$) and absorbed energy till fracture ($E_{abs\#}$) in the two impact velocity conditions were compared by means of Mann-Whitney U tests. Prior to this analysis six outliers were discerned. The results of the comparisons are summarised in Table 2. In general, $F_{\#}$ and $Eabs_{\#}$ were higher in the high impact velocity condition than in the low impact velocity, while $d_{\#}$ was slightly lower or remained stable. In the frontal impacts $F_{\#}$ was significantly higher in case of a high impact velocity, while the increase of $E_{abs\#}$ was much smaller and non-significant. In the parietal and occipital tests both $F_{\#}$ and $E_{abs\#}$ were significantly higher in the high velocity condition.

Table 2. $F_{\#}$, $d_{\#}$ and $E_{abs\#}$ summary for each impact location and impact velocity. M=Mean, SD=Standard Deviation, p-values from Mann-Whitney U test comparing velocity conditions

	$F_{\#}$ [N]		$d_{\#}$ [mm]		$E_{abs\#}$ [J]	
Frontal	Low v	High v	Low v	High v	Low v	High v
	N = 8	N = 6	N = 8	N = 6	N = 9	N = 6
	M 3242	M 6114	M 1.69	M 1.44	M 3.49	M 4.52
	SD 699	SD 2436	SD 0.39	SD 0.53	SD 0.96	SD 1.91
	p = 0.0014		*p = 0.37*		*p = 0.18*	
Parietal	Low v	High v	Low v	High v	Low v	High v
	N = 9	N = 11	N = 9	N = 11	N = 9	N = 11
	M 2841	M 4417	M 2.69	M 2.77	M 4.40	M 7.72
	SD 534	SD 1229	SD 0.79	SD 0.80	SD 1.33	SD 2.74
	p = 0.0008		*p = 0.88*		*p = 0.0043*	
Occipital	Low v	High v	Low v	High v	Low v	High v
	N = 4	N = 4	N = 4	N = 4	N = 4	N = 4
	M 4921	M 9027	M 2.03	M 2.10	M 6.32	M 10.60
	SD 833	SD 2032	SD 0.29	SD 0.62	S 0.94	SD 1.75
	p = 0.03		*p = 1.00*		*p = 0.03*	
Temporal	Low v	High v	Low v	High v	Low v	High v
	N = 6	N = 5	N = 6	N = 5	N = 6	N = 6
	M 1142	M 1433	M 3.82	M 3.46	M 2.56	M 3.08
	SD 531	SD 355	SD 1.50	SD 0.98	SD 1.20	SD 1.40
	p = 0.54		*p = 0.80*		*p = 0.69*	

Two additional occipital impact tests (mass 75.9 kg and impact velocity 3.00 m/s) were performed and resulted in a $F_{\#}$ of 10,961 N and 9,887 N and in an $E_{abs\#}$ of 11.38 J and 10.01 J respectively. When compared with the four occipital impact tests performed with a mass of 34.7 kg and a similar impact velocity, these values for $F_{\#}$ and $E_{abs\#}$ were higher. The results for the group of occipital impact tests (excluding the two outliers) are shown in Figure 7. An increase of the impact energy, either by increasing the impact velocity or the mass of the impacting pendulum, resulted in an increase of $E_{abs\#}$.

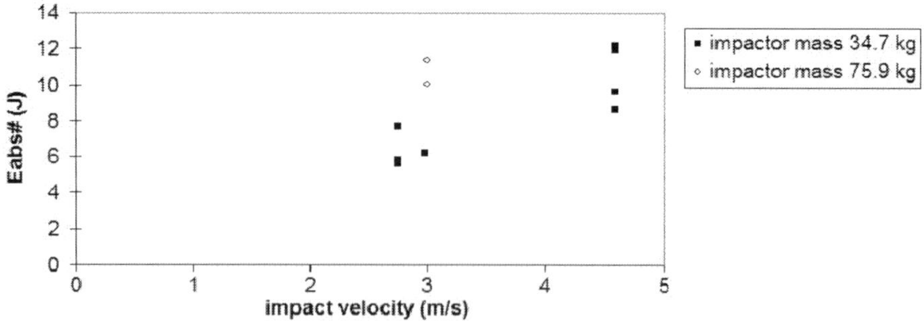

Figure 7. $E_{abs\#}$ vs. impact velocity for the occipital impact tests.

For the bicycle helmet effectiveness, only the model B helmet prevented contact between the impactor and the temporal squama and/or the zygomatic arch. In one case, with a model A helmet, the contact even led to a linear fracture, extending from the parietal bone cranially to the temporoparietal suture into the ala major.

4. DISCUSSION

The clinical survey showed that skull fractures and contusions are the most frequent lesions in bicycle accidents. Also sub- and extradural haematomas were frequently observed. Further research should therefore aim at reducing these types of lesions.

Concerning the performed accident reconstructions, some tentative trends were seen in terms of correlations between the mechanical output and the occurrence of specific lesions. The number of cases included in the study however was limited and the results should therefore be interpreted with some caution. Still, in particular with respect to skull fractures, a clear relation was found with the calculated linear head accelerations. This relation is known and has been described by the HIC **[4]**. It can however not be reliably applied to brain injuries caused by impacts generating contact forces above the limits for skull fracture **[5]**. Possibly, a criterion that also takes into account the rotational head acceleration would be more suited **[6, 7]**. Indeed, for frontal and temporal contusions, results of the accident reconstruction seem to point out that the occurrence of this type of lesions is related to the peak value of rotational acceleration and the duration of the acceleration pulse.

In the dynamic skull impact experiments, the effect of the impact velocity and impact location on $F_{\#}$, $d_{\#}$ and $E_{abs\#}$ was investigated. $F_{\#}$ increased with increasing impact velocity, which can be attributed to both the visco-elastic properties of the skull bone and to the influence of inertial resistance of the skull. $E_{abs\#}$ also increased with increasing impact velocity. This means that no fixed energy threshold can be defined, predicting the occurrence of fractures. The tolerance depends on the impact velocity. Although only two tests were performed to investigate the effect of the mass of the impacting object, the results suggest that it has an influence on the energy to failure. Thus it is more correct to state that the skull tolerance in terms of maximum energy absorbable to failure depends on the impact energy. It was also observed that the energy failure level depends on the impact location: it was higher for occipital impacts, followed by frontal and parietal. It is strongly suggested that for temporal impacts the failure level is the lowest. A weak relation between the impact site thickness and the energy to fracture was found, while the other skull properties were of no importance. Based on the results it could be hypothesised that a critical deformation level (dependent of the impact location) determines the initiation of a fracture. However, it can be seen that the critical deformation decreased with increasing impact velocity in the frontal (-15%) and temporal impacts (-10%), although this observation is not statistically significant. From impact tests, Yoganandan et al. **[8]** found that the energy to failure remained constant for impact velocities from 0.02 to 8 m/s. A similar conclusion could not be discerned from the present experiment. Conversely, the energy absorbed by the skull till fracture was showed to be correlated with the impact energy.

Finally, current bicycle helmet effectiveness was found inadequate as all helmets intended for adults failed to prevent direct contact loading in the region anterior to the ear, while 4 of the 6 helmets met even the most demanding standard (Snell's B95 standard) with respect to head coverage. In a study by Williams it was found that 63% of a total 84 impacts had occurred below the border delineated in the standard **[9]**. The potential hazard of such contact loadings was established by the fact that a linear skull fracture was induced in one of the tests. The only helmet type to prevent direct contact loading (e.g. model B intended for children), was also the only helmet that provided a better coverage of the temporal area. Lowering the temporal rims of adult helmets with as little as 20 or 30 mm may be enough to provide substantially better temporal protection and thereby improve the helmet's protective effectiveness.

REFERENCES

1. Gilchrist A. and Mills N.J., 1996, Protection of the side of the head, *Accid. Anal. Prev.*, 28: 525-535.
2. Collins B.A., Langley J.D. and Marshall S.W., 1993, Injuries to pedal cyclists resulting in death and hospitalisation, *N Z Med. J.*, 106: 514-517.
3. Oström, M., Bjornstig, U., Naslund, K. and Eriksson, A., 1993, Pedal cycling fatalities in Northern Sweden, *Int. J. Epidemiol.*, 22: 483-488.
4. Lissner H.R., Lebow M. and Evans F.G., 1960, Experimental studies on the relation between acceleration and intracranial pressure changes in man, *Surg. Gynaecol. Obstet.*, 111: 329-338.
5. Stålhammar D.A., 1990, The mechanism of brain injuries, in *Handbook of Clinical Neurology*, Vol. 13 (57): Head Injury, R. Braakman (ed.), Elsevier Science Publishers, Chapter 2, 17-41.
6. Hirsch A.E. and Ommaya A.K., 1970, Protection from brain injury: the relative significance of translational and rotational motions of the head after impact, 14th STAPP Car Crash Conference Proceedings, SAE Paper 700899, Ann Arbor, 144-151.
7. Curnow W.J., 2003, The efficacy of bicycle helmets against brain injury, *Accid. Anal. Prev.*, 35: 287-292.
8. Yoganandan N., Pintar F.A, Sinces A., et al., 1995, Biomechanics of skull fracture, *J. Neurotrauma*, 12(4): 659-668.
9. Williams M., 1991, The protective performance of bicyclists' helmets in accidents, *Accid. Anal. Prev*, 32: 119-131.

APPLICATION OF FINITE ELEMENT ANALYSIS TO HELMET DESIGN

Terry Smith, John Lenkeit and Jim Boughton
Dynamic Research Inc., Torrance, California, USA

Abstract. This paper describes an alternative method for helmet design using finite element analysis (FEA). While previous research has illustrated the use of FEA as a tool to simulate standard helmet tests, the objective of this research was to use FEA as a tool to optimize the performance characteristics of the energy absorbing (EA) liner of an existing Navy helicopter pilot helmet (i.e., the HGU-84/P). Three dimensional finite element models (FEM) of the helmet components and the test headform were developed using MSC software and material properties were estimated from published data and physical tests. A 6.0 m/s flat anvil impact was then simulated at three different sites on the helmet using MSC.Dytran software and these simulations served as the baseline. Full scale impact tests were performed with an examplar helmet in order to confirm the validity of the simulations as well as the helmet shell and liner material properties.

A group of candidate EA materials were obtained and physical impact tests were conducted in order to quantify the stress-strain characteristics of each candidate material. These stress-strain characteristics were then used to define the material properties of the existing HGU-84/P helmet liner FEM, effectively replacing the existing helmet EA liner with one fabricated from the candidate material. Whenever possible, a given EA liner material was evaluated over a range of densities. This allowed the researchers to study the effect of manipulating the EA liner density properties of the helmet without the necessity of conducting physical tests.

A total of 13 EA materials in different densities were evaluated by conducting over 120 FEM impact simulations at each of the three impact sites. The performance of a given EA liner material was evaluated by comparing simulation results with the peak headform acceleration values obtained from physical tests of the actual HGU-84/P helmet. The simulation results effectively predicted trends in liner material performance and making it

Michael D. Gilchrist (ed.), IUTAM Proceedings on Impact Biomechanics: From Fundamental Insights to Applications, 255–262.

possible to identify a small group of EA liner materials with a good likelihood of providing improved impact protection.

The results of this research clearly illustrate that FEA can be an effective tool for the analysis and design of both new and existing helmet designs. Simulation of various helmet liner materials allows for the evaluation of multiple material configurations, without the expense associated with production and testing of physical prototypes. This research and development tool can be particularly useful in the evaluation of new energy absorbing materials in applications where production and testing of individual helmet samples is quite costly (e.g., advanced composite helmet designs) and for applications where the geometric properties of the helmet are fixed (e.g., existing helmet designs).

Key words: helmet, Finite Element Method (FEM).

1. INTRODUCTION

Helmet design has traditionally taken an applied experimental approach. Overall performance of the final product cannot be known at the initial design stage; therefore, in order to develop a new or improved helmet several prototypes are produced, incorporating a wide range of material properties. These prototypes are then tested by means of standardized drop tests to determine the resulting head accelerations and thus the relative protective capability of the design with respect to applicable standards. If the desired results are not obtained (e.g., the peak headform acceleration is above the performance criteria), a change of design may be called for. This would typically involve producing additional prototype helmets in alternate material densities or shell constructions for further testing. Typically, the weight, cost, comfort and health performance are addressed after suitable impact performance is obtained; and this may require further variation of materials. While this approach has proven to be historically successful, it involves relatively high costs, long development times, and a great deal of trial and error that may or may not result in an optimized helmet design.

The application of finite element models to the simulation of helmet impacts has been reported previously [1-4]. Most of this research has been focused upon the simulation of the response of existing helmet designs and analysis of the effect of a significant design change to the helmet (e.g. ventilation holes). This research follows the same strategy in that the objective of the project was to identify replacement energy absorbing (EA) liners that would increase the level of impact protection without changing the existing geometry of the helmet and liner. A further constraint was that the helmet liner material had to be commercially available and had to have a cost similar to the existing EA liner.

2. METHODOLOGY

An exemplar large size HGU-84/P helmet was obtained and all helmet components, including the shell, liner, and ear cups, having a possible effect on impact characteristics, were digitized using a coordinate measuring machine,. From these digital point clouds, surface models were developed and then meshed with finite elements using MSC.Patran software. The helmet shell was modelled with 1018 shell elements while the liner was modelled with 1368 hex elements and the thermoplastic liner was modelled with 684 hex elements. In addition to the helmet components, a model of an ANSI Z90.1 test headform and the ANSI Z90.1 flat anvil were also developed for use within the MSC.Dytran simulation software.

The FE models of the individual helmet components were integrated into a complete finite element model of the HGU-84/P, and boundary conditions between component interfaces were identified (e.g., contact between the helmet shell and EA liner). The ANSI Z90.1 headform was modeled as a rigid, infinitely stiff material and was added to the helmet and rigid anvil components in order to complete the model used for finite element simulation of the ANSI Z90.1 impact tests [5].

Upon completion of the model development phase, ANSI Z90.1 standard helmet testing was simulated using the MSC.Dytran software package on a PC based computer system. This helmet impact simulation was initiated by imparting an initial velocity of 6.0 m/s to the helmeted test headform prior to striking the flat rigid impact anvil. The output of the model was time-based acceleration data measured at the center of mass of the impact test headform, in a manner analagous to the physical test. This acceleration data were then compared to data obtained from equivalent physical tests conducted with an exemplar helmet in the DRI impact test laboratory and with other published test data [6]. Peak headform acceleration values obtained from the center of mass of the headform and acceleration-time characteristics from the two sources were compared.

The shell of the subject helmet is constructed of proprietary materials, hence the material properties were not readily available. To enable model development, isotropic characteristics were assumed. A range of initial shell stress-strain properties were selected from typical helmet shell materials (e.g., polycarbonate and fiberglass), and by systematically varying the shell properties, good correlation with the baseline drop test results was achieved. When agreement was obtained between the baseline simulation and full scale tests at the rear and left side locations, no further manipulation of the shell properties occurred. Figure 1 shows a comparison of the results of the dynamic simulation and the full scale HGU-84/P impact tests for the rear impact locations.

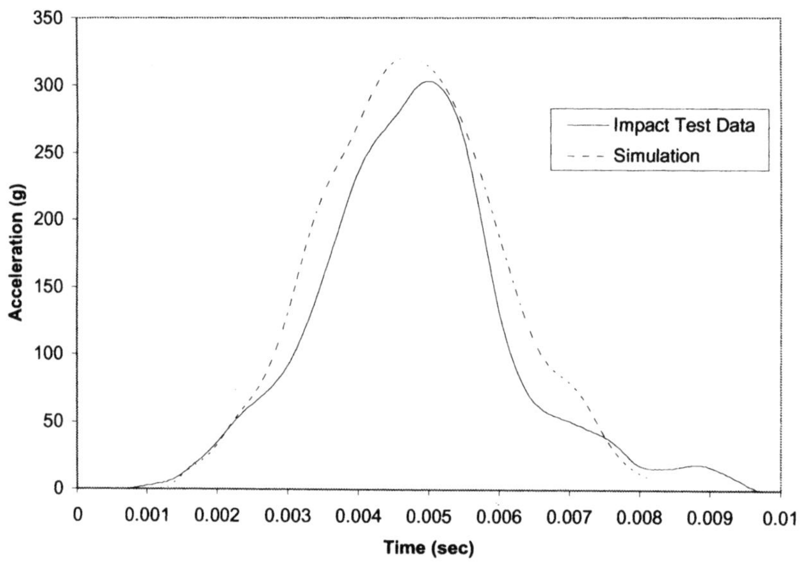

Figure 1. Comparison of HGU-84/P impact test data and HGU-84/P simulation data.

A review of the properties of existing and candidate EA materials was conducted for the purpose of developing an up-to date compendium of these materials. A relative lack of published detailed specifications, specifically stress-strain properties, of available candidate materials led to a decision to conduct in-house testing of material samples to obtain these necessary parameters. A complete list of materials tested is presented in Table 1. Since there are many different products or tradenames available within a given EA product category (e.g. expanded polystyrene), each individual EA material was given a unique identification code (e.g. EPS1, EPS2, etc.) known only to the researchers. If possible, different densities of a particular product were obtained in order to evaluate the effect of EA material density upon impact performance.

The sample materials were tested dynamically by means of the monorail drop test rig. Each sample was attached to a 5.0 kg mass with a 7.6 cm X 7.6 cm (3" X 3") flat impact face which was mounted to the drop arm assembly, and dropped from various heights. The stress on the sample during a drop test was calculated from the known contact area, the drop mass, and the acceleration-time history. The material strain was calculated from the known initial sample thickness, the known impact velocity and double integration of the acceleration data. Drop heights were selected with the goal of obtaining the maximum strain of the material within the allowable acceleration for the test measurement accelerometer (i.e., 500 g). An advantage of this method for obtaining stress-strain properties is that the material is tested under dynamic conditions that are very similar to those of actual helmet tests. Structural dynamics and any rate dependencies are embedded in the data. The EA material was modelled with the FOAM2 constitutive model in MSC.Dytran. This model used a table to define the stress-strain curve of the material.

Table 1. List of Materials Tested.

Sample ID	General Product Category	Nominal Density Kg/m^3 (pcf)
EPS1	Expanded Polystyrene	64.1 (4)
EPS2	Expanded Polystyrene	32.0, 80.1 (2.0, 5.0)
EPS3	Expanded Polystyrene	56.1, 96.1 (3.5, 6.0)
EPS4	Expanded Polystyrene	72.1 (4.5)
EPS5	Expanded Bead Blend	32.0, 56.1 (2.0, 3.5)
EPU1	Poured Blend Foam	103.0 (6.4)
EPP1	Expanded Polypropylene	67.3 (4.2)
EPP2	Expanded Polypropylene	32.0, 67.3, 97.7, 128.0 (2.0, 4.2, 6.1, 8.0)
EPE1	Polyethylene Laminate	89.7 (5.6)*
EPS6	Rubberized Expanded Polystyrene	20.0, 24.0 (1.25, 1.5)
EPP3	Expanded Polypropylene	33.6, 48.1, 88.1 (2.1, 3.0, 5.5)
XPE1	Expanded Cross-Linked Polyethylene	33.6, 51.3 (2.1, 3.2)
EPE2	Expanded Polyethylene	43.2 (2.7)

* Estimated, based on sample weight of polyethylene laminate

The validated HGU-84/P helmet model was used for all simulations, with the stress-strain properties of the EA candidate material incorporated into the liner model. Five impact locations were studied, the crown, the left side, the front, the rear and the left ear cup. The simulations were typically run with a data output step size of 0.1 ms, and a run length of 20 ms. The data for each simulation run were post-processed to determine the maximum vertical acceleration of the test headform.

3. RESULTS

More than 100 simulations were performed for 13 different candidate material categories and 23 density configurations across three different impact locations. Maximum vertical (i.e., uniaxial) headform acceleration values obtained from the simulations are presented in Table 2 for each candidate material.

Table 2. Performance of Different Candidate Energy Absorbing Materials.

Sample ID	Density Kg/m^3 (pcf)	Peak Headform Acceleration (g)			Total	% Change
		Crown	Rear	Left		
EPP2	32.0 (2.0)	162	191	241	594	+ 30
EPS2	32.0 (2.0)	184	204	236	623	+27
EPS5	32.0 (2.0)	168	195	263	625	+27
EPS1	72.1 (4.5)	197	211	223	630	+26
EPS5	56.1 (3.5)	213	223	222	655	+23
EPS6	24.0 (1.5)	144	229	302	674	+21
EPP3	88.1 (5.5)	206	248	255	708	+17
EPU1	103.0 (6.5)	211	247	258	715	+16
EPP2	67.3(4.2)	219	254	245	716	+16
EPE1	89.7 (5.6)	217	263	253	731	+14
EPS3	56.1 (3.5)	217	264	252	732	+14
EPS2	80.1 (5.0)	217	262	255	732	+14
EPP1	67.3 (4.2)	209	254	271	734	+14
EPP3	48.1 (3.0)	138	253	352	742	+13
EPS6	20.0 (1.25)	132	269	365	765	+10
EPS4	72.1 (4.5)	223	293	286	804	+6
EPP2	97.7 (6.1)	220	306	307	832	+2
EPS3	96.1 (6.0)	230	313	307	848	+1
HGU-84/P	115.0 (7.2)	223	308	322	853	0
XPE1	51.3 (3.2)	140	348	433	919	-8
EPP2	128.0 (8.0)	223	358	356	937	-10
EPE2	43.2 (2.7)	143	360	440	942	-10
EPP3	33.6 (2.1)	132	390	470	992	-16
XPE1	33.6 (2.1)	142	400	499	1041	-22

All candidate materials were compared against the response of the HGU-84/P helmet. The data presented in Table 2 shows that of the 23 combinations of material and density considered, 9 produced a 10-12% improvement over the current HGU-84/P EA liner, and 6 produced improvements in the range of 20-30%. There were also five EA liner candidates that did not perform as well as the current HGU-84/P EA liner.

4. DISCUSSION

It is important to note that these simulations were all conducted based upon the results of the ambient temperature tests that were used to develop the stress-strain properties of each EA material. Test results may be different for the different environmental conditions under which helmets may be expected to perform (e.g., hot, cold, and wet conditions). Furthermore, the samples that were tested represent only those materials that were readily available and could be supplied by the foam manufacturers given the time available for completion of this project. Poor performance relative to the HGU-84/P helmet does not suggest that this material is not a suitable energy absorbing material. It merely suggests that in this particular density, this product would not be likely to perform as well as the current helmet liner. Further impact testing and simulation would be necessary to determine the performance of these same materials at other densities.

The results shown in Table 2 indicate that there are a number of currently available materials from which EA liners could be fabricated that could improve the impact performance of the existing HGU-84/P helmet in the impact tests simulated. The results also identify several materials which may further improve the impact protection of this helmet if the EA liner were to be fabricated from other densities (e.g., EPP, EPS). It would not be expected that the results of the simulations would be exactly duplicated in full scale impact testing, but the trends predicted provide useful guidance in the selection of alternate materials and sufficient evidence to identify potential candidates for further research.

The results also demonstrate the usefulness of using computer simulation to evaluate a large number of energy absorbing liner materials in a relatively short period of time. This particular approach can be quite cost effective in those cases where there is a short product development time and the cost of performing destructive testing on an individual helmet prohibits testing of multiple different material samples. This approach also suggests that optimum material density may vary by impact region, suggesting that it may be useful to investigate the incorporation of multiple density liners as a means of improving the protective capabilities of a helmet.

The methodology developed as a result of this research also has application in other areas of head protection, and could be particularly useful in the design stage of any contemporary form of head protection, including other military helmet designs. Current helmet designs are commonly developed using CAD systems for the purpose of developing molds produced by numerically controlled machines. These CAD designs can be converted into FEA models and appropriate material properties for the shell and liner can be studied to optimize impact performance. Application of this process during product development could allow greater contingency for mold modifications or material selection prior to prototype fabrication, potentially reducing costs and development time for the helmet manufacturer.

ACKNOWLEDGEMENTS

This research was supported by a grant from the United States Small Business Innovative Research Program (SBIR) Contract N000421003-P-0742. The authors would like to thank NAVAIR for their assistance as well as all the manufacturers that supplied sample products for testing and evaluation.

REFERENCES

1. Liu, D.S., Fan, C.M., Lee, M.C. and Yen, C.Y., Development and application of a 3D finite element simulation model of impact of motorcycle helmet, *International Journal of Crashworthiness*, vol. 3, no.5, 1998, 319–327.
2. Yettram, A.L., Godfrey, N.P.M. and Chinn, B.P., Materials for motorcycle crash helmets – A finite element parametric study, *Plastics, Rubber and Composites Processing and Applications*, vol. 222, no. 4, 1994, 215–221.
3. Kormi, K. and Etheridge, R.A., Finite element analysis – Application of the finite-element method to simulation of damage to the human skull as a consequence of missile impact on a multi-layered composite crash helmet, *Journal of Biomedical Engineering*, vol. 14, 1992, 203–208.
4. Brands, D., Thunnissen, J. and Wismans, J., Modelling head injury countermeasures: a 3D helmet model. In *Proceedings of the Aerospace Medical Panel Specialists' Meeting*, Mescalero, New Mexico, USA, November 7–9, 1996, pp. 26-1 – 26-8.
5. American National Standards Institute, ANSI Z90.1 – Protective Headgear for Vehicle Users. American National Standards Institute, New York, New York, 1973.
6. McEntire, B.J., Murphy, B. and Mozo, B.T., Performance Assessment of the HGU-84/P Navy Helicopter Pilot Helmet, USAARL Report No. 96-04, 1996.

FINITE ELEMENT MODELING FOR THE PREDICTION OF BLAST TRAUMA

A. Greer[1], D. Cronin[1], C. Salisbury[1] and K. Williams[2]
[1]*Department of Mechanical Engineering, University of Waterloo. 200 University Ave. W., Waterloo, ON, Canada, N2L 3G1. 519.888.4567 ext. 2346. adgreer@lagavulin.uwaterloo.ca;*
[2]*Weapons Effects Section, Defence Research and Development Canada – Valcartier. 2459 Boul. Pie XI Nord, Val-Belair, QC, Canada, G3J 1X5. Tel: 418.844.4000 ext. 4238*

Abstract. A simplified finite element model of a human torso has been developed to investigate and predict primary blast injury to the lung. The motivation for this approach was to understand the basic origins of blast trauma to the lungs and to create a predictive model for the evaluation of injury and future development of blast protection. The model consists of a two-dimensional slice of the torso at the mid-sternum level with blast loading applied via a coupled Arbitrary Lagrangian-Eulerian approach, allowing for a variety of loads to be considered. In parallel, a simplified model of a sheep torso has been developed for direct comparison to published experimental data on blast injury. Blast loads were applied to the models based on threshold lung damage, and various lethal dose quantities for comparison to expected injury levels based on the Bowen curves. The predicted injury levels based on relative lung pressure correlated well to existing experimental data. Further, the predicted peak chest wall velocities in the model compared well to an existing trauma model and with the expected severity of trauma. Future research will focus on the prediction of lung injury in a complex blast environment, and the development of blast protection.

Key words: Finite Element Method (FEM), primary blast injury, human thorax, sheep, shock wave, lung trauma, impact biomechanics.

1. INTRODUCTION

A fundamental understanding of the thresholds and mechanisms of trauma are required for the development of improved blast protection systems.

Michael D. Gilchrist (ed.), IUTAM Proceedings on Impact Biomechanics: From Fundamental Insights to Applications, 263–271.

Advanced finite element modeling and fluid-structure coupling can be used to complement experimental blast testing, and to investigate human body response and trauma in more complex environments.

The primary objective of this study was to develop and validate a simple numerical representation of a sheep torso and a human torso subjected to five representative blast loadings. The sheep model was considered for direct comparison to experimental data and as a link to the prediction of trauma in humans. Although the solid model is simplified, the significant aspects of the geometry have been captured while focusing on appropriate material models and an injury criterion based on local lung pressure.

2. INJURY CRITERIA AND EXPERIMENTAL DATA

Blast injuries are typically classified as primary, secondary or tertiary [1]. Primary blast injury (PBI) occurs when the incident shock wave impacts and transmits through the body, and generally results in trauma to the air-containing organs of the body. These injuries are related to the tearing or failure of very sensitive biological materials. Injuries that arise from the impact of debris and fragments are classified as secondary while tertiary injuries include those sustained due to whole body displacement. The current research focuses on the prediction and protection of PBI. A more detailed description of PBI, trauma mechanisms and the initial development of related numerical models are provided by Cronin et al. [2].

The experimental data and models outlined below have been used to evaluate the response of the finite element models for simple blast loading so that they may be extended to complex blast loading in the future.

2.1 Bowen Curves for the Prediction of Air Blast Injury

The Bowen curves were developed from experimental blast data for a range of animals, and can be used to predict the lethality of free-field blast waves [3]. Given the peak overpressure and positive phase duration of a Friedlander type curve, it is possible to estimate the level of injury ranging from threshold lung damage (TLD) through to 99 % lethality. Alternately, by following an iso-injury curve, it can be shown that different blast waveforms can be created that result in similar levels of trauma. Experimental blast data can be scaled to account for differences in mass and differences in the tolerance of individual mammalian species. The original curves based on animal injury were scaled to represent a 70 kg man in various positions in either the free-field or in front of a reflecting surface. The simplicity of the curves, as well as the wealth of data used in their development, makes them a valuable reference and validation tool.

2.2 Chest Wall Velocity as a Predictor of Blast Injury

In order to predict non-auditory blast injury from complex waves, Axelsson et al. [4] correlated results from a mathematical model to their own experimental data (for sheep) and to the Bowen iso-injury curves. Given an applied pressure history, the model is capable of predicting the human chest wall response (displacement, velocity and acceleration) and intrathoracic pressures for simple and complex blast loads.

Axelsson et al. found that the peak inward chest wall velocity predicted by the mathematical model was a good indicator of the level of trauma sustained. The results compared well with the Bowen curves for TLD, LD_1 and LD_{50}, where LD_1 and LD_{50} are defined as the loadings that would result in 1% and 50% mortality during the 24 hours following exposure.

3. FINITE ELEMENT MODEL DEVELOPMENT

3.1 Thoracic Geometry and Materials

In order to predict the wave transmission and coupling between the air blast and the torso, it was necessary to create a model using a fine mesh of solid elements representing the significant geometric features of the human and sheep torsos. A mid-sternum view from The "Visible Human Project" [5] provided the geometric data required to create a simplified model of the human torso, seen in Figure 1. Geometry for the sheep model was derived in a similar manner using CT scans from the body of a 56.5 kg ewe [6]; this compares well to the sheep used in Bowen et al's experiments which ranged in mass from 48.9-59.2 kg [3]. A mid-sternum image and the finite element model can be seen in Figure 2.

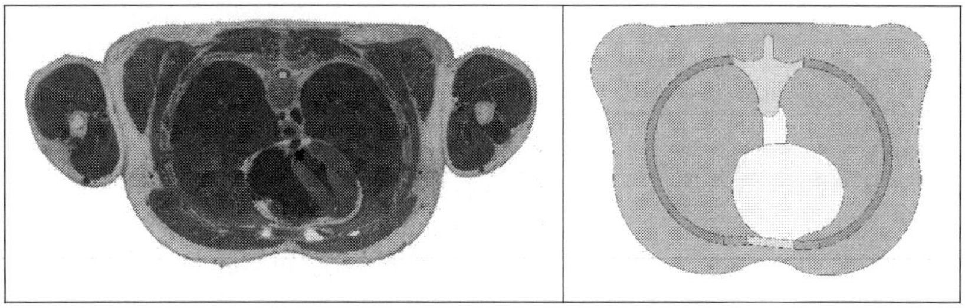

Figure 1. Axial view of the human torso at mid-sternum level [5] (left), and finite element model of the human torso (right).

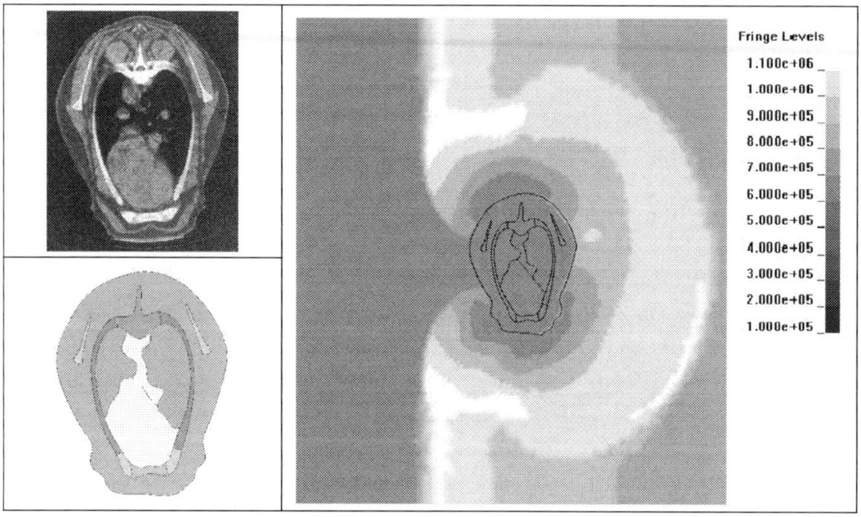

Figure 2. Axial view of a CT image of the sheep at mid-sternum level (top left) [6], and finite element model of the sheep torso (bottom left). Sheep model subject to blast loading showing contours of pressure in Pascals (right).

The sheep was chosen for this research because of the extensive experimental test data available and because it is one of the larger animals used by Bowen et al. Sheep belong to the 'high tolerance group' specified by Bowen, and man's tolerance is defined as the average of the high tolerance group. Since the blast tolerance for sheep of a given mass is stated as being higher than that for humans [3], but the 70 kg human is heavier than the sheep being studied, their relative tolerances should be comparable.

Based on a significant amount of experimental test data for individual soft and hard biological tissues, the sheep and human models were designed to include materials representing the ribs, sternum, vertebrae, scapulae (sheep), costal cartilage, muscle, fat and lungs [7,8]. In order to accurately model the wave propagation, the soft materials have been modeled as bulk materials using a Mie-Gruneisen equation of state [2].

3.2 Applied Blast Loading

A mesh of air elements was created, bounded on three sides by non-reflecting boundaries; these boundaries permit the flow of material out of the mesh, and are able to prevent reflections from traveling back into the mesh by using an impedance matching algorithm. The fourth side of the air mesh is bounded by explosive elements capable of producing a blast wave of a given peak pressure and positive phase duration.

A coupled Arbitrary Lagrangian-Eulerian (ALE) finite element technique in the LS-DYNA hydrocode was used to apply blast loading to the deformable body. The Lagrangian torso mesh is inserted directly into the air mesh of ALE elements. Coupling forces between the two substances are calculated using a penalty-based method, similar to the contact mechanism for two Lagrangian parts. Such an arrangement allows the torso deformation to influence and interact directly with the blast flow.

The peak pressure and duration of the five blast cases represent selected points from the modified Bowen TLD, LD_1 and LD_{50} curves (Table 1). The modified Bowen curves are a more recent interpretation of the original data studied by Bowen et al. [9]. The blast was directed toward the anterior surface of the human model while the right side of the sheep was oriented toward the blast, representing a typical experimental testing orientation.

Table 1. Characteristics of the modeled blast waves.

Blast Intensity	TLD	TLD	TLD	LD_1	LD_{50}
Peak Overpressure (kPa)	125	200	250	500	700
Positive Phase Duration (ms)	10.0	2.0	0.4	2.0	2.0

4. NUMERICAL MODEL RESPONSE AND PREDICTION OF INJURY

The sheep and human numerical models were subjected to five simple, free-field blast loadings of varying severity in order to compare the response to accepted experimental data and validated mathematical models. Comparisons were made using pressure data from representative elements in the lungs, and velocity data from the anterior surface of the human torso model and the right side surface of the sheep model.

4.1 Lung Pressure for Injury Prediction

Although the actual mechanism of injury at the alveolar level is not fully understood, the transmission of stress waves and the pressure history within the lungs is of great interest. Some researchers have suggested the existence of a threshold lung pressure level above which trauma will occur [10, 11].

When subjected to threshold injury blasts of varying overpressures and positive phase durations, the lung overpressures remained reasonably constant, as seen in Figure 3. These results indicate that, while the blast loadings represent durations of three different orders of magnitude, the model correctly predicts a similar level of trauma in agreement with the modified Bowen curves. When subjected to lethal dose blasts, the lung overpressures were approximately four to eight times higher than the threshold pressures, indicating a severe increase in trauma.

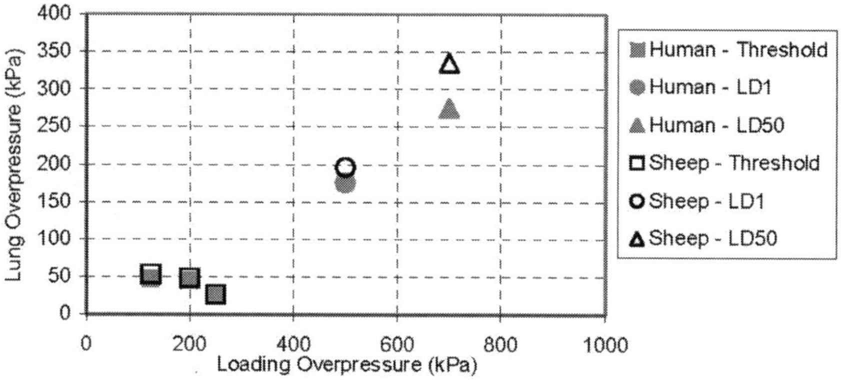

Figure 3. Predicted peak lung overpressure for the human and sheep finite element models subject to blast loading of various intensities. Values were averaged from the same three representative elements for each model.

Also of note is that the regions of highest predicted pressure coincide with locations where blast injury has been observed, such as along the ribs and around the heart [1]. In the human model, a significant region of high pressure is created around the vertebrae at the rear of the lungs and is due to the geometric focusing of pressure waves.

Although the sheep and human models were exposed to the same blast loading, the pressure in the lungs of the sheep was somewhat higher compared to the human when subjected to blasts of one and fifty percent mortality. After studying the pressure wave propagation it becomes evident that the high pressures found in the right lung of the sheep are due primarily to its orientation normal to the blast and its unprotected location. The ventral section of the left lung is protected by the heart, whereas pressures in the left-dorsal section are higher, a result of the pressure wave flowing easily from the right to the left lung.

Damage to anterior portions of the human lungs would be severe as the lungs are compressed between the dense tissues of the heart and thoracic cage. The lower pressures in the human lungs as compared to the sheep are likely due to the anterior-posterior orientation of the lungs presenting a lower cross-sectional area for the blast wave, as well as the shielding effect of the heart. Overall injury levels to the lungs, however, appear to be quite comparable between the two models.

Figure 4 shows the pressure history of three elements in the lungs of the sheep model subject to a threshold level blast. Due to the inherent difficulty of measuring internal lung pressure experimentally, esophageal pressure has typically been recorded during blast testing [10, 11]. Although the pressure magnitudes cannot be directly compared because of differences in blast environment, location within the body and vast differences in the properties of the surrounding materials, the general wave shapes and frequencies compare reasonably well.

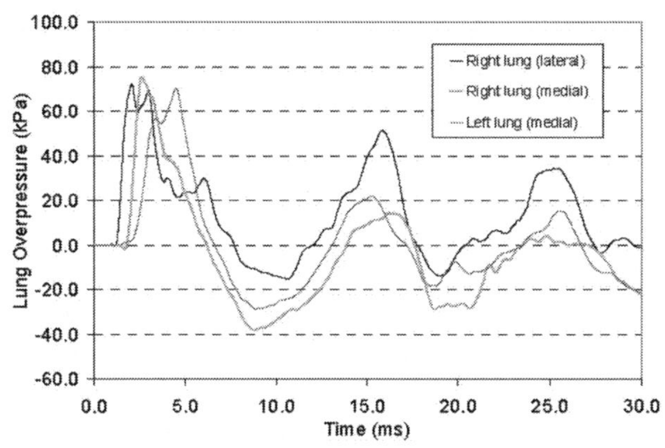

Figure 4. Predicted lung overpressure from three elements in the lungs of the sheep, subjected to a threshold level blast of 2 ms positive phase duration and 200 kPa overpressure.

4.2 Chest Wall Velocity

The chest wall velocities predicted by the sheep and human torso models subject to blast loading correlate well to the injury index scale proposed by Axelsson et al. [4] and the corresponding calculated velocities of the mathematical model.

Axelsson correlated a peak inward chest wall velocity range of 3 – 4.5 m/s for TLD, 8 – 12 m/s for LD_1 and 12 – 17 m/s for LD_{50}. As can be seen in Figure 5, these velocity ranges correspond well to the sheep and human chest wall velocity ranges predicted in the finite element models.

Once again, TLD blast loading of varying duration resulted in a nearly constant peak inward chest wall velocity, and hence similar resultant trauma. The LD_1 and LD_{50} blast loading shows a significant increase in chest wall velocity in relation to the threshold curves. In a general sense, this would indicate that peak inward chest wall velocity predicted by the finite element model can be used to predict the sustained level of trauma.

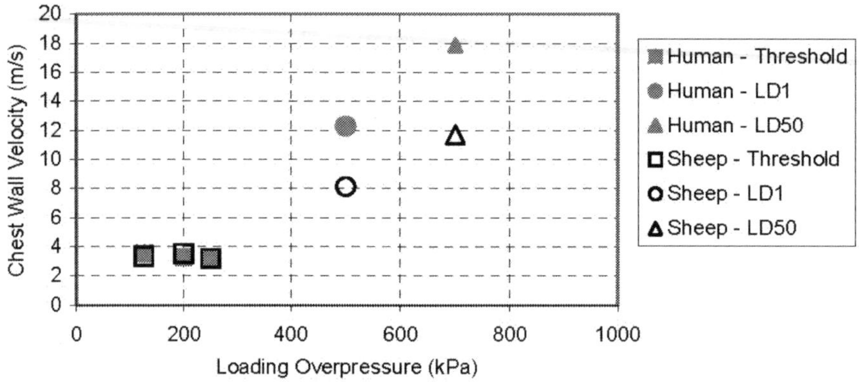

Figure 5. Predicted peak chest wall velocities for the human and sheep finite element models subject to various blast intensities. Values were averaged from the same two representative nodes for each model.

Although the sheep model predicted higher lung pressures for a given loading, the human model shows a tendency towards higher inward chest wall velocities. The difference between the two torsos, however, appears to be overstated, and is partly a result of the chest wall velocity being reported only as an average of two or three nodal velocities. A measurement technique will be implemented that more accurately reflects the velocity of the chest wall as a whole. It is also suspected that the significant mass of soft tissue toward the dorsal region of the side of the sheep resists acceleration and hence does not permit as high a velocity in that area.

5. CONCLUSIONS

Simplified finite element models of a sheep and human thorax have been developed and subjected to blast loading of varying severity. Peak local lung pressures were found to remain relatively constant when the thoraces were loaded with threshold level blasts, and increased markedly subject to lethal blast loading, in agreement with the modified Bowen injury curves. The pressure-histories of elements within the lung resemble available experimental data of esophageal pressure in terms of shape and frequency.

The peak inward chest wall velocity was found to correlate well with the mathematical model by Axelsson et al. and accurately predicted the severity of injury when related to that author's experimental testing.

The sheep and human torso models reacted similarly under varying loading conditions indicating that appropriate scaling of blast data from sheep to humans is reasonable. Future research will focus on prediction of trauma in complex blast situations, and the development of protective equipment to mitigate injury.

REFERENCES

1. Stuhmiller, J., Phillips, Y. and Richmond, D., Conventional warfare: ballistic, blast and burn injuries, In: *Textbook of Military Medicine,* Part I, Vol. 5, Zajtchuk, R. (eds), Office of the Surgeon General, Department of the Army, Washington DC, 1998.
2. Cronin, D., Greer, A., Williams, K. and Salisbury, C., Numerical modeling of blast trauma to the human torso, *Proceedings of the Personal Armour Systems Symposium (PASS) 2004,* The Hague, The Netherlands, Sept. 2004, pp. 323-332.
3. Bowen, I.G., Fletcher, E.R. and Richmond, D.R., Estimate of man's tolerance to the direct effects of air blast, Technical progress report no. DASA-2113, Department of Defense, Defense Atomic Support Agency, Washington, D.C., 1968.
4. Axelsson, H. and Yelverton, J.T., Chest wall velocity as a predictor of non-auditory blast injury in a complex wave environment, *7th International symposium of weapons traumatology and wound ballistics*, St Petersburg, Russia, Sept. 1994.
5. National Library of Medicine (NLM), *The Visible Human Project*, National Institutes of Health, www.nlm.nih.gov/research/visible/visible_human.html.
6. Davies, A., Garden, K., Young, M. and Reid, C., *An Atlas of X-ray Tomographical Anatomy of the Sheep,* Science Information Publishing Centre, Wellington, New Zealand, 1987.
7. Yamada, H., *Strength of Biological Materials,* Evans, F.G., ed. Williams and Wilkins, Baltimore, 1970.
8. Goss, S., Johnston, R. and Dunn, F., Comprehensive compilation of empirical ultrasonic properties of mammalian tissues, *J. Acoustical Soc. Am.*, vol. 64, no. 2, 1978, 423-457.
9. Williams, K., personal comm., Weapons Effects Section, DRDC Valcartier, June 2004.
10. Stuhmiller, J., Chuong, C., Phillips, Y. and Dodd, K., Computer modeling of thoracic response to blast, *The Journal of Trauma,* vol. 28, no. 1 Supplement, 1988, S132-S139.
11. Josephson, L.H. and Tomlinson, P., Predicted thoraco-abdominal response to complex blast waves, *The Journal of Trauma,* vol. 28, no. 1 Supplement, 1988, S116-S124.

EVALUATION OF HUMAN HEAD INJURY IN TRACKED VEHICLE SUBJECTED TO MINE BLAST

F. Wang, H.P. Lee, C. Lu and Q.H. Cheng
Institute of High Performance Computing, Singapore Science Park II, Singapore 117528

Abstract. Head injury of crew members in tracked vehicles subjected to mine blast is investigated. The crew members may suffer various impacts during a mine blast, for instance a collision of the head with the roof of the vehicle. The acceleration of the vehicle is derived from the finite element simulation of a tracked vehicle subjected to a mine blast. The collision of the ceiling of the vehicle with the head is modeled as a contact of the head with a rigid body. A morphologically sophisticated finite element head model is adopted. The head injury is assessed in view of various head response quantities such as the intracranial pressure, the stress in the brain and the skull, as well as the maximum strain rate in the brain.

Key words: Finite Element Method (FEM), head injury, mine blast, tracked vehicle.

1. INTRODUCTION

Tracked vehicles are usually used as military fighting vehicles and designed for ballistic and mine blast protection of crew members as well as the vehicle itself. Extensive experiments and numerical simulations have been performed to assess the safety of the tracked vehicle and the crew members subjected to a ballistic attack or mine blast. In all the experiments and numerical simulations the crew members are replaced by simplified rigid bodies (dummy body) connected with dashpots and springs. For example, when a tracked vehicle is subjected to mine blast, measurement of the linear accelerations at different locations in the vehicle and different parts of the dummy body are used to evaluate the protection function of the vehicle. The

Michael D. Gilchrist (ed.), IUTAM Proceedings on Impact Biomechanics: From Fundamental Insights to Applications, 273–280.

linear acceleration for human body reflects the loading severity and thus the likelihood of injury. However human body injury, especially the injury of human head, is a very complicated phenomenon, linear acceleration alone is not adequate for make a very reliable judgment.

There are several head injury criteria based on different kinematics parameters or response quantities in the literature. One category is the linear acceleration criteria [4, 5, 8], in which the peak value and the duration of linear acceleration are used for the injury judgment. Suspicions of the effectiveness of this category of criteria arise from experiments indicating that angular acceleration alone produces intracranial damage [9], which, thus, yields another category of angular acceleration criteria. In practice, the effects of angular and linear accelerations can not be differentiated completely. Another category of combination of the linear and angular acceleration is then formed [10]. Further arguments arise from the fact that different patterns of injuries are caused by different injury mechanisms. It is the head responses instead of the acceleration inputs that register the injury pattern and severity. Other categories of injury criteria based on head response quantities, such as the Von Mises stress, shear stress, intracranial pressure gradient, negative pressure, cavitation, peak shear strain rate etc., are established [2].

The present study focuses on head injury assessment of crew members in a tracked vehicle subjected to a mine blast using finite element (FE) method. The linear acceleration of human head from previous internal research on a tracked vehicle and crew member complex subjected to mine blast (Cheng, 2003) is adopted as the input for the FE head model. Head responses to the input accelerations are analyzed against various head injury criteria to make a more reliable head injury assessment.

2. ACCELERATION INPUT FOR HEAD

The acceleration input for the head comes from the numerical simulation of a tracked vehicle and crew member complex subjected to mine blast with different charge weights and detonation positions. Simplifications of the vehicle structure are modeled with solid, beam, shell and spring elements and 17 rigid body element connected with dashpots and spring elements at joints are used for the dummy model. Air-soil-vehicle interactions are also accounted for. Responses at different locations in the vehicle and different parts of the dummy body are recorded. One of the simulation cases (case-0) is verified against field test data and good agreement is found between the simulation and the test results. The other four cases are simulated with different charge weights and detonation locations, i.e., case-1: 5Kg charge weight under slipring; case-2: 7Kg charge weight under the right bench; case-3: 7Kg charge weight under front belly; and case-4: 10Kg charge weight under the left track. The linear accelerations for the head of the crew member nearest to the detonation location are derived as shown in Figure 1. The peak linear accelerations are summarized in Table 1.

Table 1. Peak linear acceleration for different simulation cases.

Loading Cases	Case-1	Case-2	Case-3	Case-4
Peak linear acceleration (m/s^2)	64.4	181.0	190.0	222.5

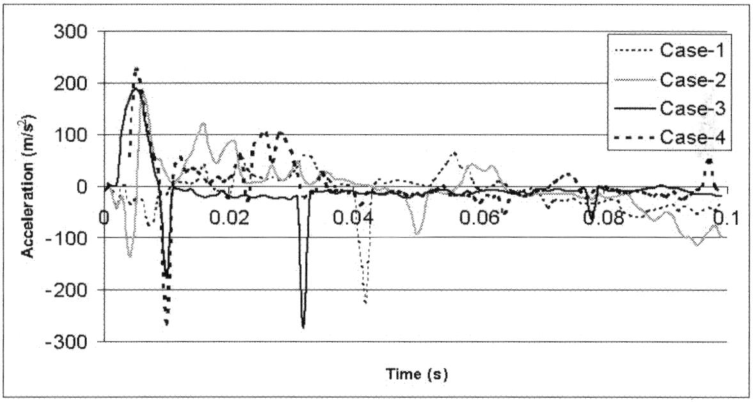

Figure 1. Vertical linear acceleration for the head.

The present study will derive through finite element simulation of the head response with a morphologically sophisticated head model actuated by these four head acceleration curves. The impact of the head to the ceiling of the vehicle is considered.

3. FE SIMULATION OF HEAD RESPONSE

The present finite element head model as shown in Figure 2 was constructed by Horgan and Gilchrist [6] and is obtained from the BEL Repository [1]. The head is modeled with solid elements for the cerebrum, cortical bone, trabecular bone, shell elements for the dura, falx, scalp, tentoriom, mandible and membrane elements for the pia. Complete connections between neighboring parts are assumed. Detailed description for this head model is seen in References [1, 6].

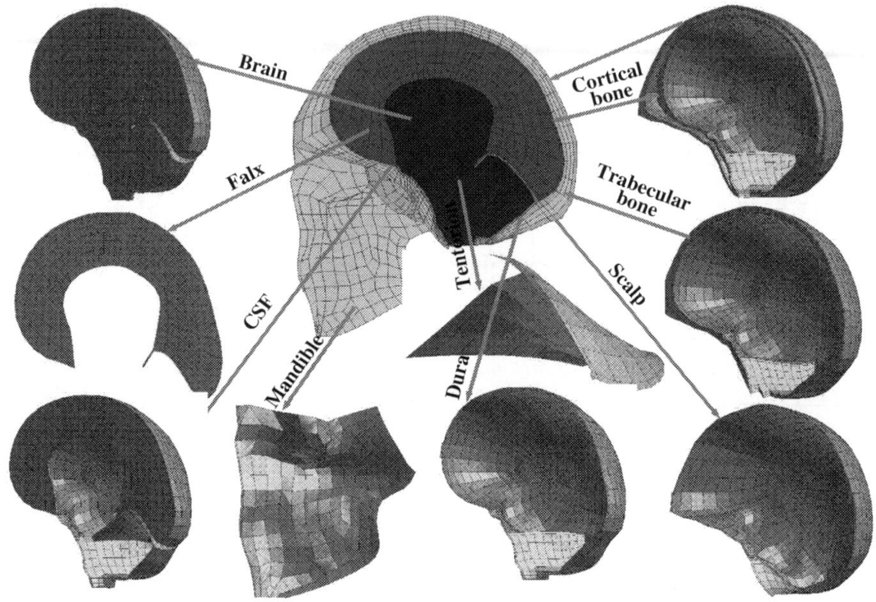

Figure 2. Finite element head model.

The vertical linear acceleration-time curve is applied at the connection between the neck and the head. Rigid plate is used to model the ceiling of the vehicle, which is 70 mm above the top of the head where the height of the crew member is assumed to be 1.73 m in height. Linear elastic material properties are adopted for all the components in the head. The material parameters are listed in Table 2 [6].

It is found from the simulation that the head in case-2 and case-4 hits the ceiling, while it does not in case-1 and case-3.

The maximum Von Mises stress and the peak intracranial pressure for different cases are listed in Table 3. The maximum cerebrum principal strains and the strain rates are tabulated in Table 4.

Very similar stress and intracranial pressure distribution patterns are found for different loading cases in spite of whether the head hits the ceiling. The reason is probably that the velocity when the head hits the ceiling in case-2 and case-4 is very low.

Figures 3 and 4 are the time history plots of strain and strain rate of selected elements in the cerebrum, where the maximum strain and strain rate occur.

Table 2. Material properties.

Materials	Young's modulus (Pa)	Mass Density (Kg/m^3)	Poisson's ratio
Cortical bone	1.5×10^{10}	2000	0.22
Trabecular bone	1.0×10^{9}	1300	0.24
Dura	3.15×10^{7}	1140	0.45
Tentorium	3.15×10^{7}	1140	0.45
Scalp	1.67×10^{7}	1130	0.42
Falx	3.15×10^{7}	1140	0.45
Pia	1.15×10^{7}	1130	0.45
CSF	1.5×10^{5}	1000	0.499989
Mandible	5.0×10^{9}	2500	0.23
Cerebrum	1.23×10^{5}	1040	0.499991

Table 3. Maximum Von Mises stress and intracranial pressure.

Loading Cases	Von Mises Stress (Pa)		Intracranial Pressure (Pa)	
	Skull	Brain	Positive	Negative
Case-1	2.8×10^{8}	–	2.932×10^{5}	-1.987×10^{5}
Casx10-2	5.523×10^{8}	1.2×10^{5}	1.143×10^{6}	-2.354×10^{6}
Casx10-3	4.17×10^{7}	–	1.823×10^{5}	-1.501×10^{5}
Casx10-4	3.6×10^{9}	1.679×10^{5}	6.59×10^{6}	-9.73×10^{6}

Table 4. Maximum cerebrum principal strain and strain rate.

Loading Cases	Maximum Principal Strain	Maximum strain rate (/s)
Case-1	0.266	72.0
Case-2	0.58	190.1
Case-3	0.0823	8.325
Case-4	1.81	381.1

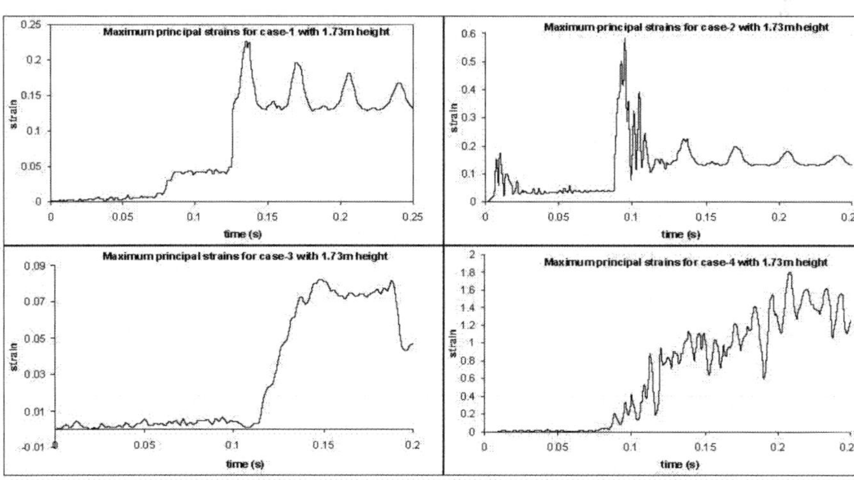

Figure 3. Maximum cerebrum principal strains.

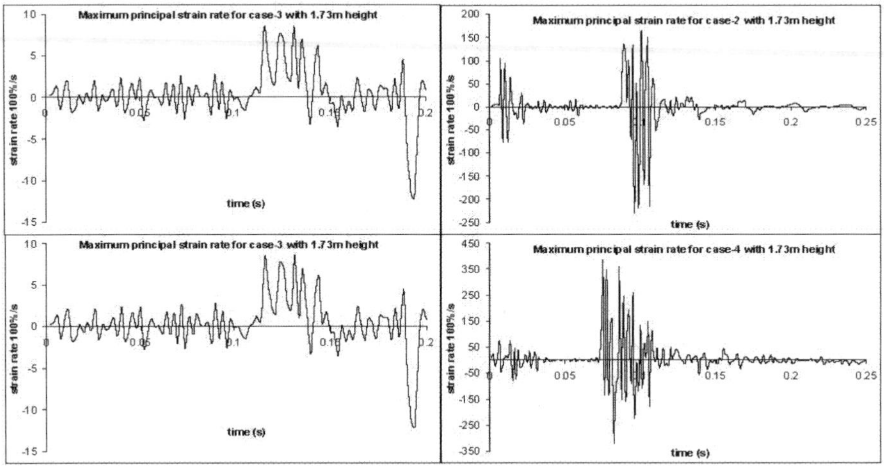

Figure 4. Maximum cerebrum principal strain rates.

4. DISCUSSIONS OF THE SIMULATION RESULTS

From the literature, it was suggested when the ratio of the peak positive pressure to the peak negative pressure is in the range of (5.3×10^4~1.3×10^5 Pa)/(-4.8×10^4~-1.28×10^5 Pa), the injury occurs, while the ratio of the peak positive pressure to the peak negative pressure falls in (4.0×10^4~1.01×10^5 Pa)/(-2.0×10^4 ~-7.8×10^4 Pa), there is no injury outcomes. In view of this, the different degrees of traumatic injuries may occur in the case-3 and case-4, where the peak positive intracranial pressures are 1.823×10^5 and 6.59×10^6 respectively, and the peak negative intracranial pressures are -1.501×10^5 Pa and -9.73×10^6 Pa (Table 3). With reference to the material failure values of 1.05×10^9 Pa and 9.75×10^7 Pa for the skull and brain, the skull failure would happen in the case-4 where the maximum Von Mises stress is 3.6×10^9 Pa. Since the maximum Von Mises stress of 1.679×10^5 Pa in the brain is lower than 9.75×10^7 Pa, the brain would not be tore up.

Some researchers believe that the strain rate is a key biomechanical cause for brain injury and concussion [7]. It is found that the strain rate ranges from 23 to 140/s for the injury outcomes. Hence, the maximum principal strain rate of 381.1/s for case-4, 190.1/s for case-2 and 72/s for case-1 may cause severe or mild brain damage.

From the above discussion, it is observed that head injury criterion proposed by various researchers is based on the various dynamic response measurement, contradictory judgments of the head injure might happen when different criteria are employed. Further comparison of the dynamic response of the head in four cases suggests that a higher response measurement such as the stress level indicates a higher probability of the injury since most of other measurements like the pressure level and strain rate are recorded higher as well. It is prudent to evaluate various response measurements before to make the conclusion of an injury.

5. CONCLUSIONS

From the analysis of a head subjected to linear acceleration, the following conclusions are derived.

(1) The ranking of the injury severity of the head in four cases is from the case-4, case-2 to case-1 and case-3 according to the injury criteria of the pressure level and strain rate developed in the brain.

(2) The measurements of the dynamic response of a head are consistent with each other, i.e. when one dynamic measurement is higher such as the stress level, the other dynamic measurements (intracranial pressure, cerebrum maximum principal strain and strain rate) are also higher.

(3) Head injury assessment using a single injury criterion may mislead the prediction, while overall evaluation of various measurements is more proper.

(4) The head injury would be assessed by the head response instead of the input of linear accelerations.

REFERENCES

1. Biomechanics European Laboratory, 2004. hppt://www.tecno.ior.it/VRLAB/ through the BEL Repository managed by Istituti Ortopedici, Rizzoli, Bologna, Italy.
2. Dawson SL, Hirsch CS, Lucas FV, Sebek BA. 1980. The contrecoup phenomenon. Human Pathology, Vol. 11, pp. 155-166.
3. Cheng QH. 2003. Numerical simulation of response of a turreted infantry fighting vehicle and occupant survivability in a mine blast. IHPC Project Report No. CSM/Q01-300201-1/CC.
4. Gross AG. 1958. A new theory on the dynamics of brain concussion and brain injury. Journal of Neurosurgery, Vol. 15, pp. 548-561.
5. Gurdjian ES. 1972. Recent advances in the study of the mechanism of impact injury of head – a summary. Clinical Neurosurgery, Vol. 18, pp. 1-42.

6. Horgan TJ. & Gilchrist MD. 2003. The creation of three-dimensional finite element models for simulating head impact biomechanics. International Journal of Crashworthiness, Vol. 8 (4), pp. 353-366.
7. King AI, Yang KH, Zhang L, Viano DC. 2003. Is head injury caused by linear or angular acceleration? Bertil Aldman Lecture, Proc 2003 International IRCOBI, Conference on the Biomechanics of Impact, Lisbon, Portugal, pp. 1-12.
8. Lindenberg R, Freytag E. 1960. The mechanism of cerebral concussions. Archives of Pathology, Vol. 60, pp. 440-461.
9. Ommaya AK, Hirsch AE, Martinez JL. 1966. The role of whiplash in cerebral concussion. SAE Paper 660804, pp. 197-203.
10. McLean AJ, 1995. Brain injury without head impact. Journal of Neurotrauma, Vol. 12(4), pp. 621-625.
11. Zhang LY, Yang KH, King AI. 2004. A proposed injury threshold for mild traumatic brain injury. Journal of Biomechanical Engineering – Transactions of the ASME, Vol. 126, pp. 226-236.

FINITE VOLUME SOLUTION TO HIGH RATE WAVE PROPAGATION THROUGH A LUNG ALVEOLI STACK

O. Alakija[1], A. Ivankovic[2] and A. Karac[2]
[1]*Mechanical Engineering, Imperial College London, London, UK;* [2]*Department of Mechanical Engineering University College Dublin, Dublin, Ireland*

Abstract. This paper presents the work on a Finite Volume model representing wave propagation through the lung. One and two dimensional cases are used for the analysis and the results are validated against measured known values of stress wave propagation velocities. The results of the one dimensional model show that the known low wave propagation velocity has a strong dependence on the build up of energy within the alveoli cells especially at the initial stages of propagation. The results of the two dimensional model show the influence of the connecting tissue structure in the generation of over and underpressures at the tissue connecting junctions and also on the overall wave propagation velocity.

Key words: impact, blast, lung, wave, injury mechanism, Finite Volume Method (FVM), fluid structure interaction.

1. INTRODUCTION

Blunt impact injury to the thorax can be caused by a number of physical forces of which the most common follow automobile accidents and blast. In the latter, unlike in the former, rib fractures are uncommon. Severe injuries in the form of haemorrhage and oedema have been observed especially when the thoracic wall velocity exceeds ~20-30m/s depending on the point of impact [1, 9].

It has been suspected that the damaging effects observed after exposure of the human thorax to blast overpressure is as a result of the pressure waves propagating through the lung [1, 2, 4]. This phenomenon is called Primary

Michael D. Gilchrist (ed.), IUTAM Proceedings on Impact Biomechanics: From Fundamental Insights to Applications, 281–288.

pulmonary blast injury (PPBI). It has also been suggested that this injury is caused by either or a combination of spallation, implosion or inertial effects in the lung micro structure [1]. However there are not many models to represent these mechanisms or those leading to injury and as such it is not clear what the real mechanisms are.

Various attempts have been made to explain the mode of wave propagation and thereby explain the resultant mechanism of injury in the lung but until now most of the models have been too simple. This may have been as a result of the problem posed by modelling the coupling between the alveoli structure and the air. Grimal et al. [4] investigated the wave propagation using a one-dimensional mass-spring chain. The lung was viewed as a one-dimensional stack of air and soft tissues layers. Bush et al. [5] modelled using the finite element method a one dimensional layered stack of air and tissue to show the pressure response across the layers.

The work presented here is an attempt to use a fully coupled solid-fluid interaction model to simulate the impact response of a two-dimensional polyhedral lung alveoli stack.

2. METHOD

Two wave propagation models are presented, a one dimensional model and a two dimensional one. The first model will be used to validate the fluid-structure interaction code by comparing the results obtained with other one dimensional models. The second model will be used to show the mode of wave propagation through a typical alveoli stack.

2.1 Numerical Procedure

To simulate stress wave propagation through the alveoli stack a two-system fluid-structure interaction (FSI) model, based on FVM, was employed [7]. Here, the fluid and solid parts of the solution domain have separate meshes, but there is a common interface between them. The solid and fluid models were combined within a single code *FOAM* (Field Operation And Manipulation, [8]). The systems of equations were solved for each mesh, and interface conditions were exchanged: tractions from fluid to solid and velocity from solid to the fluid. Both meshes remain fixed during the calculations. Small-strain analysis was performed for the solid and mesh distortions were neglected. An Eulerian frame of reference was used for the fluid, and the information about the motion of the neighbouring solid wall was passed to the fluid boundary via wall interface velocity. The scheme uses two sets of 'inner' iterations; one for each mesh, and implicit coupling in time was achieved through a series of "outer" iterations, which solves the total system to convergence within each time step [7].

2.2 One Dimensional Modelling

The one dimensional model is based on the assumption that the Lung parenchyma can be modelled locally as a bi-periodic stack of tissue and air layers as shown in Figure 1. The tissue is modelled as a solid and the air is modelled as a fluid in the solution domain.

The impact on the "top" surface is modelled by prescribing a step change in the displacement i.e. velocity. In this case 10 m/s was used as the velocity parameter. The "bottom" surface is fixed i.e. no displacement and the remaining left and right faces are prescribed with a *symmetry* boundary condition.

Two cases are considered in this paper. The geometric dimensions and material properties used are shown in Table 1.

Table 1. Material properties and Geometric dimensions for Case A and B.

	Case A	Case B
Young's Modulus of Tissue (*GPa*)	2.2	2.2
Density of Tissue (kg/m^3)	1000	1000
Poisson's ratio	0.485	0.485
Dynamic velocity of air (*kg/ms*)	1.85×10^{-5}	1.85×10^{-5}
Static density of air (kg/m^3)	1.18	1.18
Atmospheric Pressure (*Pa*)	1×10^5	1×10^5
Tissue layer thickness, y_t (μm)	5	5
Air layer thickness y_a (μm)	100	300
Impact velocity, v (m/s)	10	10

To simplify the model, the effects of other constituents like the surfactants and blood vessels on the surface of the alveoli are not included. The alveoli surface is assumed not to contain any pores for simplicity as well. This assumption only holds for high rate wave propagation through the lung.

The solid part (Tissue) is assumed to be linearly elastic, isotropic and homogenous, and therefore modelled using a simple Hookean constitutive model. The fluid part (Air) is in turn modelled as a compressible fluid.

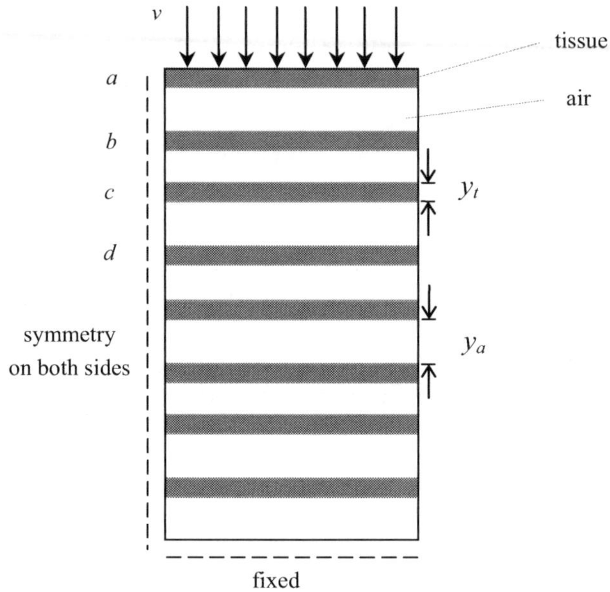

Figure 1. One dimensional bi-periodic layered model for the lung parenchyma. (*a, b, c, d…* are labels for each tissue layer).

2.3 Two Dimensional Modelling

Taking a step further we now model the parenchyma as a two dimensional polyhedral array. This geometry is a true representation of the alveoli structure as seen in the micrograph in Figure 2. Similar boundary conditions are implemented as with the one dimensional model. Again we adopt a symmetry boundary condition to the left and right faces as the nature of the problem is axisymemertic.

The alveoli wall thickness is the same throughout the geometry and is 5μm. The air space thickness is 300 μm face-to-face as the diameter of majority of the alveoli cells in the lung vary between 75 – 360 μm. We employ a step change in displacement to simulate impact at a velocity of 10m/s. We use the same assumptions as we did in the previous model.

3. RESULTS AND DISCUSSION

Firstly we present the results from the one dimensional analysis of the wave propagation through the stack. As a result of the heterogeneity of the lung it is fair to say that the wave velocity upon impact is not constant as

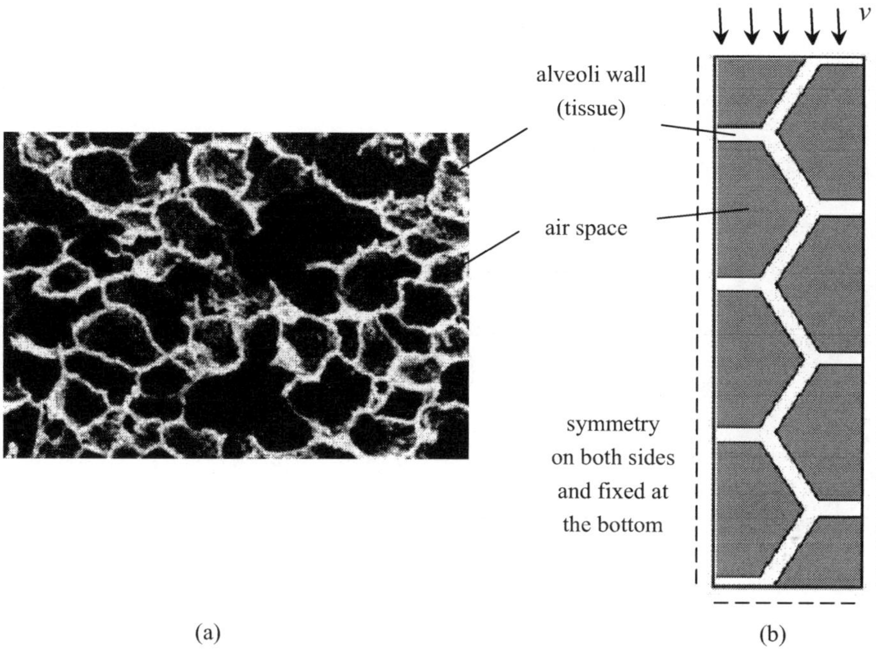

Figure 2. (a) Micrographic representation of a cross-section through an alveoli duct and sac system (b) Two dimensional mesh representation of the alveoli stack

represented in Figure 3. The wave velocity presented in the graph is calculated at each subsequent tissue layer (*a, b, c, d...*). From these results it can be seen that as the wave propagates, the overall velocity stabilises to what is know as the *wave speed.* It starts off at a very high value and reduces with distance. Figure 3a shows that for a tissue layer separation of 100μm the wave speed will eventually be about 70 m/s and Figure 3b shows for a layer separation of 300 μm, 100 m/s. This is in very good agreement with the phase velocity measurements in the Grimal Model [4].

From our model we can improve the understanding on how a stress wave would propagate through a porous weakly coupled system. From our observation the reduction in speed at the initial stages is mainly as a result of a build up in energy within the first few cells necessary for the continuation of the propagation of the wave. The further away the location considered the more energy required for propagation, this causes the reduction in speed. It is also possible if need be to determine the pressure differentials between the alveoli cells that may lead to high tensile strains causing of damage with our model though this analysis is not presented here.

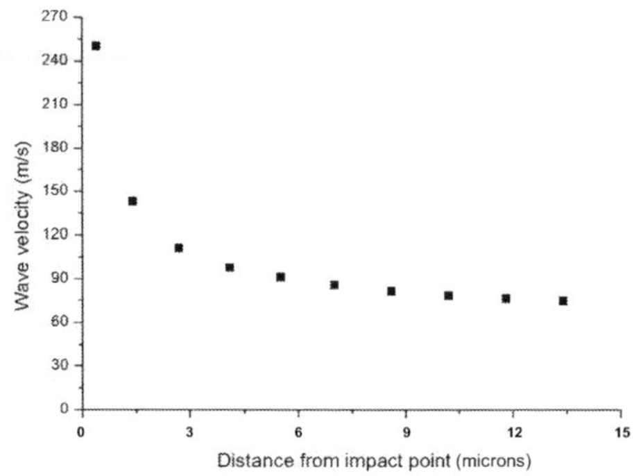

Figure 3a. Case A: Wave velocity representation across one dimensional stack.

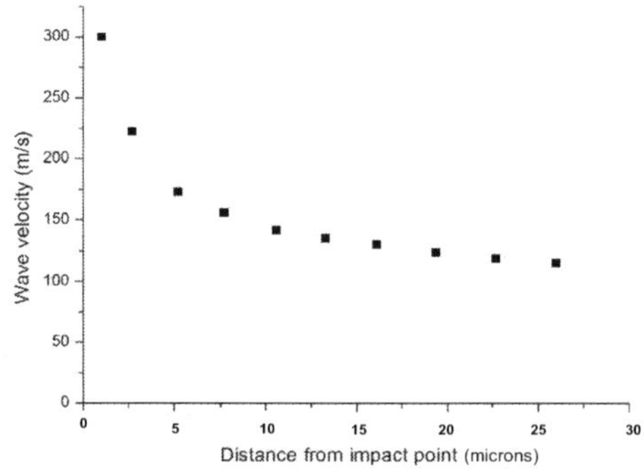

Figure 3b. Case B: Wave velocity representation across one dimensional stack.

Secondly we present the results from the two dimensional analysis of wave propagation. The analysis here is discussed purely in a qualitative sense. The overall wave propagation speed is to a certain extent more complicated to determine with our model as the overall wave in this case propagates not only by the same mode as discussed in the one dimensional case (i.e. propagation through the air and tissue sequence) but is also influenced by the stress wave propagating through the tissue structure as can be seen in Figure 4. In Figure 4a, the dark areas are regions of compressive stresses whilst in Figure 4b, the dark areas are regions of overpressure. Of interest are the regions close to the nodes (where three alveoli walls confluence), at these region we observe the fluctuation of overpressures and underpressures around this region of the alveoli sac by the compressive and shear stresses within the walls. This analysis can shed some light on how the tensile forces mentioned in [1] can lead to damage.

Applying the analysis carried out on the one dimensional case on a two dimensional 6 alveoli stack case (with the same material properties and parameters) we observe a propagation speed of 900 m/s on the sixth alveoli

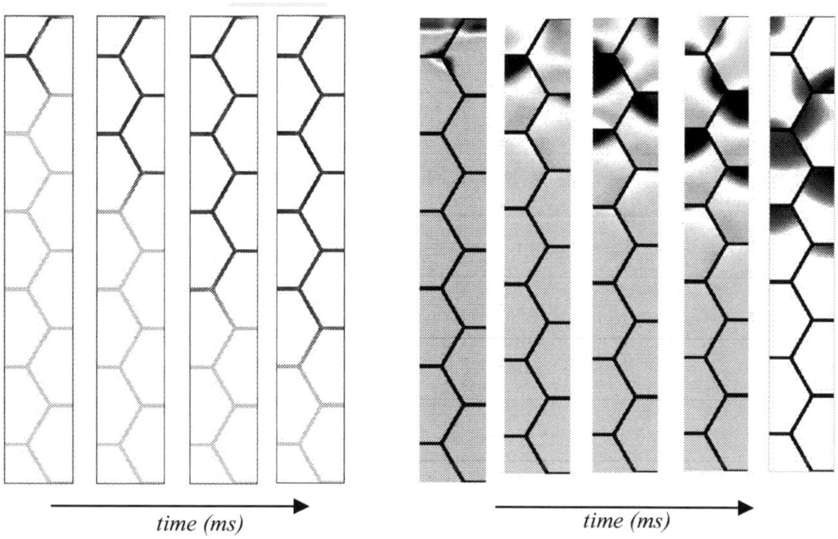

(a) (b)

Figure 4. Finite Volume analysis of 2D alveoli stack showing wave progression (a) Tissue analysis (b) Air analysis.

wall. This value is thought to being expected higher due to the influence the wave propagating through the tissue structure has on the air pressure in this case which is not exhibited in the one dimensional case.

4. CONCLUSIONS

We are not only able to predict the wave propagation speed but also shed light on why this speed is low using our model. We also have the capability of producing frequency response data from our model which is useful in determining the cut off frequencies for amour protection to the chest. It must be noted that the wave propagation speed (900 m/s) obtained from the two dimensional analysis was based on a 6 stack model, further analysis with a longer alveoli stack would need to be carried out in order to see the stabilised velocity of the wave.

REFERENCES

1. Fung, Y.C., Yen, R.T., Tao, Z.L. and Liu, S.Q., A hypothesis on the mechanism of trauma of lung tissue subjected to impact load, *Journal of Biomechanical Engineering* **110**, 1988, pp. 50-56.
2. Yen R.T., Fung Y.C. and Liu S.Q., Trauma of lung due to impact load, *Journal of Biomechanics* **21(9)**, 1988, pp. 745-753.
3. Ho, A., A simple conceptual model of primary pulmonary blast injury, *Medical Hypotheses* **59(5)**, 2002, pp. 611-614.
4. Grimal, Q., Watzky, A. and Naili, S., A one-dimensional model for the propagation of transient pressure waves through the lung, *Journal of Biomechanics* **35,** 2002, pp. 1081-1089.
5. Bush, I. and Chaneller, S., Finite element modelling of non-penetrating thoracic impact, *International IRCOBI Conference on the Biomechanics of Impacts*, 1988, Bergish-Gladbach.
6. Karac, A. and Ivankovic, A., Modelling the drop impact behaviour of fluid-filled polyethylene containers, *Fracture of Polymers, Composites and Adhesives II*, B.R.K. Blackman, A. Pavan and J.G. Williams (eds), ESIS Publication **32,** Elsevier, 2003, pp. 253-264.
7. Greenshields, C.J., Weller, H.G. and Ivankovic, A., The finite volume method for coupled fluid flow and stress analysis, *Computer Modeling and Simulation in Engineering* **4**, 1999, pp. 213- 218.
8. http://www.nabla.co.uk
9. Cooper, G.J., Townend, D.J., Cater, S.R. and Pearce, B.P., The role of stress waves in thoracic visceral injury from blast loading: modification of stress transmission by foams and high-density materials, *Journal of Biomechanics* **24**, 1991, pp. 273–295.

SESSION 6

CELLULAR BASIS FOR THE NONLINEAR CONSTITUTIVE PROPERTIES OF BRAIN TISSUE

Michael DeRidder and David F. Meaney
University of Pennsylvania, Department of Bioengineering, 120 Hayden Hall, 3320 Smith Walk, Philadelphia, PA 19104-6392, USA; Phone: 215.573.3155, Fax: 215.573.2071, email: dmeaney@seas.upenn.edu

Abstract. Brain tissue is characterized as a nonlinear viscoelastic material. One intriguing property of brain tissue is material softening at finite strains. In this study, we examine the individual motion of neural and glial cells in slices of brain tissue subjected to a controlled equibiaxial deformation. We compare the cellular motions to the kinematics of nuclei movement in astrocyte monolayers exposed to the same strain field, and to the patterns of bead movement for beads entrapped within a silicone gel surrogate. We find (a) not all cell nuclei in tissue move in response to an applied deformation, (b) the relative fraction of displaced nuclei is sensitive to the applied deformation, but not the distance of the cell from the stretched membrane, and (c) the strains computed from tracking the nuclei of moving cells is not significantly different than the strains computed from astrocyte monlayers or silicone gel surrogates. Together, these suggest a cellular mechanism underlying the strain softening behavior measured for brain tissue. Separately, these developed relationships will help determine the relative number of cells experiencing a deformation under an applied macroscopic strain field.

Key words: mechanical properties, slice culture, kinematics, brain.

1. INTRODUCTION

Defining human constitutive properties remains one of the central activities in understanding the tolerance of the human body to impact. With the emerging use of computational models for human skeletal and soft tissue

Michael D. Gilchrist (ed.), IUTAM Proceedings on Impact Biomechanics: From Fundamental Insights to Applications, 291–304.

structures, constitutive models represent a vital link to predict stress conditions within both hard and soft tissues during impact, and ultimately contribute to defining the potential injury risk of different blunt trauma situations.

Defining the complete mechanical properties of brain tissue under conditions that occur during injury has been the focus of studies that span nearly four decades (for review, see [1]). Studies reporting brain constitutive properties grow each year, and the focus has shifted noticeably from the use of small strain constitutive models [2–4] to formulations that apply for the finite strain regime [5–8]. Most models account for the potential nonlinear behavior of the brain tissue at these finite strains [9, 10]. Much less frequently, potential nonlinear time dependencies are examined [11]. The direct use of these nonlinear constitutive models in commercial finite element codes, though, is not always available. For this reason, these new nonlinear material data are not yet well integrated into computational models of the human impact response.

In this study, we examine a potential mechanism for the observed nonlinear properties of brain tissue. In recent years, several different groups independently reported properties for the brain tissue. Differences in the finite strain constitutive properties between gray and white matter occur for both in vivo and vitro tissue [7], and also appear similar across different species. Most studies show an intriguing softening behavior for brain tissue at higher strains [7, 8, 11]. The tissue softening behavior, reflected by a decreasing slope in the stress-strain curve under shear, is repeatable across consecutive tests and extends across species.

Although the brain tissue nonlinear properties are now better characterized, there is no study yet that examines how these properties emerge from the cellular components of the material. The correlation between motions of cellular structures and the resulting macroscopic material properties may be useful for studying situations that cause injury, since it would provide a method to link the macroscopic mechanical parameters (e.g. tissue stress, strain, etc.) to individual cellular damage. In this study, we explore the kinematics of cellular movement under controlled deformation protocols. We use measurements from both cell monolayers and in vitro brain tissue to study if the kinematics can provide a mechanism for understanding the nonlinear properties measured for brain tissue.

2. MATERIALS AND METHODS

Overview: The focus of the study is to characterize the motion of neurons and glia in brain tissue subjected to controlled mechanical loading. As a first step, we measure and report the motion of fluorescent spheres in a silicone gel surrogate, and the motion of nuclei in a monolayer culture of astrocytes. We use these data to interpret measures of nuclear movements in brain slice cultures exposed to a known deformation field. We examine only a quasistatic loading condition, and track the fluorescently labeled beads or cell nuclei throughout the thickness of the slice during the deformation. Using a new analysis algorithm, we define the fraction of cells that displace under the applied loading, and compute the deformation measured between cell nuclei within a cross-section of the slice culture.

Silicone gel surrogate fabrication: As an alternative to a brain tissue sample, many investigators have used a silicone gel material that can undergo finite strains without failing [12–14]. Moreover, the silicone gel can be formulated to have mechanical properties that are in the range of the properties measured for brain tissue. We used this material for comparative purposes, examining if the motion of rigid beads incorporated within the material would follow the underlying substrate deformation more closely than the nuclei tracked in brain slice samples. Gel samples approximately 3–4 mm thick were poured in a ratio to provide mechanical properties in the range measured for brain tissue, and samples were mixed with 2 μm polystyrene beads before curing. Once cured, samples were self-adhered to the elastic membrane and placed in the same cell stretching system used for tissue samples. The position of the rigid polystyrene beads at different sections was recorded under different levels of substrate deformation.

Astrocyte monolayer culture preparations: Brain slice cultures are a mix of neurons and glia. To better interpret the data from this complex tissue architecture, we examined the motion of a confluent monolayer of astrocytes cultured on a flexible elastic membrane. Astrocyte cultures were obtained from whole brains/cortices dissected from E17-19 Sprague-Dawley rats (Charles River). Brains/cortices were dissected, the meninges were removed, and the brains/cortices were placed in trypsin/DNAse for 20 minutes at 37° C. The cells were homogenized by pipetting, centrifuged for 5 minutes, and then resuspended in DMEM (BioWhitaker) + 5% Fetal Bovine Serum (FBS) (Hyclone) + 10% Ham's F12 (Sigma). Cells were filtered, diluted to $5x10^5/4x10^5$ cells/mL and placed in T75 tissue culture flasks that had been treated with poly-l-lysine (PLL) (Sigma). The media was changed every 3-4 days. Cytosine β-d-arabinofuranoside (AraC) (Sigma) was added on day 10 to minimize contamination by fibroblasts. After 13 days in vitro the flasks were shaken overnight at 240 rpm to remove loosely adherent cells, such as microglia or oligodendrocytes. After shaking, trypsin/EDTA was added for 3-5 minutes. The cells were removed from the flasks, centrifuged, and resuspended in Neurobasal Media (Gibco) + 5% FBS. Astrocytes were plated onto the membrane at $1.5 \times 10^5/1.0 \times 10^5$ cells/mL. Media was changed every 3–4 days. Experiments were performed 13–15 days after plating onto the membranes, at which point the cultures had become confluent.

Organotypic Culture Isolation: The method for isolating organotypic cultures is similar to those described previously, with some modifications. Briefly, the brains were aseptically removed from 6 six-day-old Sprague-Dawley rat pups (Charles River Laboratories, Wilmington, MA) and sectioned coronally into 350μm sections on a McIlwain Tissue Chopper (Brinkman Instruments, Westbury, NY). The brain was then transferred to ice-cold Gey's salt solution (Sigma, St. Louis, MO) supplemented with 6.5% glucose. The slices were carefully separated each slice was plated onto a treated, flexible silastic membrane (Dow Corning, Midland, MI) stretched across the bottom of a polysulfone well. Neurobasal-A media (Invitrogen, Carlsbad, CA) supplemented with 6.5% glucose, B-27 supplement (Invitrogen), and 1μg/ml L-glutamine (Sigma) was added to the well, which was incubated on a rocker for 10 days with media changes occurring every 2–3 days. On day 10 *in vitro*, cultures were labeled with SYTO-23 dye (Molecular Probes, Eugene, OR) to highlight cell nuclei throughout the tissue sample. The cultures were placed on the sample stretching system described below, and the position of cell nuclei across different sections were tracked during a controlled deformation of the underlying elastic substrate. Two regions from each of three brain locations in the slice culture – the hippocampus, the cortex, and the thalamus – were examined.

Applying an equibiaxial strain to samples: Each sample was subjected to the same deformation protocol. Cultures or silicone gel samples were placed in a device designed to statically displace a membrane in distinct intervals [15]. The tissue stretching device consisted of a polysulfone ring to which a silicone membrane is attached across the bottom, providing a well to grow the organotypic culture. The polysulfone ring was placed over a second polysulfone component mounted on a microscope stage. This second component contained an annular indenter large enough for a microscope objective to fit in the middle and image the membrane. However, the opening prevented imaging of any area of the membrane other than the center. The membrane was deformed by moving the well down over the indenter, which was done by screwing the ring down into the stage mounted portion of the device (Figure 1). With this device, the samples were imaged under different stretch conditions since the tissue remains in the same focal plane while the membrane is stretched over the sides of the indenter. Placing a small amount of grease on the underside of the membrane minimizes friction between the membrane and the indentor.

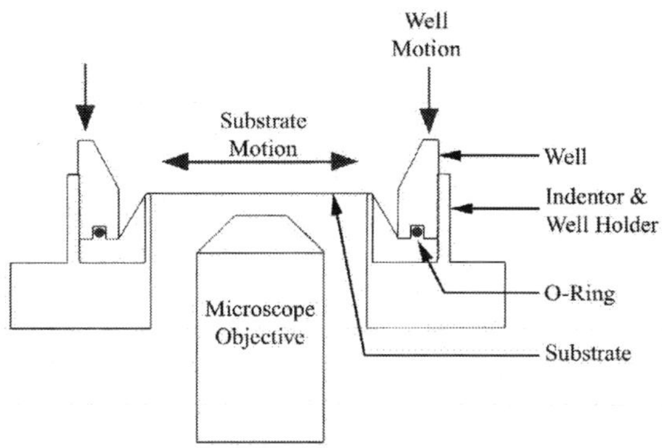

Figure 1. Schematic of a cross-section through the static cell-stretching device. The device is mounted on a microscope stage and the well is screwed down over the indenter. The membrane substrate is stretched biaxially and can be imaged continuously because it is held at a constant distance from the objective.

Stacks of images were collected for each sample type following each step increase in membrane strain using a Nikon Eclipse TE300 scanning confocal microscope equipped with Lasersharp 2000 software (Bio-Rad Laboratories, Hercules, CA) Strains up to a level of 25.5% area strain. For astrocyte monolayer cultures, only a single image plane was collected. Image stacks were analyzed to track the motion of specific cell nuclei across the deformation stages. Nuclei position was measured using Scion Image (Scion Corp, Frederick, MD). Care was taken to ensure that coordinates of the same cell were collected in each image. The number of points generated from any one image ranged from 8 to over 30.

Images were analyzed first to determine the relative fraction of moving nuclei at each level of applied strain. Coordinates of the cell nuclei from the undeformed (u) and deformed (d) image were compared, rigid body rotations were subtracted, and the net displacement of each tracked nuclei was computed. The threshold precision error (σ_{prec}) for locating an individual nucleus position was used to generate a threshold level ($\sigma_{thresh}>3\sigma_{prec}$) to separate moving nuclei from the nuclei that did not move significantly following an applied deformation. The fraction of displaced nuclei (χ) was computed for each level:

$$\chi = \frac{n(\delta > \delta_{thresh})}{n_{total}}$$

Following this analysis, the stretch ratio between line segments between displaced nuclei (n_i, n_j) were calculated according to:

$$\lambda_{ij} = \frac{\sqrt{\left(\left(x_{ni,d} - x_{nj,d}\right)^2 + \left(y_{ni,d} - y_{nj,d}\right)^2\right)}}{\sqrt{\left(\left(x_{ni,u} - x_{nj,u}\right)^2 + \left(y_{ni,u} - y_{nj,u}\right)^2\right)}}$$

where ($x_{ni,d}$, $y_{ni,d}$) and ($x_{nj,d}$, $y_{nj,d}$) denote the coordinates of the nuclei (i,j) in the deformed image, and ($x_{ni,u}$, $y_{ni,u}$) and ($x_{nj,u}$, $y_{nj,u}$) are the nuclei coordinates in the undeformed image. Average stretch ratios for moving nuclei/beads were computed for each level of applied deformation. For each level of strain, the average (μ), standard deviation (σ), and coefficient of variance ($COV=\sigma/\mu$) were calculated. In addition, linear regression was used to at each strain level to determine if the stretch ratio decreased significantly with distance from the deformed substrate. Separately, analysis of variance was used to determine if there were changes in the stretch ratios across different levels of applied strain.

3. RESULTS

Rigid polystyrene beads entrapped with silicone gel samples show a consistent kinematic pattern under equibiaxial deformation

As a first step, we examined the motion of rigid beads within an elastic material with properties approximating the mechanical properties of brain tissue measured under small strains. Above applied strains of 2.5%, all beads displaced following equibiaxial deformation of the underlying substrate (Figure 2). Below this level of deformation, nearly (96%) all of the beads moved measurably. The motion of the beads was not influenced by distance from the deforming substrate; all beads positioned approximately 100 μm from the substrate moved in the same pattern as beads close (10-12 μm) to the membrane surface ($P<.05$).

Astrocyte monolayers cultures show a consistent pattern of nuclear motion under applied strain

Cultures of confluent astrocyte monolayers showed a pattern of nuclear movement that was similar to the substrate deformation, increasing in proportion to the applied strain. In comparison to the silicone gel surrogate, though, a slightly larger fraction of nuclei did not displace significantly at the lowest deformation level (Figure 2). At higher applied deformation levels, all tracked nuclei moved in response to the substrate deformation.

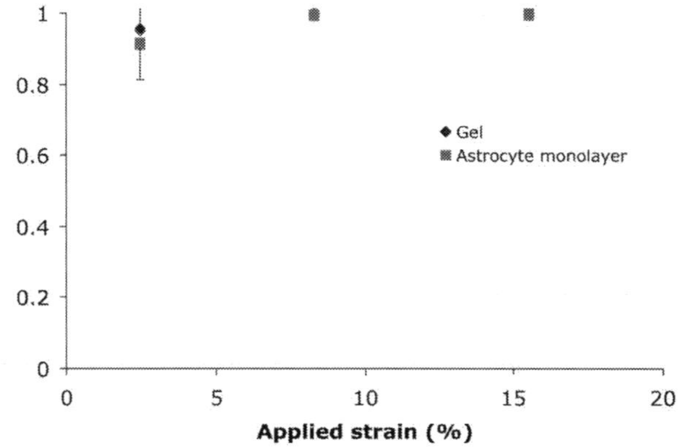

Figure 2. A. Rigid beads within a silicone gel material move with the applied substrate displacement, with nearly all beads showing uniform movement even at the lowest applied deformations. B. Confluent astrcoyte monolayers show similar movement pattern, with most of the nuclei moving in response to even the lowest applied strains.

The relative fraction of moving nuclei in brain tissue slices increases with applied strain

In brain slice cultures, the fraction of displaced nuclei showed an increase in proportion with the applied strain. However, not all nuclei would displace noticeably, even at the highest deformation level. Regression analysis showed that pattern of nuclear movement was not sensitive to the distance from the deforming substrate. On average, approximately 90% of the nuclei would move in response to strains of greater than 8.5% (Figure 3).

Segment strains among beads in silicone gel surrogates show small variation

Given that all beads in a silicone gel surrogate displace at the three highest deformation levels, we examined if the deformation measured between bead pairs changed with increasing applied strain. The average deformation between segments followed closely the applied strain to the substrate, and did not change significantly with distance from the deforming substrate (Figure 4). The variance of the deformation between bead pairs increased slightly at higher strains, but was not significantly different among the three different strain conditions.

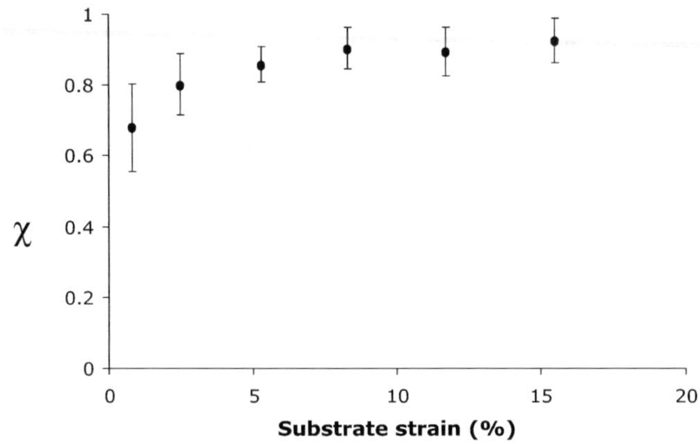

Figure 3. The fraction of nuclei displacing in response to an applied strain in brain slice cultures increases with substrate strain. No significant difference in the fraction of moving nuclei was observed for cells at different distances from the substrate.

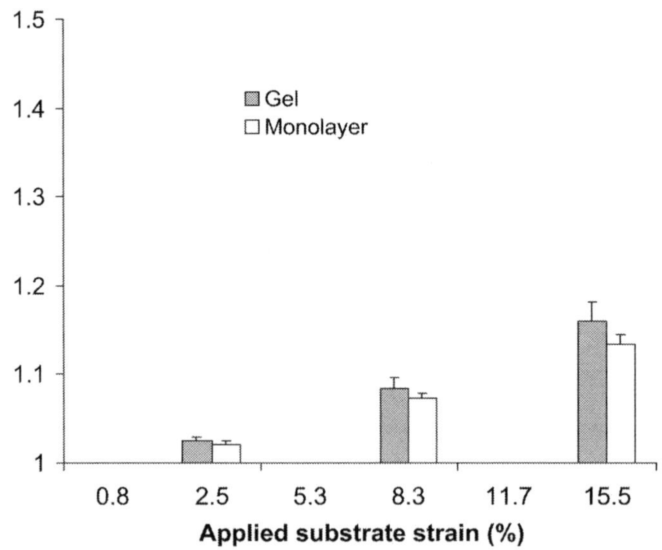

Figure 4. The average deformation measured between tracked beads in silicone gel and between nuclei in astrocyte monolayers increase in proportion to the applied strain. The variance in the measure segment deformations does not differ significantly, nor does the average strain.

Strains between nuclei in monolayer culture show increased variance at higher deformations.

In astrocyte monolayer culture, nuclei do not move in a pattern that creates a uniform strain in line segments connecting nuclei. Rather, there is an increase in the variance between line segments compared to similar measures derived from bead pairs in silicone gel samples. The variance grows in proportion to the applied strain (Figure 4).

The average deformation of line segments between moving nuclei in tissue does not change with distance from the substrate

Interestingly, the average deformation between line segments in nuclei tracked throughout different planes of brain slice cultures does not differ significantly from the average measures derived from beads in silicone gel samples. The average deformation among nuclei does not change significantly with distance from the substrate. In addition, the variance in the deformation between nuclei is increased significantly over the variance measured for beads entrapped within silicone gel samples ($p<.05$). Beyond an applied equibiaxial strain of 8.5%, the coefficient of variance for the segment strains increased more rapidly with applied strain (Figure 5).

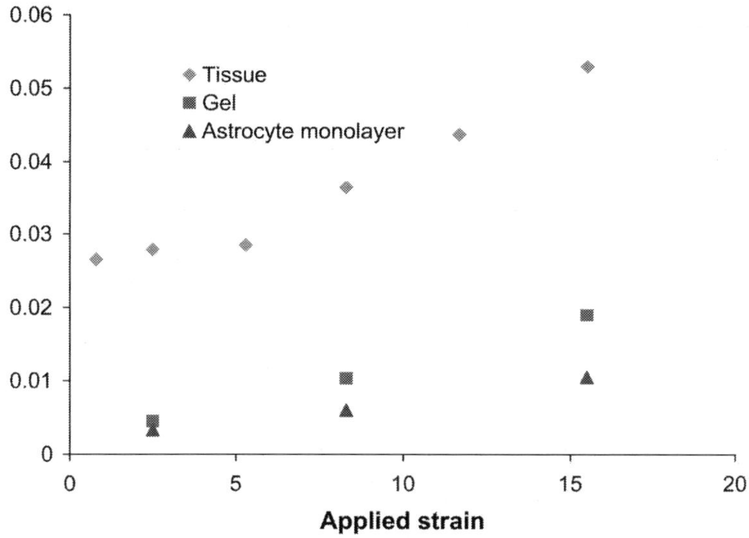

Figure 5. The coefficient of variance in the segment strains measured between tracked nuclei changed in proportion to the applied strain for both astrocyte monolayers and beads entrapped within a silicone gel. For segment strains measured for nuclei tracked within tissue, the variance increased markedly beyond applied strains of 5.5%.

4. DISCUSSION

In this paper, we studied the motion of cells within cultured brain tissue slices exposed to increasing levels of equibiaxial deformation. We were interested in examining of how the motion of nuclei – an approximate marker of cell motion – would correspond to the change in bulk mechanical properties of brain tissue as it progresses from a linear viscoelastic material under small strains to nonlinear behavior at finite strains. We use alternative approaches to study how the pattern of rigid beads within an elastic material move in response to the same applied deformation, as well as the pattern of nuclei movement in astrocyte monolayers. We show that the motion of both nuclei within tissue and in cell monolayers increases in proportion to the applied strain, and that beads within a surrogate brain material also show similar displacement patterns. Importantly, motion of nuclei within tissue suggests that most cells initially do not displace at low levels of stretch, but the number of cells showing displacement increases at higher levels of applied stretch – i.e., these cells are gradually 'recruited' into the bulk deformation state with increasing applied strain. In both cell monolayers and silicone gel samples, motion of markers is measurable at even the smallest applied strains and this recruitment phenomenon does not apply. The heterogeneity in nuclei movement is most apparent in tissue, is reflected in an increased variance of strains calculated between nuclei, and suggests a complex kinematics occurs in these brain slice cultures.

We designed this study to minimize potential complicating factors in the analysis. First, we developed measures of nuclei displacement patterns in known elastic materials, as well as a more complex single layer of cells (astrocytes) on an elastic membrane, as two comparative measures of nuclei displacement patterns in tissue. Developing these additional measures allowed us to understand variability in displacement patterns in homogenous elastic materials, as well as a single layer of cells. Second, we developed precise estimates of the errors in locating the position of nuclei beads in each preparation, and used these as a guide in defining the kinematics of movement across samples. Third, we randomized the selection of location across the brain cultures to minimize the bias that could occur if only one region was selected for analysis. Similarly, separate cultures were examined for each region.

Generally, the kinematics of cells within biological tissues under deformation is not well characterized, even though numerous studies attempt to define how these cells respond to biological forces in the in vivo environment. To our knowledge, only Sah and colleagues [16] have examined the motion of nuclei in tissue (cartilage) under a known loading (compression), showing that a nonuniform compaction of the nuclei appear closer to the loading platen for compression. We have recently studied how axons within the central nervous system elongate in white matter tissue subjected to simple extension, finding that the individual elongation pattern shows a gradual coupling of the axons to the glia/vascular matrix [17]. However, we did not examine the motion of glial cells surrounding the axons to confirm this coupling process. The current findings showing a gradual 'recruitment' of cells into the pattern of displaced nuclei suggest there is an increased coupling among cells in brain slice tissue at higher levels of deformation. The coupling among cells is more complex than simple elastic coupling, since the variance of the segment strains between nuclei is nearly four times higher than the variance observed among beads entrapped within an elastic silicone gel material. The data on segment strains between nuclei in astrocyte monolayer cultures show that, in a single layer of cells, the variance is not significantly different from beads in an elastic matrix. This finding suggests that perhaps the mixture of cells in the CNS, rather than the potential nonlinear properties of individual cell types, may give rise to the more complex displacement patterns observed in tissue.

The significance of cellular kinematics in the overall constitutive relationship is only beginning to be understood. Zahalak and colleagues developed a formulation of the constitutive behavior for tissue based on the cell mechanical properties [18, 19], and updated recently that the underlying cellular kinematics can significantly influence the resulting stress-strain behavior of the tissue. In earlier formulations of the relationship, the cellular structure is assumed to move in concert with the macroscopic deformation state. Recently, the implication of this assumption was more completely examined, and it was shown that a departure from this assumption would underestimate the predicted cellular mechanical properties by nearly an order of magnitude [19]. The progressive recruitment of nuclei movement into the bulk deformation field observed in this study indicates that the transformation between cellular motions and bulk kinematics is not exactly 1:1, and one would therefore need to account for the modifications made to the Zahalak relationship. Our data for equibiaxial deformations > 5% also point out another important factor to consider when developing a cellular based constitutive relationship – the relationship should account for a small fractions of cells that do not displace even at the highest level of strains. We expect that the recruitment pattern, combined with the mechanical behavior of these cells, would together contribute to the macroscopic properties of the tissue.

The implications of the kinematics measured in this current study are twofold: the data not only provide a potential contributing mechanism to the transition point from linear to nonlinear brain tissue mechanical behavior, but also point out the potential heterogeneity of cellular deformation that occurs under finite strains. The point of transition (5%) where rapid recruitment of cells into the deformation field approaches an equilibrium value is strikingly the same strain transition point separating linear from nonlinear material behavior in brain tissue [6, 7]. These data suggest the nonlinear constitutive behavior that occurs under finite strains is potentially due to the increased participation of the cellular tissue component. In comparison, stresses generated during the gradual recruitment process are likely from the extracellular matrix in the material. The increased segment strain heterogeneity at finite strains suggests that the deformation of the cell population may be equally mixed. Neighboring cells may experience different levels of deformation, leading a complex pattern of cellular damage even in highly localized regions of the brain. Neuropathological data from experimental models of traumatic brain injury suggest this heterogeneity at the microscale may play a role in patterns of injury in vivo, since commonly injured regions of the brain show a remarkably diverse pattern of cell death. Not only is the cell death delayed and progressive, but the activation of separate cell death pathways often occurs in neighboring cells [20].

An important next step in this study is to examine how these patterns of nuclei movement in tissue will change at higher rates of loading associated with impact. In this study, we examined only quasistatic loading conditions. Although the softening behavior of brain tissue remains even at quasistatic loading rates, the macroscopic properties of brain tissue are clearly influenced by loading rate [1]. In addition, we have shown recently that the function response of neurons is also sensitive to the loading rate [21]. If the pattern of nuclei movement will change at higher loading rates, this will have important implications on matching the cellular scale kinematic patterns to the patterns of cellular injury observed at these higher loading rates.

The measured cellular kinematics also opens up new questions on the mechanism(s) of how material softening occurs for brain tissue exposed to finite strains. The gradual recruitment of cells reaches an equilibrium point at the level where softening begins to occur in shear tests (see [7, 11]), and therefore the recruitment of cells into the deformation response would not play a role. Although the softening of cellular components presents a plausible reason, the mechanical behavior of astrocytes or neurons under such deformation conditions is not well described. However, studies of gels composed of common cytoskeletal proteins reconstituted in vitro show that the structural proteins within these cells to have reasonable linear properties even up to very large shear strains (>100%). One potential reason for softening lies in the pattern of nuclei movement at increasingly higher strains. The segment strains among neighboring nuclei show a growing variance beyond the transition point for material softening, suggesting that the cellular deformations becomes much more varied. If these cells form an appropriately interconnected network, the most compliant cells will influence more the resulting macroscopic mechanical behavior at increasing strains. This may provide a mechanism for the overall change in behavior measured for brain tissue at finite strains, and is the subject of current study in our laboratory.

5. CONCLUSIONS

By measuring the motion of beads in an elastic material, nuclei in cell monloayers, and nuclei in tissue slices under the same levels of applied equibiaxial deformation (<15%), we examined how applied bulk deformations transformed to displacement patterns measured at the cellular length scale. From these data, we conclude that beads entrapped within a homogeneous elastic material closely follow an applied equibiaxial deformation. The correspondence between macroscopic deformations and motion of individual nuclei in cell monolayers is attenuated slightly, and the average strains between nuclei is slightly smaller than the applied bulk deformation field. The motion of nuclei in tissue is most complex, with a majority (~90%) of nuclei moving in proportion to the applied equibiaxial stretch, while the remaining fraction not moving even at the highest levels of stretch. These data suggest that there will be a marked heterogeneity in cellular deformations from macroscopic deformations, and may be an underlying reason for the patterns of cellular damage frequently observed in traumatic brain injury.

ACKNOWLEDGEMENTS

Funds for this work were provided by the National Institutes of Health grants NS 312752 and NICHD 41699 and funds from the NHTSA through the Southern Consortium for Injury Biomechanics (SCIB).

REFERENCES

1. Margulies, S. and Meaney, D., Physical properties of brain tissue, in: Handbook of Biomaterials, J. Black (ed.), Springer Verlag, 1998, pp. 70-80.
2. Shuck, L.Z. and Advani, S.H., Rheological response of human brain tissue to shear, Journal of Basic Engineering, Transactions ASME, Ser. D, 94(4):905-911, 1972.
3. Galford, J.E. and McElhaney, J.H., A viscoelastic study of scalp, brain, and dura, Journal of Biomechanics, 3(2):211-221, 1970.

4. Pamidi, M.R. and Advani, S.H., Nonlinear constitutive relations for human brain tissue, Journal of Biomechanical Engineering, Transactions of the ASME, 100(1):44-48, 1978.
5. Mendis, K.K., Stalnaker, R.L. and Advani, S.H., A constitutive relationship for large deformation finite element modeling of brain tissue, Journal of Biomechanical Engineering, 117(3):279-285, 1995.
6. Arbogast, K.B. and Margulies, S.S., Material characterization of the brainstem from oscillatory shear tests, Journal of Biomechanics, 31(9):801-807, 1998.
7. Prange, M.T. and Margulies, S.S., Regional, directional, and age-dependent properties of the brain undergoing large deformation, Journal of Biomechanical Engineering, 124(2):244-252, 2002.
8. Miller, K. and Chinzei, K., Mechanical properties of brain tissue in tension, Journal of Biomechanics, 35(4):483-490, 2002.
9. Nicolle, S., Lounis, M. and Willinger, R., Shear properties of brain tissue over a frequency range relevant for automotive impact situations: New experimental results, Stapp Car Crash Journal, 48:239-258, 2004.
10. Brands, D.W.A., Bovendeerd, P.H.M. and Wismans, J.S.H.M., On the potential importance of non-linear viscoelastic material modelling for numerical prediction of brain tissue response: Test and application, Stapp Car Crash Journal, 46:103-121, 2002.
11. Takhounts, E.G., Crandall, J.R. and Darvish, K., On the importance of nonlinearity of brain tissue under large deformations, Stapp Car Crash Journal, 47:79-92, 2003.
12. Margulies, S.S., Thibault, L.E. and Gennarelli, T.A., Physical model simulations of brain injury in the primate, Journal of Biomechanics, 23(8):823-836, 1990.
13. Meaney, D.F., Smith, D.H., Shreiber, D.I., Bain, A.C., Miller, R.T., Ross, D.T. and Gennarelli, T.A., Biomechanical analysis of experimental diffuse axonal injury, Journal of Neurotrauma, 12(4):689-694, 1995.
14. Ivarsson, J., Viano, D.C., Lovsund, P. and Aldman, B., Strain relief from the cerebral ventricles during head impact: experimental studies on natural protection of the brain, Journal of Biomechanics, 33(2):181-189, 2000.
15. Tschumperlin, D.J. and Margulies, S.S., Equibiaxial deformation-induced injury of alveolar epithelial cells in vitro, American Journal of Physiology, 275(6 Pt 1):L1173-1183, 1998.
16. Schinagl, R.M., Ting, M.K., Price, J.H. and Sah, R.L., Video microscopy to quantitate the inhomogeneous equilibrium strain within articular cartilage during confined compression, Annals of Biomedical Engineering, 24(4):500-512, 1996.
17. Bain, A.C., Shreiber, D.I. and Meaney, D.F., Modeling of microstructural kinematics during simple elongation of central nervous system tissue, Journal of Biomechanical Engineering, 125(6):798-804, 2003.
18. Zahalak, G.I., Wagenseil, J.E., Wakatsuki, T. and Elson, E.L., A cell-based constitutive relation for bio-artificial tissues, Biophysical Journal, 79(5):2369-2381, 2000.
19. Marquez, J.P., Genin, G.M., Zahalak, G.I. and Elson, E.L., Thin bio-artificial tissues in plane stress: The relationship between cell and tissue strain, and an improved constitutive model, Biophysics Journal, 88:765-777, 2005.
20. Conti, A.C., Raghupathi, R., Trojanowski, J.Q. and McIntosh, T.K., Experimental brain injury induces regionally distinct apoptosis during the acute and delayed post-traumatic period, Journal of Neuroscience, 18(15):5663-72, 1998.
21. Lusardi, T.A., Wolf, J.A., Putt, M.E., Smith, D.H. and Meaney, D.F., Effect of acute calcium influx after mechanical stretch injury in vitro on the viability of hippocampal neurons, Journal of Neurotrauma, 21(1):61-72, 2004.

CEREBRAL BRIDGING VEIN RUPTURE IN HUMANS

An Experimental Evaluation

B. Depreitere[1], C. Van Lierde[2], P. Verschueren[2], H. Delye[1], D. Berckmans[3], I. Verpoest[4], J. Vander Sloten[2], G. Van der Perre[2] and J. Goffin[1]
[1]*Division of Experimental Neurosurgery and Neuroanatomy, K.U. Leuven, Belgium*
[2]*Division of Biomechanics and Engineering Design, K.U. Leuven, Belgium*
[3]*Laboratory for Agricultural Buildings Research, K.U. Leuven, Belgium*
[4]*Department of Metallurgy and Materials Engineering, K.U. Leuven, Belgium*

Abstract. A well known cause of death and disability after head trauma is the occurrence of an acute subdural haematoma (ASDH) due to bridging vein rupture. In the past, damage to the bridging veins and ASDH has been shown to be related to angular acceleration of the head in the sagittal plane. The objective of the present study was to establish critical peak angular accelerations in short duration impacts on the head (<15ms), typical for falls and collisions. 18 tests were performed. The results complement impact data from Löwenhielm. Moreover, the tolerance criteria for bridging vein disruption suggested by Löwenhielm are confirmed for short pulse durations. The first states that the peak angular acceleration cannot exceed 4500 rad/s². The second constitutes a limitation of the change in angular velocity to 40 rad/s. If both of these conditions are fulfilled, the integrity of the bridging veins is said to be assured.

Key words: bridging vein rupture, acute subdural haematoma, head impact.

1. INTRODUCTION

A well known cause of death and disability after head trauma is the occurrence of a haemorrhage inside the cranial cavity and the subsequent formation of a haematoma **[1]**, e.g. acute subdural haematoma (ASDH). These are characterized by large volumes of clotted blood in the subdural space i.e. the space between the dura mater and the arachnoid. According to Maxeiner **[2]**, two thirds of the ASDH cases are caused by large contusions bleeding into the subdural space. The remaining third of ASDH cases are

Michael D. Gilchrist (ed.), IUTAM Proceedings on Impact Biomechanics: From Fundamental Insights to Applications, 305–312.

related to disruption of parasagittal superficial cerebral veins bridging the subdural space to reach the venous sinuses. Damage to the bridging veins in ASDH has been shown to result from sudden head movement in the instant after head loading. The bridging veins that originate from the brain surface run anteriorly towards the superior sagittal sinus with angles ranging from 10 to 85° (Figure 1). The superior sagittal sinus is firmly attached to the skull. Consequently, when the brain lags behind the skull following an occipital impact on the head, a longitudinal strain will be produced in the veins and this strain can lead to rupture **[3--5]**.

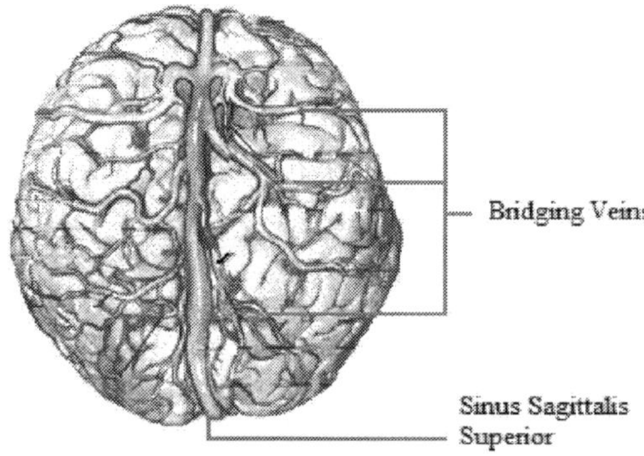

Figure 1. Top view of the brain showing the course of the bridging veins.

Results from Finite Element Analysis by both Lee **[4]** and Huang **[5]** indicate that a sudden pure rotation of the brain causes the highest strains in the bridging veins. Given the importance of angular acceleration with regards to the production of ASDH, it is likely that the occurrence or non-occurrence of the lesion can be expressed in terms of a critical value for this physical parameter.

2. MATERIALS AND METHODS

The principle of the tests was to subject the heads of intact cadavers to a rotational acceleration in the sagittal plane and forward direction induced by an occipital head impact and to subsequently investigate the integrity of the bridging veins.

Based on a technique for the detection of bridging vein leakage at autopsy described by Maxeiner **[2]**, a method was developed to check the integrity of the bridging veins using radioscopy. After creating a small bone flap in the skull near the external occipital protuberance, the method involved introducing a balloon catheter into the superior sagittal sinus and fixing it into its position by inflating its proximal end. Through this tube a radio opaque contrast fluid was injected into the venous canal and the bridging veins at low pressure while performing real time imaging. In cases where extravasation was noticed, rupture had occurred (Figure 2).

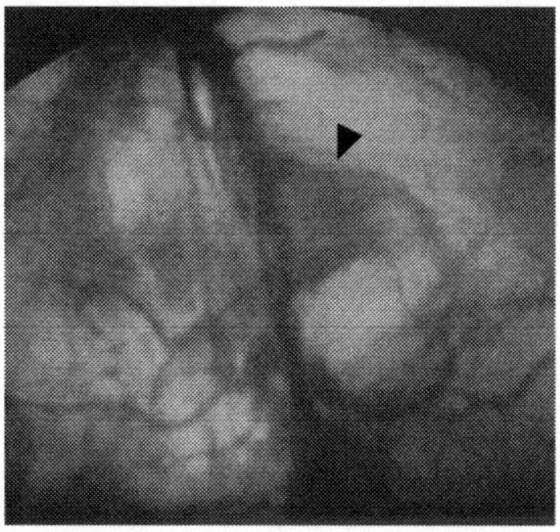

Figure 2. Post-impact radioscopic image of the head, showing leakage of contrast fluid in the subdural space, indicative of a ruptured bridging vein.

Ten intact unembalmed cadavers were obtained. The mean age and the mean interval between decease and experiment were 79.2 years and 3.1 days respectively. Each specimen was fitted with a 16 cm curved aluminum profile screwed on the left side of the head behind the ear on a vertical line projecting through the lower cervical vertebrae (Figure 3-A). Three uniaxial accelerometers were mounted on this and the distances between the accelerometers were recorded. Due to the light-weight profile, no significant interference of the head impact response was assumed. They were positioned in such a way that their axis was in the sagittal plane and perpendicular to the profile. A small craniectomy was performed immediately above the external occipital protuberance and the superior sagittal sinus was opened and canulated with a 10 French tube to make rupture detection.

The impacts were delivered by a steel pendulum of 148 cm length and weighing 14.3 kg, mounted over a steel bar (Figure 3-B). The impact energy could be varied by altering the pendulum release angle or by adding/removing mass by means of steel weights. The cadaver was placed in front of the pendulum in an upright sitting position with the back of the cadaver facing the pendulum and the head sagittal plane in line with the plane of motion of the pendulum. The head was held in position by means of a string attached to the nose that did not impede the forward rotation of the head. A sandbag was put on the body to minimize the movement of the trunk, and thus maximize the portion of the energy input that was transformed into head rotational movement. The lower end of the pendulum stayed above the cadaver's shoulders. The impactor was a steel block with a surface of 8 x 22 cm and a thickness of 8 cm attached to the front side of the pendulum that was fitted with a polystyrene padding to prevent contact injuries. Moreover, by altering the thickness and density of the padding the acceleration pulse duration could be varied. Prior to each impact 10 ml saline was injected into the superior sagittal sinus to create a physiological pressure inside the sinus of 6 – 8 mmHg. For the present series of tests, the pendulum release angle was varied between 30° and 180° and the total mass of the pendulum between 43.4 and 63.0 kg.

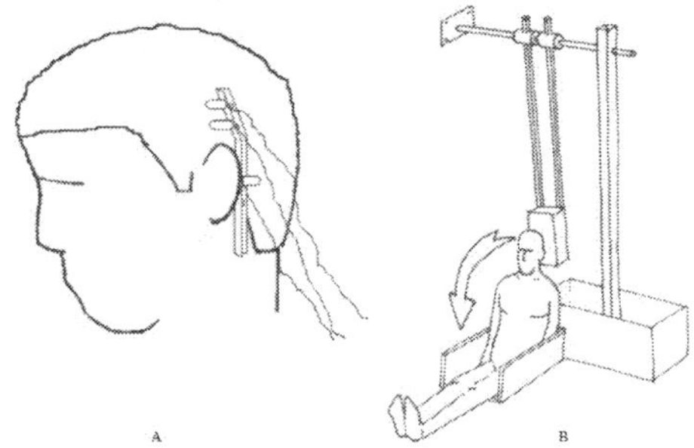

Figure 3. A) Aluminum profile with three uniaxial accelerometers screwed on the left side of the head. B) Rotational acceleration of the head in the sagittal plane induced by an occipital impact delivered by the pendulum.

Testpoint® and Matlab™ software were used for data acquisition and manipulation. The pulse duration was estimated from the linear acceleration signals that were filtered at 250 Hz, an experimentally established value to remove signal noise. Each set of two uniaxial accelerometers enabled the calculation of a peak rotational acceleration from the difference in peak linear acceleration and from their relative distance. Using the combinations of the uppermost and lowest and of the middle and lowest accelerometers, two peak rotational accelerations were obtained, of which the mean was calculated. The linear acceleration measured by the middle uniaxial accelerometer was considered an estimation of the tangential acceleration of the head's center of mass. Data is saved to be used for future FE modeling.

In order to compare the peak rotational accelerations obtained from the impact tests with those in a simple backward fall of a standing person, a supplementary test was performed in which a cadaver was tied to a shelf from the legs to the shoulders with the head freely mobile and this assembly was allowed to fall backwards on a tiled floor. Three uniaxial accelerometers were mounted on the head as described above. The magnitude and duration of the rotational acceleration of the head in the sagittal plane were calculated. The test was performed twice.

3. RESULTS

A total number of 18 impact tests were performed in 10 cadavers. The linear acceleration signals showed a typical pattern in 10 of the 18 tests, characterized by an acceleration phase in which the acceleration magnitude was higher in accordance with the higher position of the accelerometer and a deceleration phase in which the absolute value of the acceleration magnitude was the highest for the lowest accelerometer (Figure 4).

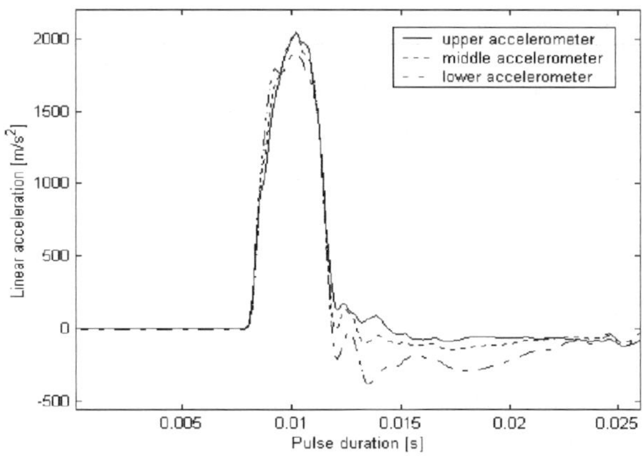

Figure 4. Linear acceleration histories obtained from impact test.

The determination of the linear and rotational accelerations failed in 3 of the 18 tests due to artifactual measurements. A bridging vein rupture was produced in 6 tests. The autopsy findings corresponded well with the radioscopy results. When a bridging vein rupture was detected at radioscopy, a collection of barium sulfate was found in the subdural space at the side of the rupture. The rupture itself could be detected at autopsy in 5 of the 6 cases. A bilateral rupture was produced in one test. The bridging vein ruptures invariably occurred in the rolandic or post-rolandic regions (frontal and parietal lobe). The two backward falls resulted in a peak rotational acceleration of the head of 9,970 and 12,314 rad/s^2 with a pulse duration of 4.8 and 4.9 ms respectively. The data is summarized in Figure 5.

4. DISCUSSION

It has been clearly demonstrated in animal experiments and finite element models that bridging vein ruptures are preferentially caused by a forward head motion in the sagittal plane and that the occurrence of ruptures is related to the rotational acceleration of the head **[5, 6]**. While bridging vein ruptures could theoretically also be provoked by purely translational head motions, it has been calculated in finite element analyses that rotational acceleration contributes approximately 2.5 times as much to the deformation of the forward-draining bridging veins as does translational acceleration **[7]**.

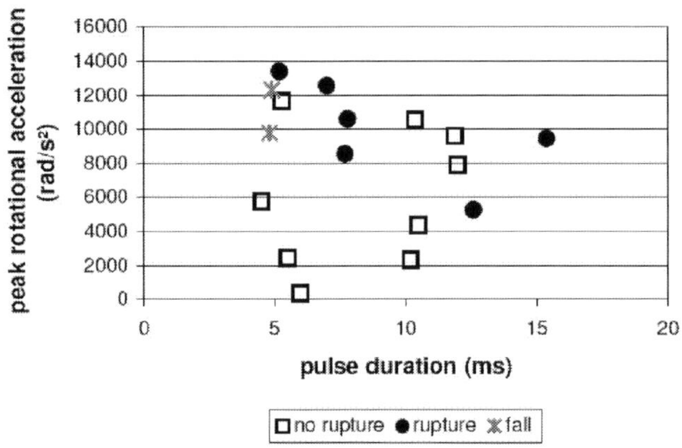

Figure 5. Peak rotational acceleration magnitude and pulse duration of the tests bridging vein rupture, the tests not leading to rupture and the two backward falls.

In the present study critical levels of rotational acceleration magnitude and pulse duration were investigated by means of impact tests in 10 human cadavers. From the results this tolerance level can be estimated to lie around 10,000 rad/s² for pulse durations below 10 ms and it seems to decrease with increasing pulse duration (Figure 5). To the best of our knowledge only one previous study has examined tolerance levels for bridging vein rupture in human cadavers **[8]**. In this study, cadavers were positioned in a car interior, containing a car seat, dashboard and windshield and accelerated backwards on a test sled. When plotted together, Löwenhielm's data is compatible with the results of the present study and could constitute a further extension of the tolerance curve for longer pulse durations (Figure 6). Based on his results, Löwenhielm proposed a critical level of rotational acceleration magnitude of 4,500 rad/s² in combination with a critical peak change in rotational velocity of 50 rad/s. In a later paper on the mathematical modeling of a single patient case of bridging vein rupture, Löwenhielm suggested to lower the critical level of the peak change in rotational velocity to 30 rad/s **[9]**.

Assuming a sinusoidal acceleration pulse, the pulse duration (t) can be calculated from the peak rotational acceleration (a_p) and the peak change in rotational velocity (v_p) which enables the plotting of Löwenhielm's tolerance levels for bridging vein rupture in Figure 6. It can be seen that all data-points associated with bridging vein rupture fall within the area defined by a curve denoting a 40 rad/s level for peak change in rotational velocity and the curve denoting the 4,500 rad/s² level for peak rotational acceleration. The present results and the tolerance levels that were proposed by Löwenhielm indicate a decrease of the critical level of rotational acceleration magnitude with increasing pulse durations. Such a course of the tolerance curve for bridging vein rupture was also suggested by Lee and Haut **[4]**.

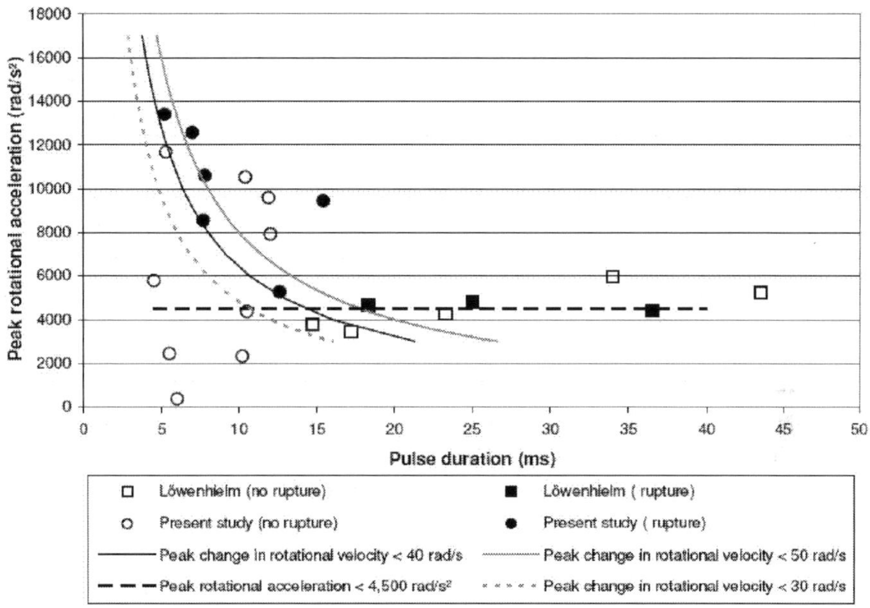

Figure 6. Peak rotational acceleration and pulse duration data of the present study and of the study by Löwenhielm on bridging vein rupture **[8]**.

The present study has some limitations that need to be addressed. The mean age of the cadavers was 79.2 years and it is likely that rotational acceleration tolerance levels are higher for younger people. Moreover, cadaver studies inevitably are an approximation of the reality. The absence of physiological intravascular and intradural pressures probably affects the relative skull-brain motion and the bridging vein tolerance. For this reason, some authors have advocated re-pressurization procedures **[10]**. Apart from the administration of a small volume of saline to mimic a physiological pressure inside the sagittal sinus immediately prior to impact, the techniques for arterial repressurization has described by these authors were not applied in the present study because the high volumes of fluid were considered incompatible with the use of intravascular contrast for radioscopy. The effect of autolysis on the vein material properties has been reported as minimal in the first five days past decease **[10]**. Invariably, the ruptured bridging veins in the present study were located in the rolandic or post-rolandic regions. These veins form angles with the superior sagittal sinus in the sagittal plane that are reported to vary between 40 and 50°. This is in close agreement with the calculations by Huang et al. in a three-dimensional finite element model of the human head that the forward draining bridging veins at angles of 50° incurred the greatest strain during occipital impact and hence, should be considered the most vulnerable **[5]**.

Although the finite element modeling efforts have been valuable in gaining insight into the mechanical pathogenesis of bridging vein rupture, they have produced tolerance levels of rotational acceleration magnitude for the human head that are much higher than the values obtained in the human cadaver experiments by Löwenhielm **[5-8]**. It is our opinion that cadaver tests, though characterized by inherent limitations as well, shall remain indispensable for the validation of mathematical models.

The rotational acceleration magnitudes in the backward fall simulations were in the vicinity of the magnitudes critical for bridging vein rupture, while it is indeed known that bridging vein ruptures may result from simple falls in the elderly **[6]**.

In conclusion, the results of a series of impact tests on 10 human cadavers with acceleration pulse durations compatible with those seen in falls support the combination of two tolerance levels, as adapted from the criteria originally proposed by Löwenhielm **[8]**: a peak change in rotational velocity of 40 rad/s and a peak rotational acceleration of 4,500 rad/s^2.

REFERENCES

1. Miller J.D., Piper I.R., Jones P.A., Pathophysiology of head injury, chapter 5 in Narayan R.K., Wilberger J.E., Povlishock J.T., *Neurotrauma*, McGraw-Hill, 1995: 61-70.
2. Maxeiner H., Detection of ruptured cerebral bridging veins at autopsy, *Forensic Science International*, 1997, 89: 103-110.
3. Bradshaw D.R.S., Linear wave propagation in traumatic brain, Doctoral Thesis Faculty of Engineering and Applied Science, Department of Electronics and Computer Science & Institute of Sound and Vibration Research, University of Southampton, 2001.
4. Lee M-C, Haut R.C., Insensitivity of tensile failure properties of human bridging veins to strain rate: implications in biomechanics of subdural hematoma, *J Biomechanics*, 1989, **22(6/7)**: 537-542.
5. Huang H-M., Lee M-C., Chiu W-T., Chen C-T., Lee S-Y., Three-dimensional finite element anaysis of subdural hematoma, *The Journal of Trauma*: Injury, Infection and Critical Care, 1999, **47(3)**: 538-544.
6. Gennarelli T.A., Thibault L.E., Biomechanics of Acute Subdural Hematoma, *The Journal of Trauma*, 1982, 22(8): 680-686.
7. Lee MC, Melvin JW, Ueno K. Finite element analysis of traumatic subdural hematoma. Paper presented at: *31st Stapp Car Crash Conference*, 1987, New Orleans (Louisiana).
8. Löwenhielm P., Strain tolerance of the Vv. Cerebri sup. (Bridging veins) calculated from head-on collision tests with cadavers, *Z. Rechtsmedizin*, 1974, **75**: 131-144
9. Löwenhielm P., Tolerance level for bridging veins disruption calculated with a mathematical model, *Journal of Bioengineering*, 1978, **2**: 501-507
10. Got C, Patel A, Fayon A, Tarrière C, Walfisch G. Results of experimental head impacts on cadavers. Presented at: *22nd Stapp Car Crash Conference*, 1978, Ann Arbor (Michigan).

THREE DIMENSIONAL PASSIVE PROPERTIES OF MUSCLE TISSUE IN COMPRESSION

Melanie Van Loocke, Ciaran Simms and Garrett Lyons
Centre for Bioengineering, Department of Mechanical Engineering, Trinity College, Dublin 2, Ireland, + 353-1-608 2978, + 353-1-679 5554, mvanlooc@tcd.ie

Abstract. Uniaxial unconstrained compression experiments have been conducted on porcine, bovine and ovine muscle samples oriented in the fibre, non-fibre and 45° from the fibre direction. The porcine and bovine samples were aged tissues. The ovine samples were fresh tissues. A transversely isotropic hyperelastic model and a model which uses the theory of transverse isotropy and strain dependent Young's moduli have been fitted to the experimental data. The results show that the hyperelastic model does not adequately fit the data in all three directions of testing. In contrast, the strain dependent Young's moduli model (SYM) gives a good fit to the experimental data in both the fibre and non-fibre direction up to 30% strain for porcine and bovine samples. The model also yields a very good prediction of the behaviour of the muscle at 45° from the fibre direction. However, preliminary data from the fresh ovine samples indicates that the SYM may not be a suitable model for description of the material properties of very fresh tissue.

Key words: skeletal muscle, compression, 3-D model.

1. INTRODUCTION

Improvements in computational power led impact biomechanics to focus more and more on virtual modelling. Finite element models of the human body can be used to predict deformations during transient loading. These models require both a description of the complex geometries of hard and soft tissues as well as the determination of their material properties under large deformations, as appearing during impacts. Very accurate geometries can be reconstructed by using medical imaging techniques. The material properties

Michael D. Gilchrist (ed.), IUTAM Proceedings on Impact Biomechanics: From Fundamental Insights to Applications, 313–320.

of bone are well documented; however biological soft tissues remain poorly classified. In particular, muscle tissue presents difficulties due to its active and passive components, non-linearity, viscoelasticity and anisotropy. Muscle tissue is subject to large deformations and is thus not represented by classical linear elasticity. Furthermore, parameters such as age, gender and species are also influential.

Three dimensional constitutive models of nonlinear anisotropic tissue exist [1–3]. Some of these have been used to model passive muscle behaviour [1] and [4]. They contain material constants which must be determined by statistical regression of test data. However, most experimental studies have considered deformation in one dimension only and significant variability between data sets is present as can be observed in Figure 1 which presents passive force-length curves obtained by tensile tests conducted variously on rat, cat or rabbit muscles [5–10]. The utility of the existing models is therefore difficult to classify.

This paper presents results of an experimental procedure to determine the quasi-static behaviour of skeletal muscle in three dimensions. Particular attention is placed on the behaviour of the muscles under compression, as this is the relevant deformation mode for impact injuries. The influence of experimental variables such as temperature and time of death is investigated and a new three-dimensional model is proposed to characterise passive muscle properties. The time-dependent muscle properties will also be investigated in future.

2. MATERIALS AND METHODS

2.1 Testing

Uniaxial unconstrained compression experiments have been conducted on porcine, bovine and ovine muscle samples. The porcine and bovine samples were obtained from a butcher and were therefore bled and hung previously (aged samples). The ovine samples were obtained immediately after death from the thoracic limbs of unbled sheep (fresh samples). This enables control of age, gender and time of death of the subject as well as method of storage of the tissue.

The compression tests, performed on a uniaxial test machine, were *ramp and hold* functions applied to cubic samples with characteristic length included between 5 and 10mm. The samples were compressed up to 30% at a strain rate of $0.05\%s^{-1}$ in order to capture only the quasi-static behaviour of the tissue. Tests were performed in three different orientations of the muscle:

the fibre, non-fibre and 45° from the fibre direction. In order to obtain samples with correct and consistent geometry and orientation, a cutting bench was developed (Figure 2). This bench is equipped with a cutting blade mounted perpendicular to a rotary table, on which samples are held in place by superficial freezing. Parallel cuts spaced as desired may be made at different orientation from the fibre direction.

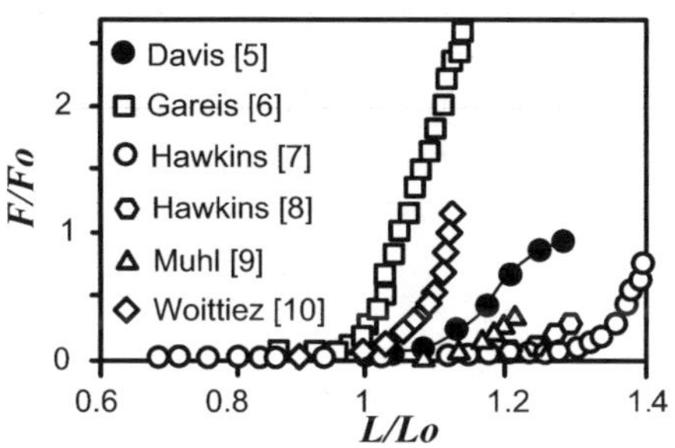

Figure 1. Comparison of published passive muscle characteristics. F_0 is maximum isometric force generated by muscle; L_0 is rest length of muscle.

Figure 2. Cutting bench. (A) blade, (B) rotary table.

The effect of temperature has been investigated by testing aged ovine samples at temperatures ranging from 3 to 36°C. Samples, kept at 3, 9, 18 and 36°C respectively, were compressed up to 10% strain. The different tests were separated by an hour to allow for temperature stabilization.

The effect of time of test after death has also been investigated by performing repeated non-destructive tests on fresh ovine tissue at regular intervals after death of the animal. Samples, kept at room temperature, were compressed up to 10% strain 2, 2.5, 3.5, 4.5, 5.5, 6.5, 7.5 and 8.5 hours after death of the animal.

2.2 Modelling

Amongst the three-dimensional constitutive models developed to represent non-linear anisotropic soft tissue behaviour, the applicability of two models to fit the experimental data were tested: a transversely isotropic hyperelastic model developed by Humphrey et al. [1] and a variant of a model by Li et al. [3] which uses the theory of transverse isotropy and strain dependent Young's moduli.

In Humphrey's model, developed to represent the behaviour of passive cardiac muscle, muscle tissue is considered as made of a matrix in which fibres are embedded. The Cauchy stress is derived from a strain-energy density function and is given by:

$$\sigma = -\,pI + 2W_1 B + \frac{W_\alpha}{\alpha} FN \otimes NF^T \tag{1}$$

p is a Lagrange multiplier enforcing incompressibility, I is the identity tensor, F is the deformation gradient tensor, $B = FF^T$, $W_1=\delta W/\delta I_1$, $W_\alpha=\delta W/\delta\alpha$, W is the strain-energy function and a is a unit vector defining muscle fibre direction in an undeformed configuration. The second term of (1) represents the matrix contribution and the third term represents the fibre contribution. W contains five parameters:

$$\begin{aligned} W(I_1,\alpha) &= c_1(\alpha-1)^2 + c_2(\alpha-1)^3 + c_3(I_1-3) \\ &\quad + c_4(I_1-3)(\alpha-1) + c_5(I_1-3)^2 \end{aligned} \tag{2}$$

These parameters are determined by fitting of the model to experimental stress-strain curves using a least square regression algorithm. Parameters c_3, c_4 and c_5 are determined by fitting of data obtained by compression of samples in the non-fibre direction. Parameters c_1 and c_2 are then determined by curve fitting of data obtained by compression of samples in the fibre direction.

The second type of model, developed to represent nonlinear transverse isotropy in porcine aortic heart valves [3], is based on classical linear elasticity extended with strain dependent Young's moduli. A transversely elastic material possesses five independent elastic constants: two moduli of extension (E_L, E_T) in the directions of isotropy, one shear modulus (G_{LT}), two Poisson's ratio (ν_{LT}, $\nu_{TT'}$). The Cauchy stress is given by:

$$\sigma = E * \varepsilon \tag{3}$$

where E is the tensor of elastic constants. To account for nonlinearity, the elastic constants are strain dependent. The Young's moduli in the fibre (longitudinal) and the non-fibre (transverse) directions take the form:

$$E_L = k_1\, exp(k_2\varepsilon) \;\; and \;\; E_T = k_3\, exp(k_4\varepsilon) \tag{4}$$

Parameters k_1 and k_2 are determined by fitting of the model to experimental data in the fibre direction; k_3 and k_4 are determined by fitting of the model to data in the non-fibre direction.

3. RESULTS

3.1 Aged Tissue

Figure 3 presents experimental stress-strain curves obtained by compression of aged porcine samples oriented in the fibre, non-fibre and 45° from the fibre direction. The cross-fibre direction is stiffer than the 45° direction, which in turn is stiffer than the fibre direction. Theoretical curves obtained by fitting of data with the hyperelastic model are also represented. The parameters of the model are determined by successive fitting of data in the cross-fibre and fibre directions. Data obtained at 45° from the fibre direction are then used to check the predictive capabilities of the model. The results show that the hyperelastic model does not adequately fit the data in all three directions of testing: good fit is obtained in the cross-fibre direction up to 30%. However the model diverges from experimental data in the fibre-direction after 20% strain and can not predict at all the behaviour of the muscle at 45° from the fibre direction.

Figure 4 presents the same experimental results fitted with the strain dependent Young's moduli model (SYM). Again, parameters are determined by fitting of data in cross-fibre and fibre directions. The SYM gives a good fit to the experimental data in both the fibre and non-fibre direction up to 30% strain. The model also yields a very good prediction of the behaviour of the muscle at 45° from the fibre direction. Similar results are obtained for aged bovine samples.

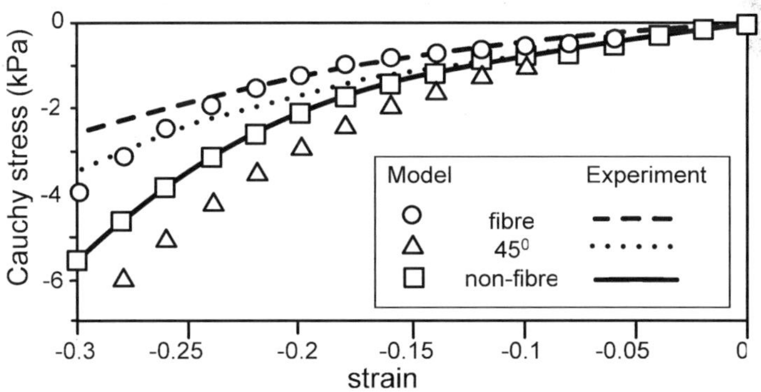

Figure 3. Compression tests on porcine samples fitted with Humphrey model. Lines denote experimental data, symbols denote mathematical models.

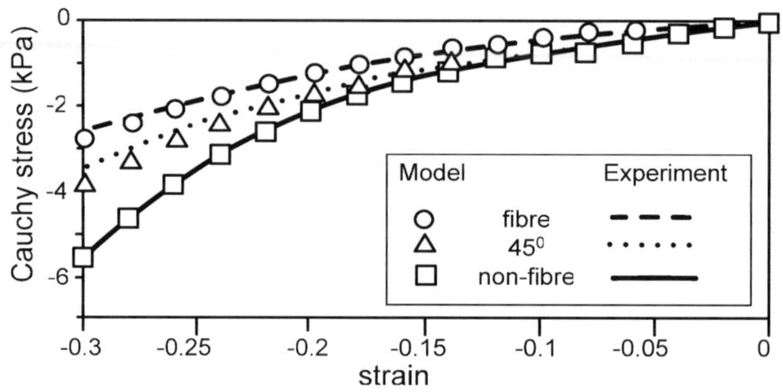

Figure 4. Compression tests on porcine samples fitted with fitted with general strain dependent elastic model. Lines denote experimental data, symbols denote mathematical models.

3.2 Fresh Tissue

Preliminary data from the fresh ovine samples show a different behaviour of the tissue under compression. As illustrated on Figure 5, the cross-fibre direction is still stiffer than the 45° and fibre directions. However, the fibre direction is this time stiffer than the 45° direction.

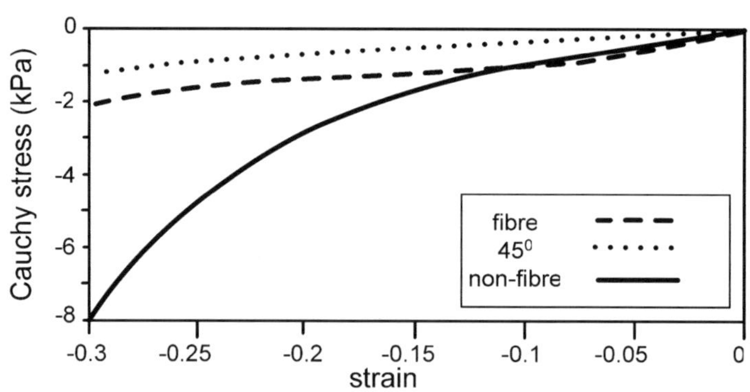

Figure 5. Compression tests on fresh ovine samples.

Furthermore, for the ovine samples, important stiffening of muscle behaviour (rigor mortis) is noted a few hours after death of the subject: the maximum compressive stress measured at 10% strain is increased approximately eight-fold 8.5 hours after death (Figure 6). Stiffening is also noted at low temperatures: the maximum compressive stress measured at 10% strain is doubled from 18 to 9°C as illustrated on Figure 7.

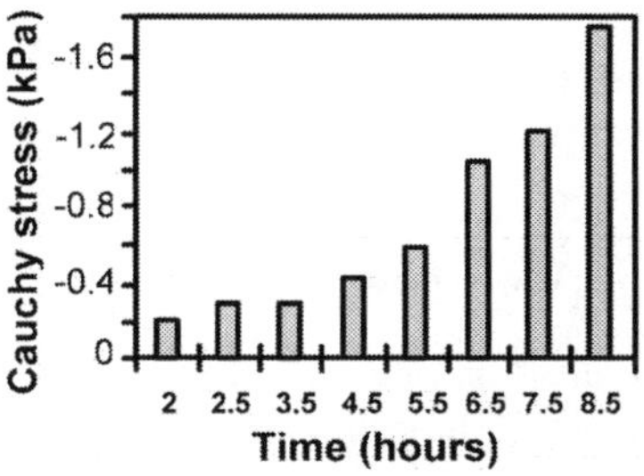

Figure 6. Evolution of the maximum compressive stress with time after death for fresh ovine samples.

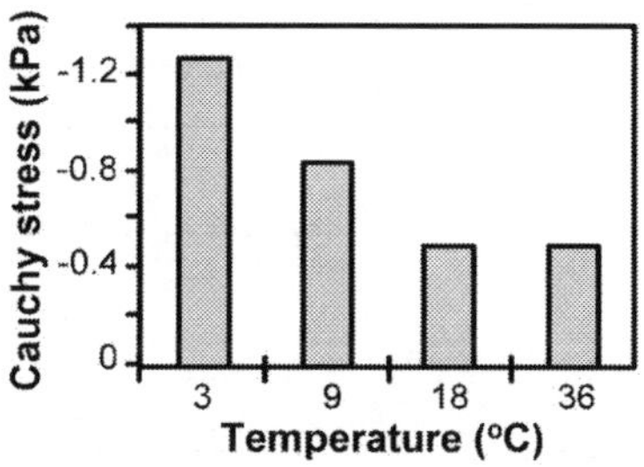

Figure 7. Evolution of the maximum compressive stress with temperature for aged ovine samples.

4. DISCUSSION

Compression tests conducted on mammalian skeletal muscles in three different orientations clearly demonstrate the non-isotropy of muscle tissue. A difference in behaviour is however observed between aged and fresh tissue. Aged bovine and porcine tissue show an increase in stiffness with an increase of angle from the fibre direction. This behaviour is adequately represented using the SYM. However, preliminary data from the fresh ovine samples shows that the weaker direction is at 45° from the fibre direction. This might be due to shearing appearing between adjacent fibre fascicles during compression. Further tests need to be performed to confirm these results.

Experiments conducted on fresh animal tissue also show that extreme care must be taken with the control of experimental variables such as temperature and time of death, which influence greatly muscle behaviour. Tissues must be tested within a few hours after death of the subject in order to obtain data representative of living tissue properties.

Buckling of muscle fibres can be observed in samples compressed in the fibre direction. To diminish the effects of this phenomenon, samples with a ratio L/D (height/transverse dimension) approximating ½ are used.

In most studies conducted to characterise skeletal muscle, only animal tissue has been tested. These data are used to develop models of human skeletal muscles but no comparison has been established between animal and human tissue. One of the aims of our work is to obtain data on human muscle tissue.

Furthermore, only the quasi-static behaviour of these tissues has been investigated. However, in order to fully characterize the mechanical behaviour of muscle its visco-elastic properties must also be determined. This is part of the future work that will be under taken in the project.

5. CONCLUSION

Experiments conducted on aged animal muscles show that the SYM represents adequately the passive quasi-static behaviour of muscle tissue submitted to compression. However, further tests need to be performed to assess the suitability of the SYM for the description of very fresh tissue properties.

ACKNOWLEDGEMENTS

This project is funded by the Programme for Research in Third Level Institutions (PRTLI), administered by the HEA.

REFERENCES

1. Humphrey, J.D., Strumpf, R.K. and Yin, F.C.P., Determination of a constitutive relation for passive myocardium, *Journal of Biomechanical Engineering*, **112**, 1990, 333–346.
2. Weiss, J.A., A constitutive model and finite element representation for transversely isotropic soft tissues, PhD Thesis, The University of Utah, 1994.
3. Li, J., Luo, X.J. and Kuang, Z.B., A nonlinear anisotropic model for porcine aortic heart valves, *Journal of Biomechanics*, **34**, 2001, 1279–1289.
4. Huyghe, J.M., Van Campen, D.H., Arts, T. and Heethaars, R.M., The constitutive behaviour of passive heart muscle tissue: a quasi-linear viscoelastic formulation, *Journal of Biomechanics*, **24**, 1991, 841–849.
5. Davis, J., Kaufman, K.R. and Lieber, R.L., Correlation between active and passive isometric force and intramuscular pressure in the isolated rabbit tibialis anterior muscle, *Journal of Biomechanics*, **36**, 2003, 505–512.
6. Gareis, H., Solomonow, M., Baratta, R., Best, R. and D'Ambrosia, R., The isometric length-force properties of nine different skeletal muscles, *Journal of Biomechanics*, **25**, 1992, 903–916.
7. Hawkins, D. and Bey, M., A comprehensive approach for studying muscle-tendon mechanic. *Journal of Biomechanical Engineering*, **116**, 1994, 51–55.
8. Hawkins, D. and Bey, M., Muscle and tendon force-length properties and their interactions in vivo, *Journal of Biomechanics*, **30**, 1997, 63–70.
9. Muhl, Z.E., Active length-tension relation and the effect of muscle pinnation on fibre lengthening, *Journal of Morphology*, **173**, 1982, 285–292.
10. Woittiez, R.D., Huijing, P.A., Boom, H.B.K. and Rozendal, R.H., A three-dimensional muscle model: a quantified relation between form and function of skeletal muscles, *Journal of Morphology*, **182**, 1984, 95–113.

A NATIONAL REVIEW OF SURGICALLY TREATED BLUNT TRAUMATIC THORACIC AORTIC TRANSECTIONS

Peter A. Naughton[1], Lars Nölke[1], Collette Shaw[1], David G. Healy[1], Aongus O'Donnell[2] and Alfred E. Wood[1]

[1]Prof. Eoin O'Malley, National Center for Cardiothoracic Surgery, Mater Misericordiae University Hospital, Eccles St, Dublin 9, Ireland; [2]Department of Cardiothoracic Surgery, Cork University Hospital, Wilton, Cork, Ireland; Tel: 00-353-1-8032164, Fax: 00-353-1-8034773, Email: freddie@woodcts.iol.ie

Abstract. Blunt traumatic transection of the aorta (BTTA) is associated with a very high mortality rate. Most victims die at the time of the injury. A review of surgically treated BTTA in Ireland between 1989 and 2002 was performed. The accident files of all road traffic deaths during 2001 and incidence data on national road traffic deaths for the study period were also reviewed. There were 30 acute and 9 chronic BTTA. Of the 30 acute transections, 29 were due to road traffic accidents (RTA) and one a crush injury. Major associated injuries included head injuries (7), fractures (13) with paraplegia in 2 and abdominal injuries requiring surgery in 4. There were 3 (10%) hospital deaths (2 arrested on induction) and no cases of postoperative paraplegia. In 2001, 18.4% of RTA victims had a BTTA on post-mortem and 8.7% of these survived to hospitalisation. 60% of BTTA victims surviving to hospitalisation have undergone successful repair. BTTA must be considered following blunt trauma and should be out-ruled by present scanning techniques.

Key words: trauma, aortic transections, road traffic accidents.

1. INTRODUCTION

Blunt traumatic transection of the thoracic aorta (BTTA) is the second most common cause of death following road traffic accidents (RTA) and carries an 85-95% instantaneous risk of death [1, 2]. Patients who survive to

Michael D. Gilchrist (ed.), IUTAM Proceedings on Impact Biomechanics: From Fundamental Insights to Applications, 321–328.

hospitalisation, usually require management of multiple injuries and prompt intervention, as more than half of the early survivors will die in the first 24 hours and 75% in the first week [3]. Approximately one percent of patients admitted with signs of blunt chest trauma will have an aortic injury [4].

In order to develop a better understanding of the management of blunt traumatic injuries of the thoracic aorta in Ireland, we reviewed the medical and transportation records of all patients treated surgically for blunt traumatic transection of the aorta over a 13-year period (1989-2002). In addition we reviewed the incidence of road traffic deaths in Ireland (1989-2002) and the accident files and post-mortem results of all road traffic deaths for a 1year period (2001) to ascertain the true incidence of traumatic aortic transection related road traffic deaths in Ireland.

2. PATIENTS AND METHODS

This is a retrospective study of all patients treated surgically for BTTA in the cardiothoracic departments of the Republic of Ireland between the 1st of January 1989-31st of December 2002. The Roads Safety and Research section of National Roads Authority (www.nra.ie) and the Environmental Research Unit reports were reviewed to obtain data on annual road traffic accidents and deaths. The Garda Síochána (Irelands National Police Service) National Traffic Bureau supplied detailed information (accident reports and post mortem reports) on all reported road traffic deaths in 2001.

3. RESULTS

3.1 Patients

Thirty patients with acute traumatic aortic transection were treated surgically in Ireland during the 14 year period. A further 9 patients presented with pseudoaneurysms of the aorta, due to chronic (more than 3 months after the trauma) blunt traumatic transection of the aorta. Twenty-nine of the acute presentations occurred as a result of RTA and one was sustained from a crush injury. Automobile accidents with frontal impacts occurred in 24 and side impacts in 2, and the other was not known. Two of the victims were ejected from their vehicle and only 10 were known to have been wearing seat belts. The driver of the vehicle died in three cases. Motorcycle accidents resulted in two victims. Both of these had multiple associated injuries including head injuries, long bone and pelvic fractures. Overall the mean age was 32.8 years, range 18-76 years and 24 (80%) were males. Additional injuries occurred in 27 patients and are summarized in Table 1. Five patients required surgical intervention for non-cardiac issues prior to repair of their aortic injury. Transection occurred at the aortic isthmus (junction of the aortic arch and the descending aorta and the insertion of the ligamentum arteriosum) in 28 patients. Two patients sustained a contained avulsion of the brachiocephalic artery from the arch of the aorta. Twenty-six patients were initially managed by trauma / orthopaedic / general surgeons and then were transferred for conformation of diagnosis and surgical management to a cardiothoracic centre.

Table 1. Co-morbidities found at presentation with Blunt traumatic transection of the aorta.

Associated surgical injuries		No.	%	
Orthopaedic	Fractures	13	43.3%	
	Sternal	2	*6.7%*	
	Limb	5	*16.7%*	
	Spinal	3	*10%*	Paraplegia- 2
	Rib	2	*6.7%*	
	Pelvis	1	*3.3%*	
Cardiothoracic	Pneumothorax	1	*3.3%*	
	Haemothorax	1	*3.3%*	
Neurological	Initial decreased Glasgow Coma Scale	7	*23.3%*	1 case required a burr hole pre BTTA repair
	Eye enucleation	1	*3.3%*	
	Intracerebral bleed requiring intervention	1	*3.3%*	
Gastrointestinal	Liver laceration	1	*3.3%*	4 cases required a laparotomy pre BTTA repair
	Splenic rupture	1	*3.3%*	
	Jejenal laceration	1	*3.3%*	
	Ruptured diaphragm	2	*6.7%*	

3.2 Diagnosis and Radiological Findings

All patients had chest x-rays performed on admission of which 2 were normal. The most consistent finding was widening of the mediastinum in 26 (86.7%) patients. Other abnormal findings on chest x-ray included pleural capping in 8 (26.7%) patients, haemothorax in 2 (6.7%) patients and deviated bronchus in 4 (13.3%) patients. Twenty-four patients underwent contrast enhanced CT of the thorax to further assess widening of the mediastinum. In 8 patients it was possible to make a definitive diagnosis and these patients proceeded to immediate surgery. This has especially been the case in recent years with the introduction of high-resolution computed tomographic angiography. Twenty-two patients had an arch aortogram to confirm the diagnosis, 16 of these patients had a previous contrast CT.

3.3 Mortality

Three (10%) patients died in hospital. One patient arrested prior to anaesthetic induction. The second mortality was a 33 year old male motorcyclist, with transection of his aortic isthmus, long bone fractures and a head injury with an initial diminished Glasgow Coma Scale. He had progressive brain injury and was diagnosed with brainstem death. The third mortality was a 68 year man with known ischemic heart disease and preoperative paraplegia due to a third thoracic vertebrae fracture and spinal cord transection. He died in the intensive care 5 hours post-operatively secondary to haemorrhage.

3.4 Morbidity

Recurrent laryngeal nerve injury resulted in hoarseness in 3 (10%) patients. One patient who sustained multiple long bone fractures developed a pulmonary embolus on the 3rd day post-operatively. There was no case of paraplegia secondary to operative repair. A further patient with a preoperative head injury has a residual partial hemiplegia and can mobilize with a walking aid.

3.5 RTA Data

During the 13 year period January 1989 – 31st December 2001, 5,692 people died and 144,839 were reported injured in Ireland as a result of road traffic accidents. There have been 39 victims with acute and chronic BTTA treated surgically during this time giving an overall incidence of 27 per 100,000 RTA victims treated successfully for a BTTA.

During 2001 there were 410 reported road traffic deaths in Ireland. The accident reports and the post mortem results of these cases were individually reviewed. Of the 341 complete files, 290 road traffic victims died at the scene or were dead on arrival into hospital. 53 (18.3%) of these deaths had traumatic aortic transection documented on their post-mortem. 51 road traffic victims survived to hospitalisation and 29 of these underwent a post-mortem examination. Traumatic aortic transection was document in two or

6.8% of survivors to hospitalisation and an additional three patients were operated upon in 2001 for BTTA. Therefore during 2001 there were 58 documented traumatic transections of the aorta, 14.8 per one million population. There were 10,633 reported hospital admissions for road traffic injuries during 2001, giving an incidence of 54.5 BTTA injuries per 10,000 reported road traffic accident victims with minor or major injuries. 91.3% (53) BTTA victims died at the scene or before hospitalisation, 3.5% (2) died within hours of admission with no intervention or diagnosis and 3 (5.2%) survived surgical intervention.

4. DISCUSSION

This report reviews all surgically treated cases of aortic transection in the Republic of Ireland over a 13 year period. As reported in other series, this is an uncommon injury associated with a very high mortality rate. This injury occurred in 16 per million of the population which is similar to the rate of 13-30 BTTA victims per 1 million population reported in other series [2, 5, 6]. BTTA was found in 18.3% of road traffic deaths in Ireland, which is consistent with the 12-29% reported by autopsy series [1, 7-9]. Although we labelled BTTA as the main cause of death, it has been estimated that at least half of the victims would have died from their other injuries [10].

This national series allows assessment of risk of rupture of a BTTA, but it does not take into account aortic transection victims as result of trauma other than road traffic accidents. It also does not include victims who died more than 30 days after the trauma, as these victims are not included in the national road traffic data and they may not have had a post mortem examination. It also cannot include those with a BTTA injury who have survived the injury and the diagnosis was missed, who live for years and die of other causes.

The mechanism of aortic transection remains uncertain, but is likely to be due to a number of anatomical and mechanical factors. Anatomically, aortic rupture classically (95% of clinical cases) occurs at the aortic isthmus [9, 11]. This may be due to the relative mobility of the ascending aorta and aortic arch, relative to the fixed descending aorta. Severe chest crushing may cause downward and leftward displacement of the heart resulting in ascending transection or evulsion of the brachiocephalic artery [9, 12]. It has been estimated that a sudden deceleration may result in a massively increased intravascular pressure, 'the water hammer effect'. The osseous pinch mechanism, where the aorta is 'pinched' between the sternum and the vertebrae [13, 14] may be responsible in some cases. It is generally accepted that aortic transection is a complex multivariate process caused by a combination of mechanical stresses including shear, torsion, stretch, compression and hydrostatic forces. Although this injury is usually produced by a high velocity impact, most commonly (95%) an automobile accident, it may occasionally be produced by low velocity impact, falls or crush injuries [15]. Aortic injury can vary in depth (intimal to transmural), length (1mm – circumferential) and number (single – multiple), depending on the type and severity of the impact, as well as the age of the victim and the presence of underlying vascular disease.

Clinical suspicion is therefore essential to establish the diagnosis and avoid the catastrophic results of a missed aortic injury. Victims frequently present following a rapid deceleration (>50km/hr) injury involving fatalities, ejection from a vehicle or as a pedestrian. Occasionally other causes including falls from heights (>3m), crush injuries or direct blows to the chest may be responsible for the injury. These details must be ascertained in the history. Suggestive signs on physical examination include chest wall deformities, steering wheel imprint on the chest, unexplained hypotension, pseudo-coarctation syndrome (mean arterial blood pressure in upper limb 10 mmHg greater than the lower limb), hoarseness or paraplegia. Four patients were found to have a pseudocoarctation syndrome on admission in this series, which was discovered using none invasive right arm and right leg pressures. Chest x-ray may suggest the diagnosis, but cannot exclude aortic injury. 6.6% of the victims in this study had initial normal chest x-rays, while other series have reported 7 – 27% normal chest x-rays on admission [6, 16]. Chest x-ray findings suggestive of aortic injury include a widened mediastinum (>8cm), left pleural apical cap, blurring of the aortic knuckle, tracheal deviation, depression of the left main bronchus (>140 degrees), deviation of a nasogastric tube, fracture of the first rib or a left sided haemothorax.

Once the diagnosis is suspected, the diagnosis can be confirmed using arch angiography, usually considered the gold standard. However, arch angiography may take significant time to organise and it is time consuming, rendering the patient less accessible during the study. Arch angiography also requires intravenous contrast, which may compromise renal function. Helical or high resolution computed tomographic angiography in comparison, allows rapid assessment of the mediastinum and other areas in these frequently multiply injured patients. Approximately 1% of blunt trauma patients have a thoracic aortic injury identified on helical CT [4]. More recently transesophageal echocardiography has been suggested as a screening modality [17]. However transesophageal echocardiography has the disadvantage that it lacks the sensitivity of arch angiography or helical CT, as it may miss the distal ascending and the proximal aortic arch due to the intervening trachea and bronchus.

Due to the significant false negativity of plain chest x-ray it may be appropriate to perform a helical CT of the thorax in all patients with a suggestive injury mechanism [18]. Helical CT has a sensitivity and negative predictive value of 100% and a specificity of 83% [4]. Helical CT is now the prime diagnostic technique for BTTA in Ireland, with aortography being reserved for equivocal Helical CT scans.

Initial management involves rapid diagnosis and careful resuscitation, with prevention of hypertension. Blood pressure control (systolic pressure <140mmHg) and wall stress reduction with beta blockers and antihypertensives (usually nitroprusside) has been advocated once BTTA is suspected or confirmed [6, 11, 19]. Anecdotal cases and small series of medically treated aortic transections suggest this is a viable option in cases were immediate repair is not possible, as may be the case where other major injuries coexist or when a patient declines surgical intervention [11, 17]. Half of the victims may have an associated head injury [6] and 29-32% of patients in some series, 16.6% in this series require a laparotomy or other operative procedure prior to aortic repair, thereby frequently necessitating a multi-disciplinary decision regarding operative timing based on the patient's cardiovascular status [11, 17]. It is important to remember that the risk of instantaneous death in the absence of treatment maybe as high as 94% in the first 24 hours. Although this high number is disputed by Pate et al when fastidious medical management of blood pressure is used [11].

In conclusion, this report summarizes a national experience with traumatic aortic transection injuries. Nationally there appear to be at least 40% of hospital survivors of road traffic accidents with blunt traumatic aortic transection not diagnosed in time for surgical intervention. Patients presenting with a history of a rapid deceleration injury or signs suggestive of BTTA warrant the exclusion of this potentially fatal diagnosis by helical CT angiography.

ACKNOWLEDGEMENTS

We would like to thank the Garda Síochána in particular Inspector Michael Brosnan of the Garda Technical unit for his assistance in reviewing the road traffic mortality data.

REFERENCES

1. Williams, J.S., Graff, J.A., Uku, J.M. and Steinig, J.P., Aortic injury in vehicular trauma, *Ann Thorac Surg,* **57,** 1994, 726-730.

2. McGwin, G., Reiff, D.A., Moran, S.G. and Rue, L.W., Incidence and characteristics of motor vehicle collision - related blunt thoracic aortic injury according to age. *J Trauma* **52,** 2002, 859-866.
3. Beel, T. and Harwood, A.L., Traumatic rupture of the thoracic aorta. *Ann Emerg Med* **9,** 1980, 483-486.
4. Fabian, T.C., Davis, K.A. and Gavant, M.L., Prospective study of blunt aortic injury: helical CT is diagnostic and antihypertensive therapy reduces rupture. *Ann Surg* **227,** 1998, 666-676.
5. Hill, D.A., Duflou, J. and Delaney, L.M., Blunt traumatic rupture of the thoracic aorta: an epidemiological perspective. *J R Coll Surg Edin* **41,** 1996, 84-87.
6. Fabian, T.C., Richardson, J.D. and Croce, M.A., Prospective study of blunt aortic injury: Multicenter Trial of the American Association for the Surgery of Trauma. *J Trauma* **42,** 1997, 374-380.
7. Hossack, D.W., The pattern of injuries received by 500 drivers and passengers Killed in road traffic accidents. *Med J Aust* **2,** 1972, 193-195.
8. Sevitt, S., Fatal road accidents in Birmingham:times to death and their causes. *Injury* **4,** 1972, 281-293.
9. Shkrum, M., McClafferty, K., Green, R., Nowak, E. and Young, J., Mechanisms of aortic injury in fatalities occuring in motor vehicle collisions. *J Forensic Sci* **44,** 1999, 44-56.
10. Parmley, L.F., Mattingly, T.W., Manion, W.C. and Jahnke, E.J. Jr., Nonpenetrating traumatic injury of the aorta. *Circulation* 1958;**17**:1086-1101.
11. Pate, J.W., Fabian, T.C. and Walker, W., Traumatic rupture of the aortic isthmus: an emergency? *World J Surg* **19,** 1995, 119-125.
12. Sevitt, S., Traumatic ruptures of the aorta: a clinico-pathological study. *Injury* **8,** 1977, 159-173.
13. Crass, J.R., Cohen, A.M., Motta, A.O., Tomashefski, J.F., Jr. and Wiesen, E.J., A proposed new mechanism of traumatic aortic rupture: the osseous pinch. *Radiology* **176,** 1990, 645-649.
14. Javadpour, H., O'Toole, J.J., McEniff, J.N., Luke, D.A. and Young, V.K., Traumatic aortic transection: evidence for the osseous pinch mechanism. *Ann Thorac Surg* **73,** 2002, 951-953.
15. Answini, G.A., Sturdevant, M., Sing, R. and Jacobs, D., Blunt Traumatic Rupture of the Thoracic Aorta: A report of an Unusual Mechanism of Injury. *Am J Emerg Med* **19,** 2001, 579-582.
16. Gammie, J.S., Ashish, S.S., Hattler, B.G., Kormos, R.L., Peitzman, A.B. and Griffith, B.P., Traumatice Aortic Rupture: Diagnosis and Management. *Ann Thorac Surg* **66,** 1998, 1295-1300.
17. Sweeney, M.S., Young, D.J., Frazier, O.H., Adams, P.R., Kapusta, M.O. and Macris, M.P., Traumatic aortic transections: eight-year experience with the "clamp-sew" technique. *Ann Thorac Surg* **64,** 1997, 384-387.
18. Exadaktylos, A.K., Sclabas, G., Scmid, S.W., Schaller, B. and Zimmerman, H., Do we really need routine computed tomographic scanning in the primary evaluation of blunt chest trauma in patients with "normal" chest radiograp? *J Trauma* **51,** 2001, 1173-1176.
19. Pate, J.W., Is traumatic rupture of the aorta misunderstood? *Ann Thorac Surg* **57,** 1994, 530-531.

LACK OF CONSISTENCY IN THREAT TO LIFE FROM SINGLE INJURY ABBREVIATED INJURY SCALE (AIS) 4 CODES IN DIFFERENT BODY AREAS

M. Woodford, O. Bouamra, A. Wrotchford, D.W. Yates and F.E. Lecky
The Trauma Audit and Research Network, The University of Manchester, Clinical Sciences Building, Hope Hospital, Salford M6 8HD, UK; 00 44 (0)161 206 4397, tarn@tarn.ac.uk

Abstract. The Trauma Audit & Research Network (TARN) collects and audits major injury data from half of all trauma receiving hospitals in England and Wales. At the time of extraction for this analysis there were 183,550 cases on the database. Included patients reach hospital alive and subsequently either die as a result of injury, require intensive/high dependency care or interhospital transfer, or stay for more than 3 days. Our aim was to determine from an established trauma registry whether single injury AIS 4 codes in different body areas result in a similar threat to life. Patients with single injuries giving an Abbreviated Injury Scale score of 4 were selected from the TARN database. Mortality rates were calculated with 95% confidence intervals (CI) by body area. This cross tabulation also included the median age and Glasgow Coma Score (GCS). There is variation in the mortality rates by body region for single AIS 4 injuries with lower extremity and external injuries having a higher crude mortality rate. In multivariate analysis there is a significantly lower odds of death associated with AIS 4 single injuries in the head and thorax areas than in other body regions. Single AIS 4 injuries present a significantly greater threat to life dependant on the body region injured.

Key words: Abbreviated Injury Scale, age, revised trauma score, trauma, mortality.

Michael D. Gilchrist (ed.), IUTAM Proceedings on Impact Biomechanics: From Fundamental Insights to Applications, 329–336.

1. INTRODUCTION

The study of trauma care requires a well defined and uniform system of measurement so that comparisons of both process and outcome can be made. The characterization of injuries is fundamental to both audit and research.

The Abbreviated Injury Scale (AIS) is used internationally to define and rank injuries by severity. The current AIS dictionary (1990 revision – Update 1998) contains over 1300 injury descriptions and each one is assigned a unique 7 digit numerical identifier including the body region of the injury and the AIS score according to the severity of the injury. This ranges from 1 (minor) to 6 (maximum) [1].

The AIS score is a consensus derived threat to life scale from which the Injury Severity Score (ISS) is derived [2]. ISS is utilized to express the overall severity of anatomical injury and in the TRISS methodology to calculate each injured patient's probability of survival [3, 4].

Survival probability calculations form the basis for individual case review and comparisons of institutional performance [5, 6].

It is therefore important that the AIS score is comparable in different body regions, i.e. AIS 4 injuries in the lower extremity region have a comparable mortality rate to those with an AIS 4 injury in the abdominal region.

The impact of the body region on the AIS score and subsequent mortality rates has not been previously studied.

We therefore examined the database of the Trauma Audit and Research Network to explore the relationship between body region and AIS 4 injury mortality rates.

2. MATERIAL AND METHODS

2.1 Patients

The Trauma Audit and Research Network [4, 6] established in 1989 as the UK Major Trauma Outcome Study [4, 6, 7]. It has its origins in a collaborative project with Howard Champion at the Washington Hospital Center, where a statistically based clinical audit system was first introduced in the early 1980s [5]. Approximately half of all trauma-receiving hospitals in England and Wales and other parts of Europe currently submit information to the Trauma Network. Trauma patients are included if they fulfill any of the following entry criteria:

- admission to hospital for more than 72 hours,
- transfer to a specialist center for extended trauma care,

- admission to an intensive care area,
- death occurring in hospital (irrespective of cause).

Patients with only isolated simple injuries (including those aged more than 65 years with fractured femoral neck or pubic rami) are excluded from the database.

Data collected by the Network include patient demographics, physiology on arrival at hospital, time and nature of medical interventions, complications, length of hospital stay, mortality within thirty days of admission and details of all injuries sustained. To ensure consistency these forms are completed only by staff who have attended a Trauma Network training course on data collection.

All patients receiving care at one hospital and those transferred into a hospital for further care were eligible for this study. Data was restricted to the 183,550 cases admitted between 1989 and 2004 inclusive.

2.2 Abbreviated Injury Scale Coding

Injuries were recorded by the participating hospitals as free text and coded, at the coordinating office of the Trauma Network, using the Abbreviated Injury Scale 1990 Revision - 98 Update [1]. The central coders have been trained and recognized by the American Association of Automotive Medicine [1].

Patients with single injuries AIS4 in any body region were selected from the Trauma Network database. Those cases where the patients had single injuries AIS1 to AIS3, AIS5 and AIS6 and those with multiple injuries were excluded. This resulted in a total of 1,961 patient records in this study. The body regions for each injury were also recorded as head, neck, thorax, abdomen, spine, extremity including bony pelvis and external.

2.3 Statistical Analysis

The mortality rates of single AIS 4 injuries in different body regions were compared. 95% Confidence Intervals (CI) were also calculated. A multiple logistic regression was performed to determine whether any body region had a significant impact on the odds of death after adjustment for age and presenting physiology. Interpretation was based on the size of effects rather than reliance purely on statistical significance.

Table 1. Patients with single AIS4 injuries presenting between 1989–2004.

Body area	No of patients	Mortality % (95% C.I.)	Median Age (yrs)	Median RTS
Head	788	9.8 (7.7-11.8)	31.0	7.84
Neck	8	25.0 (0.0-55.0)	24.0	7.23
Thorax	534	7.7 (5.4-9.9)	52.6	7.84
Abdomen	168	2.4 (0.1-4.7)	26.4	7.84
Spine	177	5.6 (2.2-9.1)	44.0	7.84
Lower Extremity	74	16.2 (7.8-24.6)**	35.7	7.84
External	212	15.1 (10.3-19.9)**	35.0	7.84

** Significantly higher mortality rate than that given for abdominal injuries $p<0.05$

3. RESULTS

Patients whose most severe single injury AIS4 totaled 1,961, 1.06% of the Trauma Network Database. Table 1 lists the distribution of those patients, the body region injured, the age, presenting physiological measurements and the variation in mortality.

There is a significantly higher rate associated with AIS 4 single injuries in the lower extremity and external area than in the abdomen. This does not appear to be explained by presenting physiology differences between these groups. However there was a significant difference in the median age of these groups of patients.

However the multiple logistic regression suggested that the increased mortality rate in extremity injuries was explained by this variation in age as there was no independent increase in the odds of death with lower extremity injuries. In contrast AIS4 injuries to the head and thorax were associated with significant reductions in the odds of death compared to those in other body regions (0.12-0.51 and 0.21-0.92 respectively, Table 2). Increasing age and reduced Glasgow Coma Score also contributed independently to the odds of death (1.05-1.08 for each increase in year above 0, and 0.67-0.76 for each unit increase in the GCS above 3).

4. DISCUSSION

We have found a variation in the mortality rates for single AIS4 injuries. There is a significantly higher rate associated with AIS4 single injuries in the lower extremity and external area than in the abdomen. This does not appear to be explained by presenting physiology differences between these groups. However the lower extremity group has a significantly greater age than patients with AIS4 abdominal injuries hence the overall effect was not significant in multivariate analysis. However the odds ratios when age and Glasgow Coma Scale on arrival are accommodated show that the head and

Table 2. Impact of body region, age and presenting GCS on odds of death in patients with single AIS4 injuries.

	B	S.E.	Wald	df	Sig.	Odds Ratio	95.0% C.I. for OR Lower	Upper
In Abdomen			24.83		.00			
In Head	-1.40	.37	14.46	6	.00	.25	.12	.51
In Neck	.81	1.02	.64	1	.42	2.25	.31	16.48
In Thorax	-.83	.38	4.74	1	.03	.44	.21	.92
In Spine	-.38	.69	.30	1	.58	.68	.18	2.65
In Lower Extremity	-.81	.51	2.55	1	.11	.45	.17	1.20
In External/Burn/Other	.21	.53	.16	1	.69	1.24	.44	3.49
GCSE	-.34	.03	115.75	1	.00	.71	.67	.76
AGE	.06	.01	122.20	1	.00	1.07	1.05	1.08
Constant	-.84	.47	3.2	1	.07	.43		

B: coefficients of regression
S.E.: standard error of coefficients of regression
Wald: statistic for testing whether B=0
df: degree of freedom
Sig: p-value significance (to be compared with 0.05)

thorax have significantly reduced odds of death compared to other body regions.

The weaknesses, as always, with analyses of this sort, relate to data quality and accurate injury information. It is not possible for us to state categorically that all injury descriptions were clinically complete however we are sure that the coding assignments are consistent in all cases. Additionally all data collectors undergo at least one training course in how to extract data from clinical notes and some are checked by clinicians.

We have not found other studies comparing the mortality rate of single AIS4 injuries. However detecting a variation in the threat to life within AIS severity codes is not entirely surprising as their initial assignment was consensus rather than data based. We suggest a potential reason for the discrepancy may be that most AIS4 head and thoracic injuries would not require invasive surgery, this may be more likely in AIS4 injuries of the abdomen, neck, spine and lower extremity. This appeared to be true in this sample with surgery occurring in 16 and 35% respectively of thoracic and head single AIS4 patients. Rates in other patients ranged from 38% (spine) to 81% (abdomen).

This work has implications for the practical applications of the abbreviated injury scale i.e. its use in biomechanics and the benchmarking of trauma systems [6]. For example, it is possible that hospitals with a low prevalence of head and/or thoracic injury their casemix will consequently have a lower "performance benchmark" which does not fairly reflect quality of care.

We recommend that further analysis of this sort within the remaining AIS coding severities and that this evidence base be taken account of in further iterations of the AIS dictionary.

ACKNOWLEDGEMENTS

With thanks to the Trauma Audit and Research Network (TARN): Clinical Staff and Data Coordinators at participating hospitals in England and Wales:

University Hospital of North Staffordshire, Addenbrooke's Hospital, St Bartholomews Hospital, Royal London Hospital, Hammersmith Hospital, Guy's Hospital, Calderdale Royal Hospital (The), Royal Liverpool Childrens Hospital (Alder Hey), University Hospital of Copenhagen, Lisbon Teaching Hospital, Morriston Hospital, Hope Hospital, Kings College Hospital, Nottingham University Hospital, Royal Manchester Children's Hospital, Royal Victoria Hospital Belfast, Selly Oak Hospital, Birmingham Children's Hospital, St George's Hospital, Southampton General Hospital, Pinderfields General Hospital, Bristol Royal Infirmary, Royal Preston Hospital, James Cook University Hospital, Leeds General Infirmary, Hull Royal Infirmary, Manchester Royal Infirmary, Booth Hall Children's Hospital, Basildon Hospital, Whipps Cross University Hospital, North Manchester General Hospital, Derriford Hospital, Newcastle General Hospital, Royal Liverpool University Hospital, University Hospital Aintree, Hinchingbrooke Hospital, Walton Centre for Neurology, Royal Hallamshire Hospital, Regional Spinal Injuries Unit Southport, Frenchay Hospital, Atkinson Morley's Hospital, St James Hospital, John Radcliffe Hospital, Sheffield Children's Hospital, Royal Devon & Exeter Hospital, Royal Sussex County Hospital, St Mary's Hospital London, Withington Hospital, Wexham Park Hospital, St Thomas' Hospital, Royal Lancaster Infirmary, Leicester Royal Infirmary, Craigavon Area Hospital, Southend Hospital, Colchester General Hospital, Ipswich Hospital, Dewsbury District Hospital, St James' University Hospital, Fairfield General Hospital, Tameside General Hospital, Harrogate District Hospital, Royal Oldham Hospital, Ealing Hospital, Ashford Hospital, Royal Cornwall Hospital, Norfolk & Norwich University Hospital, Burnley General Hospital, Blackburn Royal Infirmary, Bedford Hospital, Nevill Hall Hospital, Royal Berkshire Hospital, Broomfield Hospital, Milton Keynes General Hospital, Huddersfield Royal Infirmary, Ormskirk & District Hospital, Airedale General Hospital, Great Western Hospital Swindon, Warwick Hospital, Northampton General Hospital, Stoke Mandeville Hospital, Southmead Hospital, Derbyshire Royal Infirmary, Royal Surrey County Hospital, Daisy Hill Hospital, Wycombe Hospital, Pilgrim Hospital, Peterborough District Hospital, St Peters Hospital Chertsey, Queen Alexandra Hospital, Whiston Hospital, Kettering General Hospital, Doncaster Royal Infirmary, Lincoln County Hospital, Chesterfield & Nth Derbyshire Royal Hospital, Leighton Hospital, Barnsley District General Hospital, William Harvey Hospital,

Pontefract General Infirmary, Taunton & Somerset Hospital, Gloucestershire Royal Hospital, Warrington Hospital, Poole Hospital, Weymouth & District Hospital, University Hospital of North Tees, University Hospital of Wales, Royal Hampshire County Hospital, Conquest Hospital, Medway Maritime Hospital, Darent Valley Hospital, Kings Mill Hospital, Wansbeck General Hospital, Worthing Hospital, Sunderland Royal Hospital, North Tyneside General Hospital, Glan Clwyd District General Hospital, Withybush General Hospital, Princess Alexandra Hospital, Jersey General Hospital, Torbay Hospital, Maidstone General Hospital, Prince Philip Hospital, Friarage Hospital, University Hospital of Hartlepool, West Wales General Hospital, Kent & Canterbury Hospital, Worcester Royal Hospital, Blackpool Victoria Hospital, Northern General Hospital, Royal Bolton Hospital, Countess of Chester Hospital, Stepping Hill Hospital, Weston General Hospital, York Hospital, Coventry & Warwickshire Hospital, Homerton Hospital, Epsom General Hospital, Wythenshawe Hospital, West Middlesex University Hospital, Hillingdon Hospital, Royal Albert Edward Infirmary, Leigh Infirmary, The Horton Hospital, University Hospital Lewisham, Chorley & South Ribble District General Hospital, Rochdale Infirmary, Scarborough Hospital, Scunthorpe General Hospital, Trafford General Hospital, Sandwell General Hospital, Diana, Princess of Wales Hospital, City Hospital Birmingham, Kent & Sussex Hospital, Princess Royal Hospital Telford, Royal Victoria Infirmary, Newcastle Upon Tyne, Southport & Formby District General Hospital, Rotherham District General Hospital, Bassetlaw District General Hospital, Bradford Royal Infirmary, West Cumberland Hospital, Arrowe Park Hospital, Mater Infirmorum Hospital, John Coupland Hospital, James Paget Hospital, Hereford County Hospital, South Tyneside District Hospital, Royal Shrewsbury Hospital, Princess Royal University Hospital Orpington, Waterford Regional Hospital, Birmingham Heartlands Hospital, Skegness & District Hospital, Eastbourne District General Hospital, Royal United Hospital, Queen Elizabeth Queen Mother Hospital, Cheltenham General Hospital, Northwick Park Hospital, St Helier Hospital, Watford General Hospital, Queen Elizabeth Hospital Kings Lynn, Furness General Hospital, Royal Gwent Hospital, Queen Elizabeth Hospital (Gateshead), Grantham & District Hospital, Prince Charles Hospital, Cumberland Infirmary, Ysbyty Gwynedd District General, Wrexham Maelor Hospital, Crawley Hospital, East Surrey Hospital, Dorset County Hospital, Toolo Hospital Helsinki University Central Hospital, Meilahti Hospital Helsinki University Central Hospital, West Suffolk Hospital.

REFERENCES

1. Committee on Injury Scaling, Association for the Advancement of Automotive Medicine. The Abbreviated Injury Scale, 1998 revision. Des Plaines, Chicago: Association for the Advancement of Automotive Medicine, 1998.
2. Baker, S.P., O'Neill, B., Haddon, W. and Long, W.B., The injury severity score: a method for describing patients with multiple injuries and evaluating emergency care, J Trauma, 1974, 14:187-196.
3. Boyd, C.R., Tolson, M.A. and Copes, W.S., Evaluating trauma care: The TRISS method, J Trauma, 1987, 27:370-378.

4. Hollis, S., Yates, D.W., Woodford, M. and Foster, P., Standardised comparison of performance indicators in trauma: A new approach to case-mix variation, J Trauma, 1995, 38:763-766.
5. Champion, H.R., Copes, W.S., Sacco, W.J., et al., The Major Trauma Outcome Study: Establishing national norms for trauma care, J Trauma 1990, 30:1356-1365.
6. Lecky, F., Woodford, M. and Yates, D.W., Trends in trauma care in England and Wales 1989-97, Lancet, 2000, 255:1771-1775.
7. Yates, D.W., Woodford, M. and Hollis, S., Preliminary analysis of the care of injured patients in 33 British hospitals: first report of the United Kingdom major trauma outcome study, BMJ, 1992, 305:737-740.

SESSION 7

MECHANICAL FACTORS IN OSTEOARTHROSIS

Contribution of Repetitive Impulsive Loads to the Development of the Cartilage Failure in Human Joints

John O'Connor
University of Oxford, Department of Engineering Science and Orthopaedic Engineering Collaboration, Botnar Centre @ Nuffield Orthopaedic Centre, Oxford OX3 7LD, England

Abstract. The paper describes experiments and calculations investigating Radin's hypothesis that repetitive impulsive loads leads to cartilage damage and is the mechanical factor involved in the development of osteoarthrosis of the joints.

Key words: cartilage failure, impulsive loads, gait analysis, heelstrike transient, gait simulations, knee anteromedial osteoarthritis.

1. INTRODUCTION

Osteoarthrosis or osteoarthritis (OA) is a degenerative disease of the joints which manifests itself usually in the seventh decade of life and for which there is no known cure [**1,** page 98]. Equally, there is no known cause although secondary forms of the disease can follow injury to the bones or to the ligaments which hold bones together at joints. Secondary osteoarthrosis is common amongst retired athletes and ballet dancers. The primary disease is said to be ideopathic in the sense that it is not caused by infection or diet, is not contagious but arises for unknown causes internal to the individual. It can affect 25% or more of the elderly population. It is a major cause of disability and pain in the elderly and a major healthcare cost.

In OA, the low-friction layers of cartilage which cover the articular surfaces of the joints and act as bearing surfaces are damaged and eventually eroded, exposing the underlying bone. Cartilage itself does not contain nerves so that the early stages of cartilage damage are not painful. The

Michael D. Gilchrist (ed.), IUTAM Proceedings on Impact Biomechanics: From Fundamental Insights to Applications, 339–346.

underlying subchondral bone is richly enervated and its exposure leads to a quite sudden onset of severe pain and limitation of activity. Only one joint, or a matching pair (the knees), can be affected.

Osteoarthrosis is unlike rheumatoid arthritis which starts as an inflammation of the joints, usually with multiple joint involvement. Destruction of the cartilage layers is then secondary to the inflammation.

Fortunately, joint replacement is a satisfactory treatment and will usually survive the elderly patient, offering perhaps fifteen years of pain-free restoration of activity. The same guarantee cannot be given to the younger patient (aged $<$ 55 years) in whom the failure rate of joint replacement is significantly higher. There is still need for a cure for OA, based on an understanding of the causes.

The bulk of research in this area has been biological in approach, as would be expected, concentrating on the microstructural nature of the tissues [**2**] and their disruption. There has been only a little work on the nature and magnitude of the loads required to cause cartilage damage.

Radin and colleagues [**3**] have been the main group to discuss the effects of load on the development of cartilage damage. In the laboratory, they found that large static loads damaged the underlying bone but appeared to have little effect on the cartilage. They showed that cartilage could be damaged by repeated impact loads, applied during joint movement. They were able to demonstrate arthrosis-like cartilage damage in the knees of live rabbits when subjected to repeated loads applied rapidly through a cam [**4**] although loads of the same amplitude applied more slowly failed to produce damage. It is not possible to perform such experiments on human subjects so that establishing a relationship between patterns of loading and the incidence of OA is more difficult and has to be done indirectly.

Radin *et al* [**5**] found a higher rate of loading at foot-contact (heelstrike) during level walking in a group of volunteers who had experienced pre-clinical knee joint pain than in an age-matched sex-matched group of volunteers free of lower-limb problems. It has been known for some time that the incidence of OA in the lower-limb joints of Chinese populations is lower than in age-matched sex-matched Caucasian populations [**6**]. Chen *et al* [**7**] found a lower rate of loading at foot-contact during level walking in a group of Chinese women than in an age-matched group of Caucasian women and related this to differences in the speed of normal level walking. Chen [**8**] also found a significantly higher loading rate at foot-contact in level walking in subjects with no lower-limb problems but whose like-sex parents had a proven history of OA of the knee than in an age-matched sex-matched group with no such family history. Loughlin *et al* [**9**] have found a genetic linkage with the incidence of OA which they attributed to the chromosome locus that could regulate the mass of subchrondral bone supporting the articular cartilage. However, the increased subchondral bone mass could be the result of repeated impulsive loading [**10**].

2. IMPULSIVE LOADING DURING WALKING

Simple level walking is an activity in which some of the population apply repeated impulsive loads to the lower limb. Force plates set flush with the floor are used in gait laboratories to record the three components of force applied through the ground to the foot during the stance phase of gait. A typical trace of the vertical component of force is shown in Figure 1.

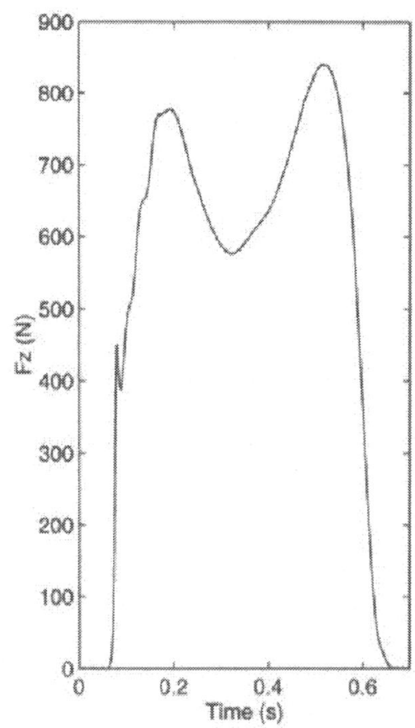

Figure 1 shows the vertical component of the force transmitted between the planted foot and the ground over the period of ground contact during level walking. After foot contact at about 0.05 seconds, the trace shows two main peaks, at about 0.2 and 0.6 seconds, corresponding to the moments of maximum downward acceleration of the mass centre of the body. In addition, the subject exhibited a sharp initial impulsive peak (called the heel-strike transient, HST) and a brief period of high loading rate, immediately after foot-contact. A maximum loading rate of about 75kN/sec occurred at 0.075sec. HST is exhibited only by a proportion of the population, most subjects exhibiting a steady slower loading rate to the first main peak at 0.2 seconds.

Figure 1. Vertical component of ground reaction during stance for a subject weighing about 700N. This load cycle is repeated with every stride. At the instant of heelstrike, the leg is straight and the knee is fully extended.

For a group of 218 healthy subjects, Chen *et al* [**7**] found a mean maximum vertical loading rate at heelstrike of 78.8 ±40.2 BW/sec, where BW represents body weight. The large standard error of the mean implies that, across the population at large, there is a very wide variation in maximum loading rate at heelstrike.

3. ANTEROMEDIAL OSTEOARTHROSIS OF THE KNEE

Contact between the femur and the tibia at the knee joint occurs in two separate compartments, lateral and medial. In each, a more or less spherical femoral condyle makes contact with a more or less flat tibial plateau. White *et al* [**11**] described "anteromedial osteoarthritis of the knee" as the early manifestation of OA in the joint. The lesion is confined to the medial compartment and affects the cartilage on the front half or two thirds of the tibial plateau and on the most distal portion of the medial femoral condyle. Erosion of the underlying bone gives rise to severe disabling pain. The cartilage at the back of the medial tibial plateau and on the back of the medial femoral condyle remains of full thickness (about 3mm). These focal lesions can persist for many years as long as the anterior cruciate ligament remains intact and functional. Once the anterior cruciate ligament ruptures, the cartilage lesions spread rapidly throughout the knee.

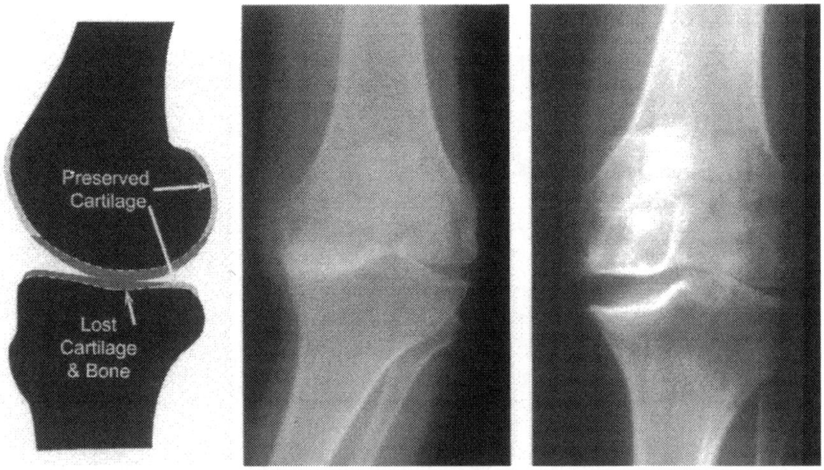

(a) (b) (c)

Figure 2. (a) Diagram of the knee seen from the side showing areas of cartilage and bone erosion and areas of cartilage preservation. (b) X-ray from the front while standing showing deep erosion on the medial side of the joint. (c) X-ray from the front with the leg straightened by the clinician showing space between the bone surfaces filled with intact cartilage on the lateral side and substantial bone-loss on the medial side.

The preserved areas of cartilage are those which make contact when the knee is bent to about 90° [**12**]. At 90°, i.e. during stair climbing or rising from a chair, the compressive forces transmitted across the articular surfaces of the knee can be very large because the lever-arms available to the surrounding muscles are much shorter than those available to the external loads and the muscle forces are therefore correspondingly larger than the loads, several times body-weight. The compressive reactions at the surface interface are larger still. Thus, the absence of cartilage damage in an area transmitting large loads confirms Radin's laboratory evidence [**3**].

The eroded areas of cartilage and bone are the regions which make contact when the knee is straight or nearly so when the intra-articular compressive forces are relatively smaller, as during the stance phase of level walking. It is therefore tempting to ascribe cartilage damage to "wear and tear" caused by the repeated reversing sliding of the articular surfaces under load or to fatigue failure [**13**] due to cyclic loading. However, the entire population experiences these loading and sliding conditions but only a proportion suffers from the osteoarthrotic consequences of cartilage failure. An hypothesis is needed to explain why only a proportion of the population suffer from OA. The wide range of biological research [**2**] has not found an explanation in terms of local pre-arthrotic deterioration of tissue structure.

Radin's hypothesis [**14**] provides such an explanation. It states that cartilage failure follows from the transmission of repetitive impulsive loads. Chen's [**7**] work confirms that of many others and demonstrates that the loading rate at heelstrike during normal level walking varies considerably across the population. The knee is straight and the areas of cartilage liable to erosion are in contact at the instant of heelstrike. The hypothesis therefore suggests that the individuals who exhibit the highest rates of loading at heelstrike during each stride are the group at risk of developing OA.

4. LOAD TRANSMISSION THROUGH CARTILAGE

The cartilage layers which transmit compressive force to the underlying subchondral bone are 2-3mm thick and consist of a mesh of collagen fibers, comprising 15-20% of the mass, with the interstices filled with proteoglycans molecules and water. The proteoglycans molecules (bottle brush-like macromolecules) offer significant frictional resistance to the flow of water through the interstices so that the effective relaxation time for cartilage uner steady load is about 1000 seconds [**15**, page 243]. On the other hand, the response of cartilage to rapidly applied loads is described by Mow *et al* [**15**] as similar to that of an incompressible elastic layer.

Kelly and O'Connor [**16**] analysed the response of an incompressible elastic layer to pressure applied over part of its free surface. They found significant opening-mode stress intensity factor values for surface-breaking cracks lying perpendicular to the free surface and for closed cracks lying near to and inclined at 45° to the bone/cartilage interface. They suggested that their analysis could explain the incidence of such cracks in cartilage observed by Meachim [**17**], Radin *et al* [**4**] and Sokoloff [**18**]. Amalgamation of crack networks could lead to loss of discrete fragments of cartilage and exposure of the underlying bone, the initiation of osteoarthrosis. The tensile stress fields leading to crack formation are less likely when slowly applied loads are accompanied by significant interstitial fluid flow.

5. HEELSTRIKE TRANSIENTS AND WALKING PATTERN

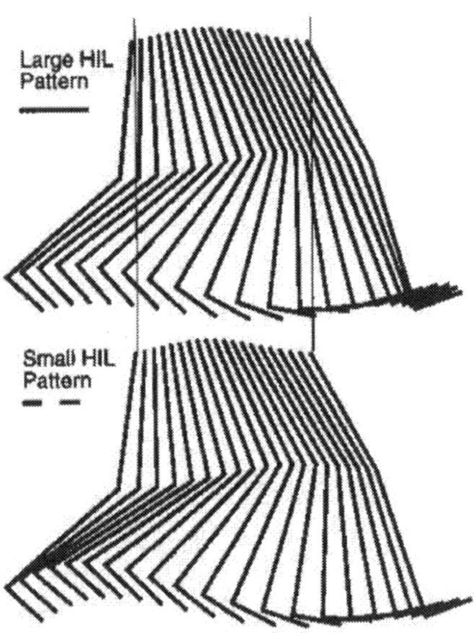

Gill and O'Connor [**19**] performed detailed analysis of the gait patterns of a group of young adults who exhibited high loading rates at heelstrike and of a similar group who did not. The impact at heelstrike is set up by the pattern of leg motion during the immediately preceding swing-phase. In a simulation [**20**], they showed that subjects with a small Heelstrike Impulsive Loading rate (HIL) start forward-swing with the ankle lifted well clear of the ground (Figure 3); the foot swings forward and touches the ground more or less tangentially, with little need for vertical impulse.

Figure 3. Simulation of the swing-phase of level walking for a subject with a high rate of loading at heelstrike and a subject with a low rate. The critical difference is the height of the ankle at the beginning of swing which determines the shape of the subsequent arc of motion of the foot. Both subjects have the same pattern of hip motion.

Subjects with a large HIL start forward swing with the ankle closer to the ground so that the foot swings through bottom-dead-centre and a sudden vertically-downwards velocity is necessary at the end of swing to achieve foot contact, necessitating the application of an impulsive force. Gill and O'Connor [**19**] found that a subset of their high rate loaders also exhibited a medially directed impulsive load at heelstrike, putting the medial compartment of the knee more at risk than the lateral. Walking speed is also a factor. Collins and Whittle [**21**] investigated 13 normal young adult male subjects walking at five self-selected speeds. They found that loading rate at heelstrike was significantly correlated to walking speed. Chen *et al* [**7**] explained differences in loading rate between Chinese and Caucasian women aged greater than 55 years by differences in their normal walking speeds. They suggested that loading rates could be reduced by the simple device of walking more slowly.

6. DISCUSSION

Many factors contribute to the development of osteoarthrosis; this paper has investigated the contribution of repeated impulsive loads to cartilage failure. The evidence has to be circumstantial but the hypothesis seems to explain well the pattern of development of the disease in the human knee [**11**] and the differences in incidence of the disease in Chinese and Caucasian populations [**7**]. There appears to be a clear genetic factor [**9**] but differences in loading rate at heelstrike have been found between families with a known incidence of the disease and families with no history [**8**].

The hypothesis has difficulty in explaining the lower incidence of primary osteoarthrosis in the ankle even though it too sees the impulsive loads applied during gait in the population thought to be at risk. It also has difficulty explaining the most common site of osteoarthrosis in the human body at the first metacarpal-phalangeal joint (the joint between the palm of the hand and the thumb). Nonetheless, the hypothesis appears to have some merit and suggests how the risk of developing the disease may be reduced by modification of gait patterns.

REFERENCES

1. Radin, E., *Orthopaedics for the Medical Student.* JP Lippincott, Philadelphia, 1987.
2. Archer, C.W., Caterson, B., Benjanin, M. and Ralphs, J.R (Eds), *The Biology of the Synovial Joint.* Harwood Academic Publishers, Amsterdam, 1999.
3. Radin, E.L., Parker, H.G., Pugh, J.W., Steinberg, R.S. Paul, I.L. and Rose, R.M., Response of joints to impact loading, III. Relationship between trabecular microfractures and cartilage degeneration. *Journal of Biomechanics* **6**, 1973, 51-57.

4. Radin, E.L., Martin, B.M., Burr, D.B., Caterson, B., Boyd, R.D. and Goodwin, C., Effects of mechanical loading on the tissues of the rabbit knee. *Journal of Orthopaedic Research* **2**, 1984, 221-234.
5. Radin, E.L., Yang, K.H., Riegger, C., Kish, V.L. and O'Connor, J.J., Relationship between lower limb dynamics and knee joint pain. *Journal of Orthopaedic Research* **9**, 1991, 398-405.
6. Hoaglund, FT., Yau, A.C.M.C., Wong, W.L., Osteoarthritis of the hip and other joints in Southern Chinese in Hong Kong. *Journal of Bone and Joint Surgery [Am]* **55A**, 1973, 545-557.
7. Chen, W-L., O'Connor, J.J. and Radin, E.L., A comparison between the gaits of Chinese and Caucasian women with particular reference to their heelstrike transients. *Clinical Biomechanics* **18**, 2003, 207-213.
8. Chen, W-L., *Impulsive Loading in Gonarthrosis.* D.Phil thesis, University of Oxford, 1998.
9. Loughlin, J., Mustafa, Z., Smith, A., Irven, C., Carr, A.J., Clipsham, K., Chitnavis, J., Bloomfield, V.A, McCartney, M., Cox, O., Sinsheimer, J.S., Sykes, B. and Chapman, K.E., Linkage analysis of chromosome 2*q* in osteoarthritis. *Rheumatology* **39,** 2000, 377-381.
10. Farkas, T., Boyd, R.D., Schaffler, M.B., Radin, E.L. and Burr, D.B., Early vascular changes in rabbit subchondral bone after repetitive impulsive loading. *Clinical Orthopaedics and Related Research* **219**, 1987, 259-267.
11. White, S.H., Ludkowski, P.F. and Goodfellow, J.W., Anteromedial osteoarthritis of the knee. *Journal of Bone and Joint Surgery [Br]* **73B**, 1991, 582-586.
12. O'Connor, J.J., Shercliff, T., Biden, E. and Goodfellow, J.W., The geometry of the knee in the sagittal plane. *Proceedings of the Institution of Mechanical Engineers Part H, Journal of Engineering in Medicine* **203H**, 1989, 223-233.
13. Freeman, M.A.R., The fatigue of cartilage in the pathogenesis of osteoarthrosis. *Acta Orthopaedica Scandanavica* **46**, 1975, 323-328.
14. Radin, E.L., Burr, D.B., Caterson, B., Fyhrie, D., Brown, T.D. and Boyd, R.D., Mechanical determinants of osteoarthrosis. *Seminars in Arthritis and Rrheumatism* **21**, 1991, 12.
15. Mow, V.C., Hou, J.S., Owens, J.M. and Ratcliffe, A., Biphasic and quasilinear viscoelastic theories for hydrated soft tissues. Chapter 9 in Mow, V.C. Ratcliffe, A. and Woo, S.L-Y, (eds) *Biomechanics of Diarthrodial Joints.* Springer-Verlag, New York, 1990, 215-260.
16. Kelly, P.A., and O'Connor, J.J., Transmission of rapidly applied loads through articular cartilage. Part I: Uncracked cartilage. Part II: Cracked cartilage. *Proceedings of the Institution of Mechanical Engineers, Journal of Engineering in Medicine* **210H,** 1996, 27-49.
17. Meachim, G., Articular cartilage lesions in osteoarthritis of the femoral head. *Journal of Pathology* **107,** 1972, 199.
18. Sokoloff, L., Microcracks in the calcified layer of articular cartilage. *Archives of Pathology and Laboratory Medicine* **117**, 1993, 191-195.
19. Gill, H.S. and O'Connor, J.J., Heelstrike and pathomechanics of osteoarthrosis: a pilot study. *Journal of Biomechanics* **36**, 2003, 1625-1631.
20. Gill, H.S. and O'Connor, J.J., Heelstrike and the pathomechanics of osteoarthrosis: a simulation study. *Journal of Biomechanics* **36**, 2003, 1617-1624.
21. Collins, J.J. and Whittle, M.W., Influence of gait parameters on the loading of the lower limb. *Journal of Biomedical Engineering* **11**, 1989, 409-412.

THE INITIAL ASSESSMENT AND MANAGEMENT OF BLUNT TRAUMA

Aidan Gleeson
Consultant in Emergency Medicine, Beaumont Hospital, Dublin 9, Ireland

Key words: blunt trauma, patient management.

1. INTRODUCTION

There 1.2 million attendances at Emergency Departments in Ireland. There is a wide variety of pathology treated covering all Medical and Surgical Specialties. A significant percentage of patients present with injuries, the majority being classified as relatively minor trauma. Major trauma, which is predominantly blunt in origin, constitutes no more than 0.5% of the normal workload of an Emergency Department.

2. FORMS OF BLUNT TRAUMA

1. Road Traffic Accidents – Vehicular, motorcycle, cyclist and pedestrian.
2. Falls
3. Assaults.

Michael D. Gilchrist (ed.), IUTAM Proceedings on Impact Biomechanics: From Fundamental Insights to Applications, 347–353.

3. MANAGEMENT OF THE INJURED PATIENT

3.1 History

The history is a vital component in the evaluation of any injured patient. A doctor will look for information regarding *when* the injury occurred, *how* it happened and what associated *symptoms* the patient has.

The mechanism of injury is of particular importance in blunt trauma. Whilst it might not give a doctor the exact diagnosis before the patient is examined, it can narrow down the differential diagnosis considerably in many cases. A good example of this is with a patient who complains of shoulder pain after a fall. If the mechanism of injury involved the patient falling directly onto the shoulder, such as when knocked off a bicycle then the most likely injury is that of an acromioclavicular joint subluxation. A fracture at the outer end of the clavicle has to be considered as well. The mechanism of injury may not be entirely helpful in narrowing down the differential diagnosis with certain trauma, such as wrist injuries. The majority of Emergency Department Clinicians are all to well aware that a fall onto the outstretched hand, causing a hyper-extension injury of the wrist, can cause a scaphoid fracture, but it can also cause the following injuries:

(a) Ligament sprain.
(b) Lunate dislocation.
(c) Perilunate dislocation.
(d) Triquetral fracture.
(e) Tear of the triangular fibrocartilage.
(f) Scapholunate dissociation or rotatory subluxation of the scaphoid.
(g) Fracture of the distal radius – displaced (Colles, Barton's) or undisplaced.

The degree of force involved during the traumatic incident can be a useful indicator of whether the patient is likely to have sustained a significant intra-cranial injury. A fall from a height of 2 meters or greater is much more likely to cause significant intra-cranial damage, particularly if the head strikes a firm surface such as concrete. Additionally, if a patient has been struck on the head, the larger and harder the object used in the assault, the more likely the patient is to suffer a severe brain injury.

Loss of consciousness and amnesia are good indicators of a significant force being used. The clinician should look for other indicators of severe brain injury such as a headache, nausea, vomiting and visual disturbance, but these can be present in patients with benign head trauma and are generally considered to be non-specific.

With chest trauma a history of chest pain is generally non-specific but, if particularly severe, may indicate that the patient has a rib fracture. One should then seek information regarding potential complications of rib fractures, such as a pneumothorax (collapsed lung) or a haemothorax (bleeding into the chest cavity) which might be indicated by the patient being short of breath.

Spinal pain is again quite non-specific and may be seen with a simple soft tissue injury. However, if the pain is particularly bad and the patient is unable to mobilize properly then there may be a bony injury. It is important to evaluate for nerve root or spinal chord injuries as well which may be evidenced by the patient complaining of paraesthesia (pins and needles), numbness, weakness or paralysis of a limb or limbs.

The presence of abdominal pain can indicate the presence of intra-abdominal injury, such as solid organ injury with intra-abdominal haemorrhage or hollow viscus performation, i.e. bladder or bowel rupture. A history of frank haematuria (blood in the urine) may indicate that there has been damage to one or both kidneys or other areas of the genitourinary tract, such as the bladder or urethra.

Other important aspects of injury evaluation which, constitute normal components of history, taking involve questions on past medical history (inclusive of injuries), what medication the patient is taking and whether or not he/she has any allergies.

3.2 Examination

The examination of the injured patients, in its totality, is best illustrated by the Advanced Trauma Life Support format. This system is used when assessing a potential case of major trauma. It comprises four phases:

(a) *Primary survey.*
(b) *Resuscitation.*
(c) *Secondary survey.*
(d) *Definitive care.*

3.2.1 Primary survey

This involves a quick assessment of the patient to look for immediate life threatening injuries. It is broken up into the following components:

(i) ***A*** *– Airway and cervical spine control*

The airway is assessed for obstruction, which could be indicated by stridor, a harsh sound heard on inspiration. One looks at the airway for evidence of oral trauma. One must ensure that the cervical spine is protected through the use of a spinal board, a cervical collar and sand bags either side of the head.

(ii) ***B*** *– Breathing*

One assesses the respiratory rate which, if increased, may indicate intra-thoracic pathology, such as a pneumothorax or a haemothorax. Alternatively, an increased respiratory rate may simply be a manifestation of shock caused by injuries elsewhere. One would also look at the chest for evidence of bruising or lacerations. A stethoscope is used to check for normal breath sounds in both lungs and the trachea (windpipe) position is evaluated as, if it is shifted to one side, it may indicate the presence of a tension pneumonthorax on the opposite side.

(iii) ***C*** *– Circulation and haemorrhage control*

If the patient's pulse is fast it can be an indicator of shock, as can alteration of the components of the blood pressure. If there is any external bleeding, then initial management consists of applying pressure to that area only and blind clamping of internal structures is not recommended.

(iv) ***D*** *– Disability*

This is a quick neurological evaluation using the AVPU scale or the Glasgow Coma Score. One should also check the pupils, as a dilated pupil that does not react to light likely reflects raised pressure within the skull from either brain swelling or an expanding haematoma.

(v) ***E*** *– Exposure*

All the patients clothing must be removed so that the whole body can be inspected, otherwise, serious injuries may go unnoticed. One should take care, however, to ensure that the patient does not become excessively cold.

3.2.2 Resuscitation

This phase runs in tandem with the primary survey and generally constitutes giving the patient oxygen and intravenous fluids. Fluid is given to replace blood lost through either internal or external bleeding, but one must be aware that the over judicious administration of fluids can itself cause further bleeding and compromise matters more.

At the end of the primary and resuscitation phases three x-rays are performed which will help to diagnose immediate life threatening problems. These are a lateral x-ray of the cervical spine, a chest x-ray and an x-ray of the pelvis. In the unconscious patient, in particular, pelvic fractures may not be obvious from the clinical examination alone and it is extremely important that these are detected early, as certain pelvic fractures can cause life threatening haemorrhage.

3.2.3 Secondary survey

If the patient is responding to the treatment in the Emergency Department and there is no evidence of uncontrolled haemorrhage, then the treating doctor moves onto the secondary survey. This is a top-to-toe examination of the patient, which incorporates the re-evaluation of areas assessed in the primary survey. The management of the injured patient is a dynamic process where patients who appear to be stable in the early phase of their care can become acutely unwell later on, as the effects of their injuries develop over time. Hence it is essential that the different body systems are continually re-evaluated.

The scalp is examined for swelling and lacerations. A gloved finger is placed in open wounds of the scalp to feel for skull fractures. The ears are examined for bleeding from the external auditory canal, which can be a sign of a base of skull fracture. Later on, bruising behind the ear, the so called Battle's sign, may appear which is also indicative of a basic skull fracture. Bleeding from the nose or the presence of Racoon eyes can indicate a fracture of the anterior portion of the skull base.

The eyes are examined in detail, paying particular attention again to pupil size and reactivity to light. The facial bones are palpated and the oral cavity, inclusive the upper and lower jaw are inspected. A sublingual haematoma (haematoma under the tongue) is pathognemonic of a lower jaw fracture.

The neck veins are inspected for distention, which may be indicative of a tension pneumothorax or a cardiac temponade (bleeding within the outer sac of the heart).

The chest is re-examined as before.

The abdomen is inspected for wounds, bruising and distention. It is palpated for tenderness. One can listen for bowel sounds, but the presence or absence of such is not a reliable indicator of significant abdominal injury. The genitalia are inspected for bruising and swelling. The urethra, in particular, is evaluated for blood, as the presence of such can be indicative of a urethral rupture.

A rectal examination is essential to check for the tone of the anal sphincter. A lax anal sphincter is seen with a spinal cord injury. Other findings of note from a rectal examination are those of a high riding prostate gland (ruptured urethra), the presence of blood on the glove (rectum has been perforated) and palpating bony fragments from a pelvic fracture.

The pelvis is "sprung" to check for instability, a procedure which should only be performed once, otherwise it can promote excessive bleeding in a patient with an unstable pelvic fracture.

All four limbs are evaluated for the presence of fractures, dislocations and tendon or nerve injuries. The presence of bruising, swelling and deformity may be indicative of a bony injury. Altered sensation and decreased power in the limb can be indicative of a nerve injury, either in the limb or in the spinal cord.

As one goes through the secondary survey, injuries that are picked up are treated accordingly.

Further evaluation of injuries is undertaken with diagnostic tests inclusive of blood tests, plain radiography, ultrasound and CT scanning. Continued resuscitation/evaluation also occurs through the insertion of a urinary catheter, where appropriate.

3.2.4 Definitive care

This involves the complete management of particular injuries. In the Emergency Department simple wounds are cared for with dressings and closed using sutures or other means. Limb fractures are splinted and dislocations are reduced where possible. Pain relief is given as appropriate, making sure not to give excessive analgesia, thus compromising the evaluation of their injuries. Definitive care of injuries beyond the Emergency Department can involve a large number of specialty teams including Orthopaedics, Neurosurgery, General Surgery, Maxillofacial Surgery, Plastic Surgery, Urology, Cardiothoracic Surgery, Anaesthetics and Intensive Care Medicine.

4. SUMMARY

The evaluation of the injured patient is an intricate process. It begins with the history, paying particular attention to the mechanism of injury. The examination should be thorough and can be indicative of likely injuries. Adjunct to the history and examination are investigations, such as x-rays, which facilitate definite diagnosis and the patient is then treated accordingly

BIOENGINEERING AND ORTHOPAEDIC SURGERY IN PRACTICE

Review of the Applicability of Biomechanical Principles to the Outcomes in Trauma and Orthopaedic Surgery

Paraic Murray
The Galway Clinic, Doughiska, Galway, Ireland
Tel.: +353 91 720160, Fax: +353 91 720161

Abstract. This paper explores where bioengineering has a role in understanding surgical pathology and treatment options.

Key words: bioengineering, surgery, orthopaedics, knee, fracture.

1. HISTORICAL BACKGROUND

Trauma and orthopaedic surgery has evolved from surgery in general over the past century. In the middle ages the learning of the ancient Greeks, Romans and Egyptians had been lost. The seat of learning was the medieval monastery with one monk to look after tonsure, the cutting of the monks hair. Presumably as the tonsure monk had a facility with sharp instruments he also lanced the abscesses. The early surgeons were the tonsure monks in these monasteries. Barbers and Surgeons with their common heritage were included together when the guilds were organised. The red and white pole symbolised blood on bandages and is still used in the College of Surgeons [1] as a symbol on conferring day.

Michael D. Gilchrist (ed.), IUTAM Proceedings on Impact Biomechanics: From Fundamental Insights to Applications, 355–360.

2. EVOLUTION OF SURGERY

As our specialty is evolving a common trend is for mystique and multiple hypotheses, some of which are over simplifications, to be applied to any poorly understood phenomena. The knee was thought to be a hinge however we now know that the exquisite anatomy, mechanics and function of the knee are coordinated like a magnificent symphony.

Although biomechanics has made great strides the pure application of engineering to biological systems is similar to examining some of the wreckage of the space shuttle and trying to find out how to get to the moon.

3. CLINICAL CORRELATION

Theories based on Bioengineering need to be modified by careful analysis of clinical results.

Knee surgery and pathology are used as examples in this paper to give some idea of where we are in this evolutionary process of understanding.

The Swiss Association for the Study of Internal Fixation [2] of fractures (AO-ASIF) has revolutionised fracture surgery. When they began fracture disease was the plague of conservative treatment of fractures. Fracture Disease is the syndrome of joint stiffness, muscle wasting and osteopenia which follows prolonged conservative treatment of limb injuries. Early aggressive surgery and rehabilitation have eliminated these problems. One of the major contributions of the AO has been the research into implant design and performance. Over the years accurate anatomical reduction of diaphyseal (midshaft of a long bone, e.g. the tibia) fractures has given way to biological fixation with restoration of length axis and rotation of the injured limb. The principles of anatomical reduction of the fragments and rigid fixation make sound engineering sense, however, lessons have been learned, and perfect reduction at the expense of blood supply and soft tissue cover is counterproductive.

Prior to the AO implants were of limited success, not only were the materials inadequate for treating biological systems, but the application of implants is much better taught and understood.

Articular fractures where the reduction is to within half the thickness of the articular cartilage are less likely to suffer post traumatic osteoarthritis than where the reduction is less accurate.

Despite this lateral tibial plateau fractures are a conundrum. Accurate anatomical reduction with restoration of the joint alignment is desired but many people do very well with up to a centimetre of joint depression.

4. KNEE BRACES

An interesting study by Kirkley et al. [3] compared the use of the G2 unloader brace and a neoprene sleeve in patients with Osteoarthritis of the medial compartment of the knee and found, despite good biomechanical principles, that there was no benefit with the more sophisticated brace. The purpose of this study was to compare a custom-made valgus-producing functional knee (unloader) brace, a neoprene sleeve, and medical treatment only (control group) with regard to their ability to improve the disease-specific quality of life and the functional status of patients who had osteoarthritis in association with a varus deformity of the knee (varus gonarthrosis).

The study design was a prospective, parallel-group, randomised clinical trial. Patients who had varus gonarthrosis were screened for eligibility. The criteria for exclusion included arthritides other than osteoarthritis; an operation on the knee within the previous six months; symptomatic disease of the hip, ankle, or foot; a previous fracture of the tibia or femur; morbid obesity (a body-mass index of more than thirty-five kilograms per square meter); skin disease; peripheral vascular disease or varicose veins that would preclude use of a brace; a severe cardiovascular deficit; blindness; poor English-language skills; and an inability to apply a brace because of physical limitations such as arthritis in the hand or an inability to bend over. Treatment was assigned on the basis of a computer-generated block method of randomisation with use of sealed envelopes. The patients were stratified according to age (less than fifty years or at least fifty years), deformity (the mechanical axis in less than 5 degrees of varus or in at least 5 degrees of varus), and the status of the anterior cruciate ligament (torn or intact). The patients were randomly assigned to one of three treatment groups: medical treatment only (control group), medical treatment and use of a neoprene sleeve, or medical treatment and use of an unloader brace. The disease-specific quality of life was measured with use of the Western Ontario and McMaster University Osteoarthritis Index (WOMAC) and the McMaster-Toronto Arthritis Patient Preference Disability Questionnaire (MACTAR), and function was assessed with use of the six-minute walking and thirty-second stair-climbing tests. The primary outcome measure consisted of an analysis of covariance of the change in scores between the baseline and six-month evaluations.

One hundred and nineteen patients were randomised. The control group consisted of forty patients (thirty-one men and nine women; mean age, 60.9 years); the neoprene-sleeve group, of thirty-eight patients (twenty-seven men and eleven women; mean age, 58.2 years); and the unloader-brace group, of forty-one patients (twenty-eight men and thirteen women; mean age, 59.5

years). Nine patients withdrew from the study. At the six-month follow-up evaluation, there was a significant improvement in the disease-specific quality of life ($p = 0.001$) and in function ($p <$ or $= 0.001$) in both the neoprene-sleeve group and the unloader-brace group compared with the control group. There was a significant difference between the unloader-brace group and the neoprene-sleeve group with regard to pain after both the six-minute walking test ($p = 0.021$) and the thirty-second stair-climbing test ($p = 0.016$). There was a strong trend toward a significant difference between the unloader-brace group and the neoprene-sleeve group with regard to the change in the WOMAC aggregate ($p = 0.062$) and WOMAC physical function scores ($p = 0.081$).

The results indicate that patients who have varus gonarthrosis may benefit significantly from use of a knee brace in addition to standard medical treatment. The unloader brace was, on the average, more effective than the neoprene sleeve. The ideal candidates for each of these bracing options remain to be identified.

5. KNEE ALIGNMENT

Undoubtedly alignment of the knee [4] is important for longevity in knee replacement and poorly aligned knees fail earlier. Much of the instrumentation in total knee replacement is dedicated to restoring the anatomy as accurately as possible in joint replacement. Careful use of these instruments will allow a satisfactory outcome in the majority of patients despite the limitations imposed by the design of the implant. Many tibial implants are unisexed; there are no left and rights. Consequently one will either fail to completely cover the medial tibial plateau or over hang on the lateral side. The articulating surface is flat to concave on both sides despite the convex nature of the natural lateral tibial plateau.

The three variables that consistently correlated with wear were patient age, mechanical axis alignment, and the shelf-age of the component. Gamma Irradiation was introduced to eliminate the storage problems associated with ethylene oxide sterilisation of polyethylene implants. This logical change has been associated with dramatic increases in the wear rates of total hip replacements. Normal wear rates are predicted to be in the order of micrometers over years, with hardly perceptible wear over the lifetime of a joint replacement. The present author has revised joints where the femoral component of a THR had worn completely through the polyethylene.

The background to this is related to the manufacture of polyethylene. There are only a few companies who manufacture this commodity and there is no such thing as a medical grade. Consequently the grade used for implants is the top of the industrial grades. Companies using this polyethylene were unable to modify its properties, for the better of patients and their own profits. Hylamer was introduced by DePuy to correct this anomaly and was expected to out perform standard polyethylene. Unfortunately this failed to materialise. Hylamer was eventually withdrawn and the company are currently defending class actions by patients. I remain to be convinced that implant companies' sole focus is on the welfare of patients.

Malalignment of the extensor mechanism is best treated with principles which respect the biomechanics of the knee. Lateral subluxation of the patella requires correction of the quadriceps (Q) angle by lateral release of the patella, with medial advancement of the medial retinaculum and possibly a Fulkerson slide of the tibial tuberosity. Accurate correction of the extensor mechanism kinematics generally gives a good result. Failure to do so may lead to recurrent subluxation.

6. MOLECULAR BIOLOGY

An area of exploding research is in molecular biology. A study we performed was in modifying scar formation in surgical incisions.

Adhesion [5] formation is one of the foremost obstacles to a reliably good outcome in tendon and joint surgery. Thermal preconditioning has been found to reduce the inflammatory response through the induction of molecular chaperone expression, a recently described family of cytoprotective intracellular proteins. The authors analysed the effect of thermal preconditioning on the inflammatory response to surgery, on tendon healing, and on the formation of peritendinous adhesions in 16 New Zealand White rabbits. Very significant decreases in adhesion formation and in the gliding and dimensions of tendons in animals that had thermal preconditioning were found. Tendons from these animals also showed a decreased level of adhesion formation and a significantly diminished inflammatory response on histological examination with no biomechanically significant deleterious effect on the strength of tendon healing on testing load to failure. These findings are consistent with induction of heat shock proteins by hyperthermic pre-treatment. Such prevention of peritendinous adhesions and the inflammatory response to injury and surgery without compromising healing are findings that have significant implications for tendon surgery and all surgery involving joints and soft tissues.

7. SUMMARY

My talk will explore where bioengineering has a role in understanding surgical pathology and treatment options. The future should involve an even greater role for Surgeon Scientists in the design of new implants. Bioengineers should be more involved in the rigorous assessment of new implants. Why is it in North America that the introduction of all new drugs are carefully controlled while new knee replacements are introduced with the grandfather clause, they are similar to previous implants?

REFERENCES

1. www.rcsi.ie Virtual Tour, Council Room.
2. AO Manual for Internal Fixation.
3. Kirkley, A., Webster-Bogaert, S., Litchfield, R., Amendola, A., MacDonald, S., McCalden, R and Fowler, P., The effect of bracing on varus gonarthrosis, J. Bone Joint Surg. Am. 1999 Apr., 81(4):539–548.
4. Gerard, A and Engh, M.D., Causes of Implant Wear with TKA/UKA, the Arthritic Knee, Dublin 2005.
5. Mulhall, K.J., McLaughlin, R., Kay, E., Kiely, P., Bouchier-Hayes, D. and Murray, P., Thermal preconditioning prevents peritendinous adhesions and inflammation, Clin. Orthop. 2002 Dec., (405):258–266.

THE IMPACT BIOMECHANICS OF SPINAL COLUMN INJURIES

M.J. Shelly and A.R. Poynton
Department of Orthopaedic Surgery and the National Spinal Injuries Unit, Mater Misericordiae University Hospital, Eccle St., Dublin 7, Ireland

Abstract. The vertebral column is the central bony pillar of the body and serves to protect the spinal cord from injury. Vertebrae show regional differences but they all possess a common pattern. Analysis of the biomechanics of individual components of the spine allows one to predict how the spinal column behaves during impact. Although many forces and moments in different directions act on the spine, they can be resolved into the basic loads – compressive/tensile forces, shearing forces, sagittal plane bending and torsion. The normal motion and function, mechanisms of injury and residual stability of the cervical spine is explored. Stabilisation techniques employed in the management of cervical spine trauma are described along with the clinical indications, advantages and disadvantages of each technique.

Key words: spine, biomechanics, force, stabilisation.

1. BASIC ANATOMY

The vertebral column is the central bony pillar of the body. It supports the skull, pectoral girdle, upper limbs, and the thoracic cage, and by way of the pelvic girdle, transmits body weight to the lower limbs. Within its cavity lie the spinal cord, the roots of the spinal nerves, and the covering meninges, to which the vertebral column gives great protection. The vertebral column is composed of 33 vertebrae – 7 cervical, 12 thoracic, 5 lumbar, 5 sacral (fused to form the sacrum) and 4 coccygeal (the lower three are commonly fused). The fact that the vertebral column is segmented and made up of vertebrae, joints, and pads of fibrocartilage called intervertebral discs, makes it a

Michael D. Gilchrist (ed.), IUTAM Proceedings on Impact Biomechanics: From Fundamental Insights to Applications, 361–377.

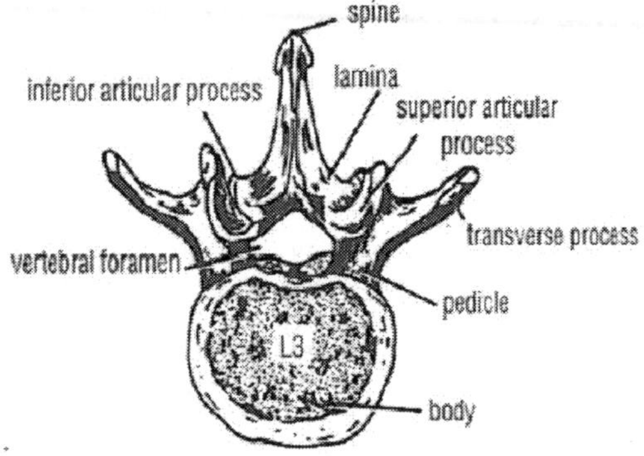

Figure 1. Typical vertebra.

flexible structure. The intervertebral discs form about one-quarter the length of the column. Although vertebrae show regional differences, they all possess a common pattern (Figure 1).

1.1 Characteristics of Atypical Cervical Vertebrae

The first cervical vertebra (atlas) does not posses a body or spinous process. It has an anterior and posterior arch with a lateral mass either side. The second cervical vertebra (axis) has a peg-like odontoid process that projects from the superior surface of the body (representing the body of the atlas that fused with the body of the axis).

1.2 Joints of the Vertebral Column

1.2.1 Atlanto-occipital joints

The atlanto-occipital joints are synovial joints that are formed between the occipital condyles (base of skull) and the facets on the superior surfaces of the lateral masses of the atlas. Two strong ligaments cross the joint, namely the anterior atlanto-occipital membrane which connects the anterior arch of the atlas to the anterior margin of the foramen magnum and the posterior atlanto-occipital membrane which connects the posterior arch of the atlas to the posterior margin of the foramen magnum. Flexion, extension and lateral flexion all occur at these joints. Rotation is not possible.

1.2.2 Atlanto-axial joints

The atlanto-axial joints are three synovial joints, one of which is between the odontoid process and the anterior arch of the atlas, while the other two are between the lateral masses of the bones. The apical ligament is a median-placed structure connecting the apex of the odontoid process to the anterior margin of the foramen magnum. The alar ligaments lie on each side of the apical ligament and connect the odontoid process to the medial side of the occipital condyles. The cruciate ligament consists of a strong transverse part and a weaker vertical part. The transverse part is attached on either side to the inner aspect of the lateral mass of the atlas and binds the odontoid process to the anterior arch of the atlas. Extensive rotation of the atlas (and thus the head) on the axis occurs at this joint.

1.2.3 Joints of the vertebral column below the axis

With the exception of the first two cervical vertebrae, the remainder of the mobile vertebrae articulate with each other by means of fibrocartilaginous joints between their bodies (the intervertebral discs) and by synovial joints between their articular processes (the facet joints) [1].

2. BIOMECHANICS OF INDIVIDUAL COMPONENTS OF THE SPINE

2.1 Basic Definitions

Although many forces and moments in different directions act on the spine, they can be resolved into the basic loads. These are compressive and tensile forces acting along the long axis of the spine, anteroposterior and mediolateral shearing forces that directly displace the vertebrae in these directions, sagittal plane bending causing flexion or extension, lateral bending, and torsion or rotation about the longitudinal axis of the spine [2].

The spine exhibits viscoelastic behaviour and under load, a spinal motion segment (two vertebrae and their interconnecting soft tissues) displace in a characteristic way. The motion segment has a neutral zone or region where it can be displaced with little force. Beyond the neutral zone, as the tensions in the tissues increases, is an elastic zone where increasing force is required to produce greater displacement in any direction. Finally, the zone of failure is reached, where tissue tearing and/or bony fracture result (Figure 2). The neutral zone has been shown to be the most sensitive

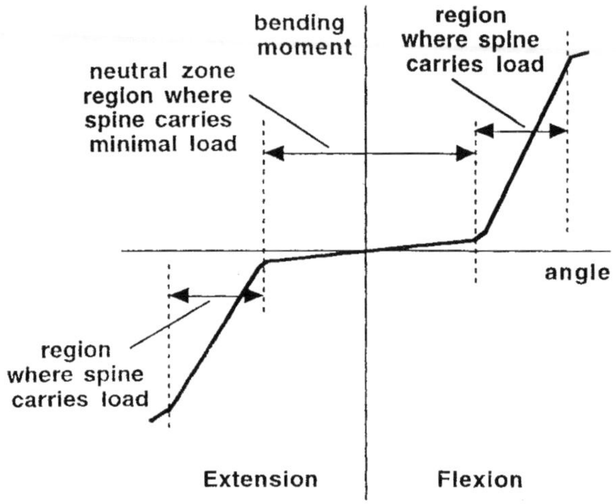

Figure 2. Typical load displacement behaviour of a spinal motion segment.

parameter in defining a spinal injury [3]. Release of the applied load after loading into the damage region results in the motion segment remaining in a deformed position because of tissue damage.

2.2 Vertebra

The vertebral body is principally formed of trabecular bone with a thin cortex and superior and inferior endplates. The vertical columns of trabecular bone are the principle load-bearing structures [4]. The cross ties (horizontal trabeculae) assist in load bearing by preventing the vertical columns from buckling. Buckling of a long slender column is a common mechanism of failure under compressive loading, and loss of cross tie support is more critical than loss of area of trabecular column.

2.3 Disc-Nucleus-Endplate

The intervertebral disc consists of three major components: the annulus fibrosis, which resists tensile stresses along with bulk compression of its material; the nucleus pulposis, which converts some of the compressive force to internal pressure that creates disc bulging or tensile stresses; and the endplates, which form a flexible seal on the vertebral body and prevent extrusion of the nucleus (Figure 3). The fibre failure tensile stress is greatest in the outer anterior ring of the annulus, and significantly, the posterior annular fibres are only about 50% as strong and considerably thinner, therefore predisposing the posterior region to fail at considerably lower load than the anterior region [5].

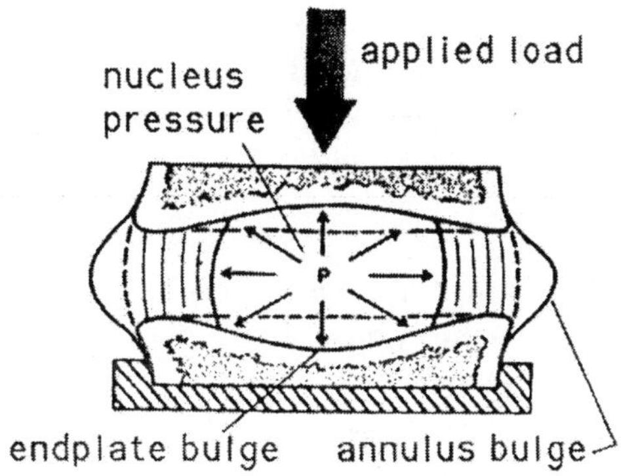

Figure 3. Structure and function of the disc-nucleus-endplate unit under load.

2.4 Facet Joints

The facets clearly limit motion in various planes by bony interaction. which vary in different regions of the spine. In the cervical spine, the facets limit anterior translocation (of the superior vertebra) and compression, in the thoracic spine they limit anterior translocation and in the lumbar spine they limit axial rotation and anterior translocation. The facets are also surrounded by a very tight fibrous capsule; therefore, loads that tend to separate the facet surfaces also contribute to loading the facets and limiting spinal motion [2].

2.5 Spinal Ligaments

The spinal ligaments have well-defined structural functions. Many posterior ligaments, based on elongation, contribute to resisting flexion, especially the supraspinous, interspinous, and facet capsules. Similarly, the anterior longitudinal and capsular ligaments resist extension, and the intertransverse ligaments and facet capsules resist lateral bending. The major strain during rotation occurs in the capsule of the facet whose faces are separating [6, 7].

3. OCCIPITO-ATLANTO-AXIAL JOINTS

3.1 Normal Motion and Function of Components

The function of the C0-C1-C2 joint complex can be appreciated by observing the arrangement of its bony surfaces (Figure 4). C0-C1 allows about 13 degrees of flexion/extension and 8 degrees of lateral bending. In contrast, C1-C2 permits a significant axial rotation (47 degrees) and some lateral bending (10 degrees), but negligible flexion and extension [8]. The centre of axial rotation is aligned with the longitudinal axis of the odontoid. The odontoid functions not only as a guiding axis for head rotation, but resists shearing (horizontal transverse forces) and bending of the head in the posterior and lateral directions, and, indirectly through the transverse ligament, flexion. Ligaments function to limit motion as their internal tension increases. The transverse ligament restrains flexion and anterior motion of the head [9] and the capsular ligaments of C1-C2 act as backups especially in axial rotation [10].

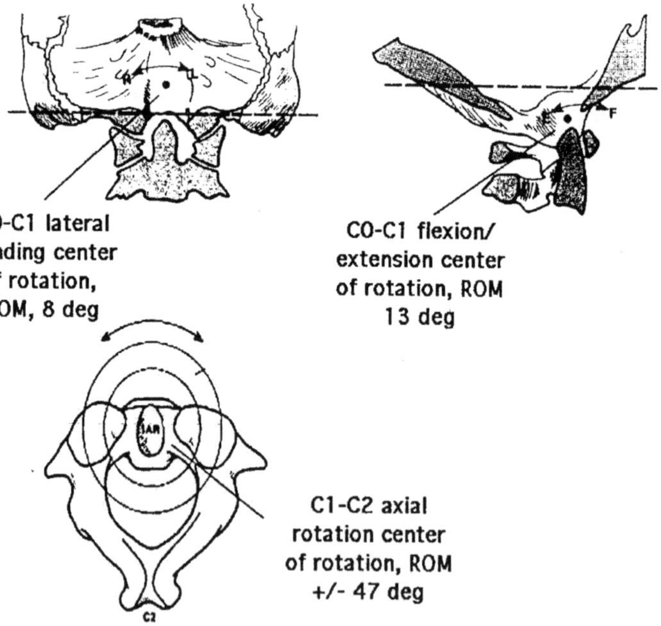

Figure 4. Anatomy and function of C0-C1-C2 complex.

3.2 Mechanism of Injury and Load Tolerance

The loads applied to the C0-C1-C2 joints are complex, typically resulting from a blow to the head or by differential motion between the head and the torso. Certain structures are typically injured due to loading in specific directions, e.g. flexion of the occiput, without compression, produces mainly type 3 odontoid fractures (inferior to the base of the odontoid, including bone of C2), lateral bending results in type 2 fractures (at the base of the odontoid), and extension creates mainly type 2 fractures [11, 12]. The odontoid has very thick cortical bone at its base, which explains its considerable and symmetric strength in different loading directions, but the mechanism of load transfer is not the same. Extension of the occiput loads the surface of C1 against the odontoid, whereas flexion loads the transverse ligament, which, because of its band arrangement, transfers load to the odontoid. The consequence of this is certain failure of the transverse ligament in flexion-type injuries [9]. The application of compression may result in fracture of the posterior or anterior arch of the atlas [13], along with other injuries such as separation of the lateral masses and ligament failure (Jefferson burst fractures).

3.3 Residual Stability

Of great importance clinically is an understanding of the mechanical stability remaining after injury. Any joint has a neutral zone or range of motion that occurs with very little load. In addition, there may be an elastic zone, depending on the structures that have been injured. The sum of these motions defines the overall range of motion of the joint [10]. The neutral zone has been found to be the more sensitive indicator of the effect of injury.

Jefferson fractures (fractures of the ring of the atlas) have been shown to increase the neutral zone motion by a mean of 27% in flexion/extension. Axial rotation and lateral bending were minimally affected. The effect of odontoid fracture has not been specifically measured. However, if the alar and transverse ligaments are damaged, there will clearly be significant increases in rotational instability as well as translation, which could cause cord occlusion.

4. CERVICAL SPINE

4.1 Normal Motion and Function of Components

The centres of rotation of the cervical spine lie close to the facet articulations, demonstrating the motion-limiting role of the facets. The measured flexion/extension, lateral bending, and rotational motions in the normal spine demonstrate fairly uniform ranges of about 10 degrees from C2/C3 to C7/T1, in each direction for each level [14]. The major stabilisers of the spine are the surrounding muscles. The vertebra is the main structure supporting compression of the spine. However, compressive load has been shown to be distributed approximately two thirds on the vertebral body itself and one third on the lateral masses [15]. Further, the uncinate processes serve mainly to limit axial rotation (removal increases rotation by 25%) and lateral bending (14%) [16]. Isolated facet capsule sectioning showed that these structures contribute about 25% of resistance in axial rotation and 30% in flexion/extension [17].

4.2 Mechanism of Injury and Load Tolerance

The cervical spine is an unstable column with a large weight (the head) on its upper end and is therefore prone to injury, typically in road traffic accidents (RTAs) and some sporting activities. C5/6 is injured most often with combined compression and flexion, followed by facet dislocation the most common mechanism Injuries to the cervical spine can be broadly grouped as tension or compression based, in both cases coupled with bending.

Differential motions between the head and the torso tend to create significant neck tensile forces combined with bending, even without direct impact. This may occur where the torso is restrained from moving by the seat and shoulder belts while high accelerations are applied to the head, causing combined bending and tension injuries to the neck. Combining tension with extension typically results in a combination of posterior element fracture, along with anterior disc and ALL damage. A common mechanism of tension/extension injury occurs during rear-end collisions in vehicles without head rests, where the torso motion is stopped by the seat back while the head continues to extend posteriorly. Tension combined with flexion can result from rear-end impacts. It is accentuated by complex interactions between the torso, seat back, head, and head rest (Figure 5). Injuries can include soft tissue damage, typically to the posterior elements combined with, at higher load levels, facet dislocation. Lateral bending combined with tension, which occurs in vehicle side impacts, can result in soft tissue injuries such as facet capsule damage and disc injury.

A second type of injury occurs by direct impact to the head. A small difference in location of the impact load vector can result in significantly different injuries to the cervical spine. When the head is flexed forward so that the cervical spine is not in its normal lordotic posture, the cervical spine tends to lock and not allow further flexion. Compression applied in this case will result in a burst injury, with vertical fracture lines in the vertebra and the possible retropulsion of bone into the canal [18]. When the impact is located progressively more anteriorly, the resulting injury changes first to a wedge compression fracture, then to a facet dislocation [2]. Trauma to the spinal cord adds a further dimension to the magnitude of injury. Spinal cord compression is usually maximal at the time of injury [18]. This observation may help to explain why cord injury symptoms frequently do not correspond to the observed post-injury canal intrusion.

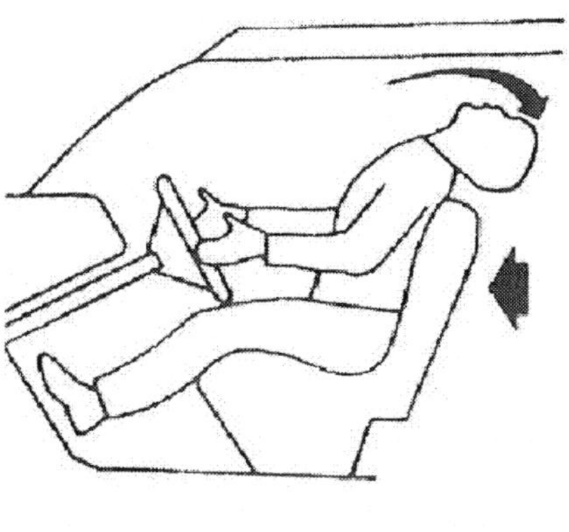

Figure 5. Hyperextension of neck in rear-end impact without a headrest.

4.3 Residual Stability

Clinical instability has been defined as the loss of the ability of the spine under physiologic loads to maintain its patterns of displacement so there is no initial or additional neurologic deficit, no major deformity, and no incapacitating pain [19]. In the case of ligamentous injuries, Panjabi et al [20] showed that loss of posterior ligaments does not significantly affect flexion or extension if the disc remains intact. However, loss of the anterior half of the disc, although not affecting flexion, does significantly increase extension motion compared to the intact spine under the same load [21]. Any injury that involves vertebral fracture results in less ability of the vertebra to carry load, but a second effect is reduction in the tensions of the surrounding ligaments, even if they are not damaged, creating increased laxity of the motion segment.

5. STABILISATION TECHNIQUES: UPPER CERVICAL SPINE

Atlanto-axial fusions may be indicated in certain cases of acute or chronic atlanto-axial instability. For most lesions in the upper cervical spine a posterior approach is preferred. However, some type-II odontoid fractures are suitable for anterior screw fixation.

5.1 Gallie Technique

This wiring technique provides a posterior stable construct for C1-2 instability particularly resistant to flexion forces. It has less resistance to extension and lateral bending forces and has been superseded by screw fixation techniques. Indications include fracture of the odontoid with anterior displacement and rupture of the transverse ligament of C1. The patient is placed prone with skeletal traction applied to facilitate reduction of the C1-2 subluxation. Lateral image intensification is used to check position and reduction. A midline incision extends from the occiput to C4. The soft tissues are cleared from occiput to C2. Lateral dissection beyond 2 cm from midline is avoided to prevent injury to the vertebral artery and venous plexus. A 1.2mm wire is fashioned into a loop with a hook configuration (Figure 6). The wire is passed from the inferior aspect of C1 cranially and looped over the superior surface of C1. The loop is carefully pulled backwards and distally sufficiently far to loop over the spinous process of C2. A cortico-cancellous rectangular bone graft is removed from the posterior iliac crest and fashioned into an H configuration. It is notched laterally. The two free ends of the wire, which are laterally placed, are then brought across to the midline after the graft has been applied to the posterior surfaces of C1 and C2 (Figure 7). The notches provide better fixation of the bone graft when tightening the wires. Postoperatively, a firm collar preventing extension of the neck is recommended for a period of 6-10 weeks.

5.2 Transarticular Screw Fixation

Indicated for acute and chronic atlanto-axial instability. This technique is biomechanically superior to a wiring technique and maintenance of reduction is possible. The patient lies in the prone position and the reduction

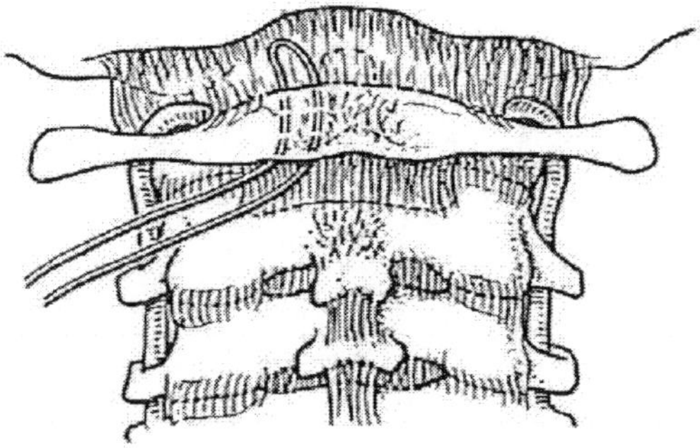

Figure 6. 1.2mm wire passed from the inferior aspect of C1 anteriorly and cranially.

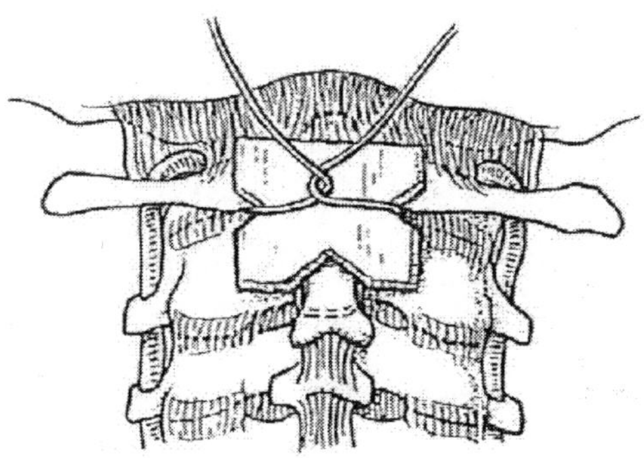

Figure 7. Bone graft held in situ by 1.2mm wires.

of C1-2 is checked using an image intensifier. Reduction is facilitated by skeletal traction. The neck is flexed as much as possible to facilitate insertion of screws. A midline incision from the occiput to the tip of the spinous process of C5 is performed. The soft tissues are dissected off posteriorly to expose the arch of C1, the spinous processes, lamina and inferior articular processes of C2 and C3. Under image intensifier control, a long 2.5mm drill is inserted in a strictly sagittal direction (Figure 8). The entry point of the drill is at the lower edge of the caudal articular processes of C2. The drill enters the lateral mass of the atlas close to it posterior-inferior edge and perforates its cortex anteriorly. Screws are inserted after tapping with a 3.5mm cortical tap across the C1-C2 joint (Figure 9). Postoperatively, patients are immobilised in a firm collar for 6-8 weeks.

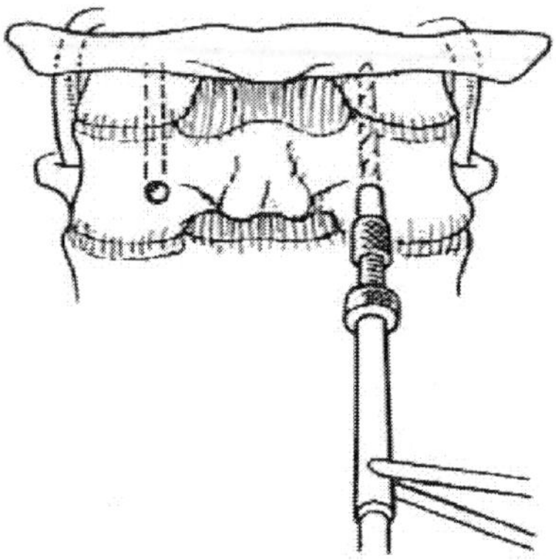

Figure 8. Entry point of the drill is at the lower edge of the caudal articular process of C2.

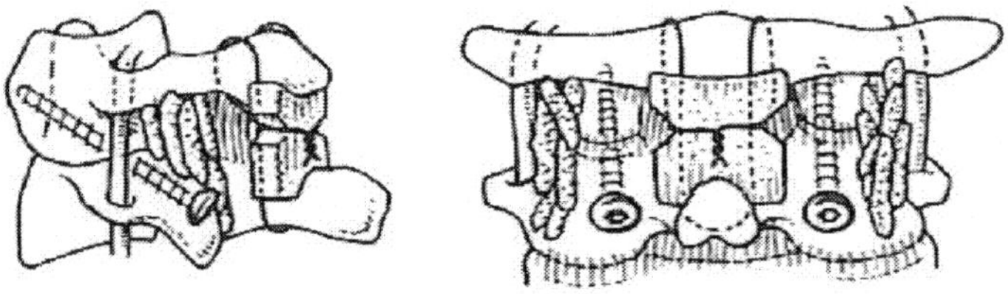

Figure 9. Configuration of the cannulated self-cutting screws.

More recently independent screw fixation of C1 (lateral mass) and C2 (pedicle) has become popular. The advantage of this technique is the ability to reduce displacement intraoperatively.

6. STABILISATION TECHNIQUES: LOWER CERVICAL SPINE

6.1 Posterior Techniques

6.1.1 Wiring technique

There are many wiring techniques for posterior fixation of the lower cervical spine. The most simple and least dangerous is interspinous wiring. It is indicated in injuries of the posterior complex involving predominantly soft tissue with insignificant injury to the vertebral body. Disadvantages of the technique include wire breakage/cut-out, poor biomechanical fixation, and a failure to maintain lordosis. A midline posterior approach is used. A hole is drilled on either side of the base of the spinous process of the upper vertebra of the injured segment. The entry point corresponds to the junction of the base of the spinous process and lamina. A pointed towel clip is placed in the holes and with a gentle rocking movement the holes are connected. A 1.2mm wire is passed through the hole and then around the base of the inferior spinous process, leaving the interspinous soft tissue intact. The two ends of the wires are tightened. Finally, the wire ends are curved around the inferior spinous process and twisted tight. Cancellous bone graft is applied to the interspinous segment (Figure 10).

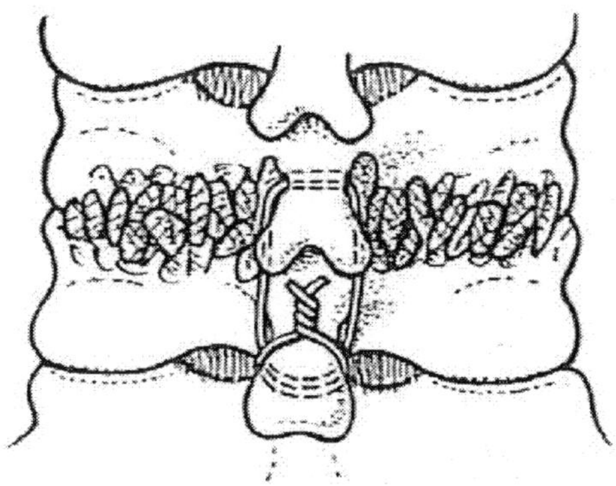

Figure 10. Posterior cervical wiring technique.

6.1.2 Plate technique

The posterior plating technique provides stable tension band fixation in flexion, increasing stability in rotation, and buttressing in extension. It is indicated in ligamentous and/or osseous lesions of the posterior complex without significant damage of the vertebral body, Uni- or multisegmental instability and instability associated with deficiency of the posterior elements from laminectomy or fractures, e.g. trauma, tumour or failed anterior fusion. It has superior stability compared to wiring techniques and can be used in the presence of lamina and spinous process fractures. The technique has the disadvantage of potential neurovascular damage. A midline posterior approach is performed. The most cranial and caudal vertebrae of the selected area of fusion are drilled; Kirschner wires inserted into the drill holes and the small titanium template is applied to assess the size and contour of the titanium plate required. The chosen titanium plate is applied to the spine. The proximal and distal holes are measured through the titanium plate; only the near cortex is tapped. Prior to screw insertion, the posterior surface of the facet joints is decorticated and packed with cancellous bone graft. The plate is fixed to the lateral masses cranially and caudally with 3.5mm cancellous screws in the drill holes. The intervening lateral masses are now drilled and screws inserted as before (Figure 11).

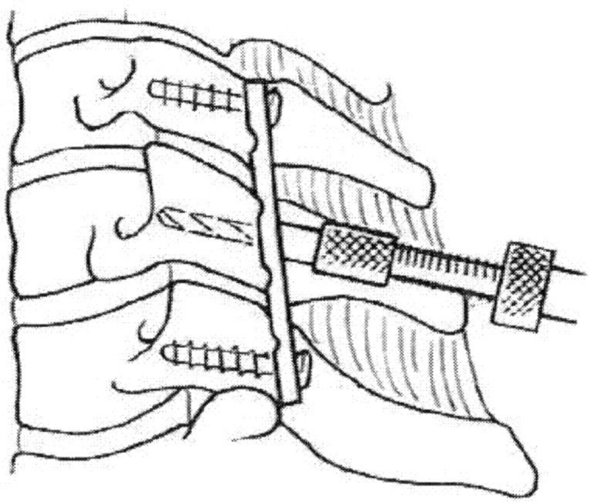

Figure 11. Posterior cervical plating.

6.1.3 Cervical spine titanium rod system

This is a modular tension band system for posterior fixation of the occipito-cervical spine, upper and lower cervical spine, and upper thoracic spine. A choice of clamps and hooks are fixed on a 3.5mm titanium rod by means of set screws. 3.5mm bone screws can be optimally positioned through the clamps in any desired direction and on each motion segment. A cross linking device may also be used to connect both rods and protect the spinal cord in cases of extensive laminectomy. Indications include occipito-cervical and upper cervical spine instabilities (e.g. rheumatoid arthritis, tumours, infections, post-traumatic conditions), instabilities of the lower cervical spine (e.g. traumatic instabilities, tumours), and degenerative conditions of the lower cervical spine. Advantages of the technique include: optimal screw insertion at all levels instrumented, optimal bone grafting, allows postoperative MRI. The technique is considerably more expensive than posterior plate fixation (Figure 12).

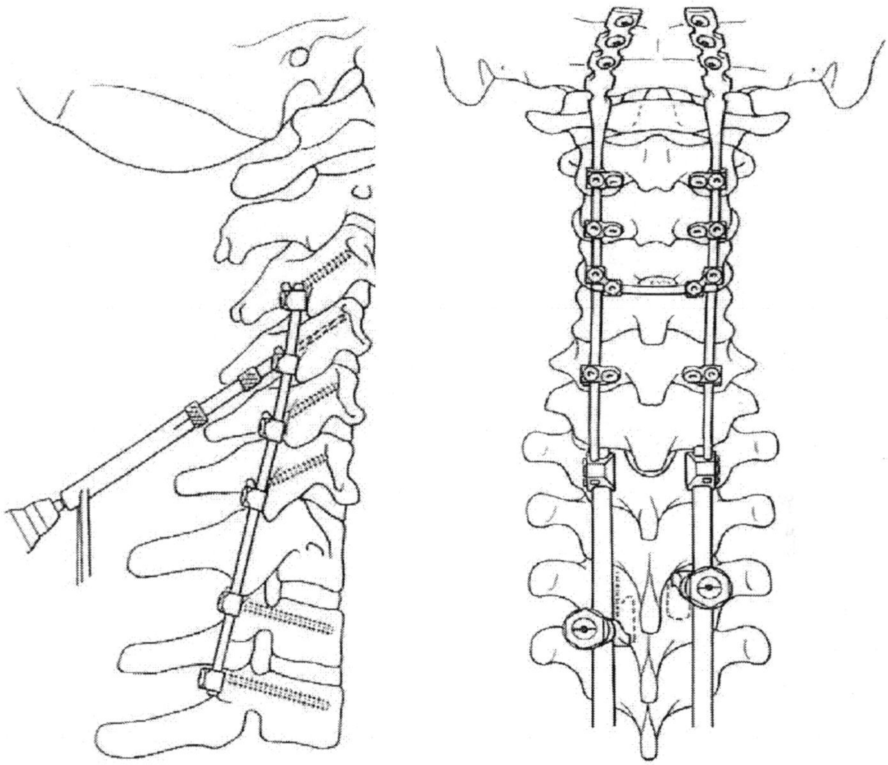

Figure 12. Cervical spine titanium rod system.

6.2 Anterior Techniques

6.2.1 Plating techniques

Anterior plate fixation is used to increase the stability of the anterior column following grafting techniques. The plate acts as a tension band in extension and as a buttress plate in flexion. It is indicated to support the anterior column when instability persists, particularly when associated with loss of height of the vertebral body following a severe wedge compression or burst fracture. It is also indicated post partial or total vertebrectomy for decompression of the spinal cord. The standard H plate is commonly used. It has the disadvantage that screws have to penetrate the posterior cortex of the vertebral body with a potential risk to the spinal cord and screw loosening with anterior migration can occur. An anterior approach is used. Before applying the plate, an anterior interbody fusion is performed using either a tri-cortical bone graft or a spacer device, such as a titanium cage. If bone graft is used, it should be wedge shaped to maintain the lumbar lordosis. A plate of suitable length is placed onto the anterior aspect of the vertebral bodies and held with a clamp. It is essential to know the sagittal diameter of the vertebral body to prevent overpenetration of the posterior cortex with the drill and thus prevent injury to the contents of the spinal canal. Only the anterior cortex is tapped with the 3.5mm tap. The appropriate length 3.5mm cortex screw is inserted. A similar procedure is performed with the other screw holes (Figure 13). The drill holes can be eccentrically drilled to allow axial compression on the vertebral graft and to enhance stability when tightening the screws [22].

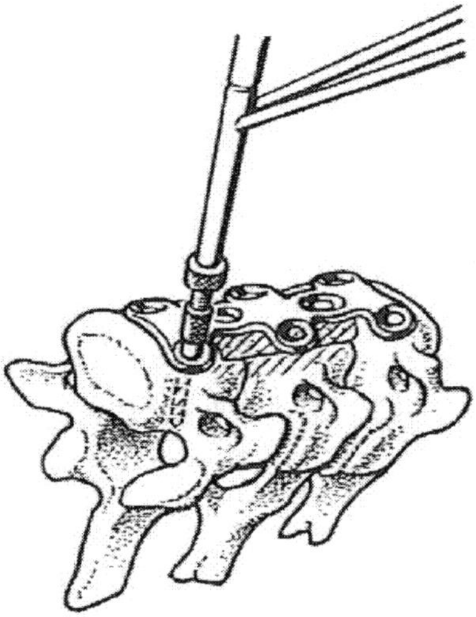

Figure 13. Anterior cervical spine plating.

REFERENCES

1. Snell, R.S., *Clinical Anatomy for Medical Students*, Fifth Ed, Little, Brown & Co., Boston, 1995.
2. Cotler, J.M., Simpson, J.M., An, H.S. and Silveri, C.P., *Surgery of Spinal Trauma*, Lippincott Williams & Wilkins, Philadelphia, 2000.
3. Oxland, T.R. and Panjabi, M.M., The onset and progression of spinal injury: A demonstration of neutral zone sensitivity, *J. Biomech*. 1992, 25:1165-1172.
4. Bell, G.H., Dunbar, O., Beck, J.S. and Gibb, A., Variation in strength of vertebrae with age and their relation to osteoporosis, *Calcif. Tissue Res.* 1:75-86.

5. Acaroglu, E.R., Iatridis, J.C., Setton, L.A., Foster, R.J. and Mow, V.C., Weidenbaum M. Degeneration and aging affect the tensile behaviour of human lumbar annulus fibrosis. *Spine* 1995, 20:2690-2701.
6. Tencer, A.F., Mayer TG. Soft tissue strain and facet face interaction in the lumbar intervertebral joint: Part 2. Calculated results and comparison with experimental data. *J. Biomech. Eng.* 1983, 105:210-215.
7. Panjabi, M.M., Goel, V.K. and Takara, K., Physiological strains in the lumbar spinal ligaments. *Spine* 1982, 7:192-203.
8. Panjabi, M.M., Dvorak, J., Duranceau, J. et al. Three dimensional movements of the upper cervical spine. *Spine* 1988, 13:726-731.
9. Fielding, J.W., Cochran, G.V.B., Lawsing, J.F. and Hohl, M., Tears of the transverse ligament of the atlas. *J. Bone Joint Surg.* 1974, 56A:1683-1691.
10. Crisco, J.J., Oda, T., Panjabi, M.M., Bueff, H.U., Dvorak, J. and Grob, D., Transactions of the C1-C2 capsular ligaments in the cervical spine. *Spine* 1991, 16:S474-S479.
11. Mouradian, W.H., Fietti, V.G. Jr., Cochran, G.V.B., Fielding, J.W. and Young, J., Fractures of the odontoid: a laboratory and clinical study of the mechanisms. *Orthop. Clin. North Am.* 1987, 9:985-1001.
12. Doherty, B.J., Heggeness, M.H. and Esses, S.I., A biomechanical study of odontoid fractures and fixation. *Spine* 1993, 18:178-184.
13. Oda, T., Panjabi, M.M., Crisico, J.J., Oxland, T.R., Multidirectional instabilities of experimental burst fractures of the atlas. *Spine* 1992, 17:1285-1290.
14. van Mameren, H., Sanches, H., Beursgens, J. and Drukker, J., Cervical spine motion in the sagittal plane:2. Position of segmental averaged instantaneous centres of rotation, a cineradiographic study. *Spine* 1992, 17:467-474.
15. Pintar, F.A., Yoganandan, N., Pesigan, M., Reinartz, J., Sances, A. Jr., Cusick, J.F., Cervical vertebral strain measurements under axial and eccentric loading. *J. Biomech. Eng.* 1995, 117:474-478.
16. Clasen, J.D., Goel, V.J. and Traynelis, V.C., Investigation of the role of the uncinate process and Luschka's joints using finite element analysis. Transactions of the 43rd Annual Meeting of the Orthopaedic Research Society, 1997, 22:359.
17. Zdeblick, T.A., Abitbol, J.J., Kunz, D.N., McCabe, R.P. and Garfin, S., Cervical stability after sequential capsule resection. *Spine* 1993, 18:2005-2008.
18. Carter, J.W., Ching, R.P., Tencer, A.F. and Mirza, S., Transient changes in cervical spinal canal geometry during wedge compression fracture. Proceedings of the 6th Injury Prevention through Biomechanics Symposium, Detroit, Michigan, 1996, 39-46.
19. Panjabi, M.M., Lydon, C., Vasavada, A., Grob, D., Crisco, J.J. and Dvorak, J., On the understanding of clinical instability. *Spine* 1994, 19:2642-2650
20. Panjabi, M.M., White, A.A. and Johnson, R.M., Cervical spine mechanics as a function of transaction of components. *J. Biomech.* 1975, 8:327-336.
21. Panjabi, M.M., Duranceau, J.S., Oxland, T.R. and Bowen, C.E., Multidirectional instabilities of traumatic cervical spine injuries in a porcine model. *Spine* 1989, 14:111-1115.
22. Aebi, M., Thalgott, J.S. and Webb, J.K., *AO ASIF Principles in Spine Surgery*, Springer-Verlag, Berlin, 1998.

BRAIN OEDEMA AND INTRACRANIAL PRESSURE

Jan Goffin
Department of Neurosurgery, Catholic University of Leuven, University Hospital Gasthuisberg, Leuven, Belgium

Key words: brain oedema, intracranial pressure.

1. INTRODUCTION

One of the most dangerous consequences of severe head injuries is the development of **brain oedema**. Brain oedema is also seen in a number of other pathologies, such as brain tumours, infections, intoxications, ischemia etc.

Brain oedema means an increase the brain volume due to an enhancement of the water content. **Vasogenic oedema** is the most common form of brain oedema encountered in clinical practice, for example with trauma and tumour. It results from increased permeability of capillaries, in which the concept of 'blood-brain-barrier' plays an important physiologic and pathophysiologic role. The endothelial cells of brain capillaries are normally connected with tight junctions that act as a barrier to the passive movement of many substances across the endothelium. In case of vasogenic oedema these tight junctions become incompetent, allowing plasma to escape into the intercellular space. **Cytotoxic oedema** is also encountered with trauma, but commonly results from hypoxia of the neural tissue (which many times accompanies traumatic conditions). The hypoxia affects the ATP-dependent sodium pump mechanism in the cell membrane, promoting accumulation of intracellular sodium and the subsequent flow of water into the cell to maintain osmotic equilibrium. Thus, this type of oedema is primarily intracellular and affects virtually all cells, including endothelial cells, astrocytes and neurons.

Michael D. Gilchrist (ed.), IUTAM Proceedings on Impact Biomechanics: From Fundamental Insights to Applications, 379–382.

The global intracranial content normally consists of 1400 gram brain tissue, 75 ml blood and 75 ml cerebrospinal fluid. These 3 compartments can not be compressed and the bony skull is a rigid structure. As a consequence, each increase of volume of one of the compartments must lead to an increase of intracranial pressure (ICP): the so-called Monroe-Kellie doctrine. Each minute about 750 ml blood enters and leaves the brain, which corresponds to a cerebral blood flow (CBF) of 50 ml/100 gr. brain tissue/minute.

Most of the cerebrospinal fluid (CSF) is produced in the ventricles of the brain. CSF circulates downwards and leaves the brain at the base of the skull. Part of the CSF circulates more downwards in the spinal canal. Eventually it migrates over the external sides of the cerebral hemispheres and is reabsorbed in large venous vessels at the inner side of the skull. As such this continuous production, migration and reabsorption of CSF reflects a dynamic equilibrium. The production of CSF is independent from the ICP, whereas its reabsorption is linearly related to the ICP.

Normal brain tissue contains 80% water, primarily located intracellularly.

2. DELETERIOUS CONSEQUENCES OF BRAIN OEDEMA

Deleterious consequences of brain oedema (and of other space-occupying lesions of the intracranial space as well): brain oedema has a **pressure effect** and a **volume effect**.

2.1 The Pressure Effect Leads to a Decrease of CBF

CBF = k.(SABP-ICP)/CVR
SABP = systemic arterial blood pressure
CVR = cerebrovascular resistance
SABP – ICP = CPP (cerebral perfusion pressure)

This formula expresses the principle of 'autoregulation' of the CBF, which in normal circumstances keeps the CBF independent from SABP, ICP or CPP. Indeed, if for example the blood pressure starts to rise, cerebral blood vessels will constrict (enhanced CVR), and the CBF remains constant.

However, in pathologic conditions, the autoregulation may be disturbed, which means that CBF inversely becomes dependent of ICP. The normal ICP is less than 15 mm Hg or 200 mm H20.

The intraventricular ICP is considered to be the golden standard and is measured with an intraventricular catheter that is inserted through a burr hole in the skull. ICP can also be measured in the brain parenchyma itself.

In case the ICP is rising, a few compensation mechanisms become effective: compression of venous thin-walled blood vessels, excessive displacement of CSF towards the spinal canal and increase of CSF absorption. After some time these compensation mechanisms become exhausted. As a consequence there is an exponential relationship between ICP and intracranial volume.

2.2 The Mass Effect Leads to Displacement and Herniation of Specific Anatomic Parts of Brain Tissue, Causing Compression of the Brain Stem

The intracranial space is divided in 3 compartments by rigid membranes that extent in central direction, starting from the inner side of the skull: the 'falx cerebri' and the 'tentorium cerebelli'. These 3 compartments respectively contain the left and right cerebral hemispheres and the cerebellum. Both the top of the tentorium and the base of the skull have openings ('incisura tentorii' and 'foramen magnum'), through which the brain stem passes. The brain stems contains vital structures, which means that survival with a damaged brain stem is not possible.

Depending on the anatomic location of brain oedema (or of another space-occupying lesion), parts of the brain may thus be displaced from one compartment to another, leading to herniation of these displaced brain structures between either the border of the incisura tentorii or the bony margin of the foramen magnum at one side and the brain stem at the other side. As such the brain stem will be compressed.

A patient with a compressed brain stem will loose consciousness and will become comatose. The regulation of his blood pressure, heart rhythm, body temperature and respiration will be disturbed. This will quickly lead to a life threatening situation and even death.

3. TREATMENT OF POSTTRAUMATIC BRAIN OEDEMA

In a clinical setting treatment of posttraumatic brain oedema is often performed in an intensive care unit and stepwise according the ICP-levels that are measured with an intracranial catheter (TIL = 'therapy intensity level'):

- The patient is sedated and curarised. The head end of his bed is elevated to about 30°.
- Hyperventilation leads to a decrease of the pCO2 level of arterial blood and, as a consequence, to constriction of cerebral vessels.
- CSF may be drained externally out of the cerebral ventricles.
- Osmotic agents such as Mannitol may displace water from brain oedema areas towards the lumen of cerebral blood vessels.
- Pharmaceutical agents such as Phenobarbital may decrease cerebral metabolism and ICP. The same probably holds true for hypothermia.
- Eventually, vasopressive agents such as Dopamin or Adrenalin, may be used for enhancing arterial blood pressure and cerebral perfusion pressure.

TRAUMATIC BRAIN INJURY
Pathological Mechanisms and Surgical Treatment

Jack Phillips
Consultant Neurosurgeon, Beaumont Hospital, Dublin 9, Ireland

Abstract. This paper outlines the pathological mechanisms and clinical presentation of traumatic brain injuries. The surgical management of patients with intra cranial pathology following head injuries is described. Prompt treatment of intra cranial haematomas following head injury is associated with a good outcome. Delayed surgical treatment has a worse prognosis. Patients presenting with a diffuse brain injury also called diffuse axonal injury not amenable to surgical treatment have a high mortality or are left with a devastating neurological injury. There is clearly room for fruitful collaboration between neurosurgeons and those professionals in the mechanical engineering and computer sciences with a special interest in the causation of traumatic brain injury.

Key words: traumatic brain injury, surgical treatment.

1. INTRODUCTION

Trauma is a leading cause of death in persons under forty years of age and head injuries contribute significantly to these deaths in seventy five percent of patients. The overall incidence of head injury is estimated to be 300 per 100,000 persons with a mortality rate of approximately 7.5%. Sixty five percent of deaths occur at the scene of the accident or on route to hospital. For every trauma death there are three to four permanent disabilities. Overall trauma accounts for more working days of lost life than cardio vascular disease and cancer combined. Days lost in the United States per year cost the exchequer approximately four hundred million dollars per day [1]. In a retrospective study of one thousand consecutive trauma deaths in England

Michael D. Gilchrist (ed.), IUTAM Proceedings on Impact Biomechanics: From Fundamental Insights to Applications, 383–391.

and Wales traumatic brain injury was the cause of death in sixty five percent of patients reaching hospital [2].

This paper outlines the pathological mechanisms and surgical treatment of traumatic brain injury.

2. INITIAL MANAGEMENT

Head injuries are categorised into mild, moderate and severe. The Glasgow coma scale has been found to be a reliable measurement of head injury severity. Patients with a severe head injury have a Glasgow coma scale of eight or lower. Moderate head injury is defined as a Glasgow coma scale between nine and twelve and those with minor head injuries have a Glasgow coma scale of thirteen to fifteen. Patients with mild head injuries may only require observation in the hospital environment while all other grades invariably require CT imaging and evaluation or consultation with a neurosurgeon.

The management of head injury patients requires the following essential information:

1. Age of patient and mechanism of injury.
2. Assessment of vital signs including respiratory and cardio vascular status.
3. Neurological examination including level of consciousness, pupil reaction and the presence of lateralised weakness.
4. Nature and type of other systemic injuries.
5. Results of x-rays or other imaging studies.

3. SURGICAL MANAGEMENT

3.1 Scalp Lacerations

Brisk haemorrhage from scalp lacerations can be controlled by placing 0 or 2-0 nylon sutures through the skin to include the galea using a mattress suture. Hair on the edge of the wound margin should be shaved and the wound well irrigated. A thorough search should be conducted for foreign bodies. Uncontaminated wound lacerations are best closed with a two layer absorbable suture through the galea and sterile clips in the skin.

Galeal sutures are responsible for tensile strength. Smaller wounds may be approximately with steri strips. Cephalohaematomas collect beneath the galea in children and frequently resolve spontaneously. A sub galeal haematoma can be mistaken for depressed skull fracture.

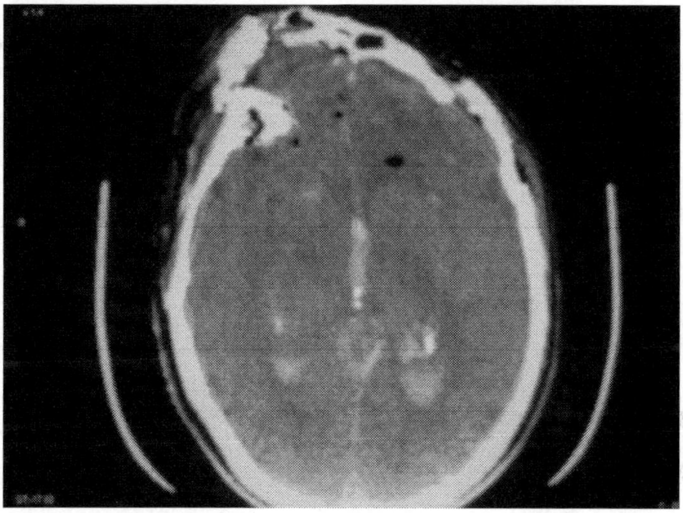

Figure 1.

3.2 Skull Fractures (Figure 1)

Linear skull fractures of the temporal region may cause injury to the middle meningeal artery resulting in an extra dural haematoma. Linear fractures which cross the sagittal or sigmoid sinus may lead to low pressure epidural collections. Linear skull fractures do not require surgical exploration.

Depressed skull fractures may be diagnosed with tangential skull x-rays or preferably by CT scanning. Open depressed skull fractures require operative debridement and repair. The underlying dura should be inspected for laceration. Depressed fractures adjacent to sinuses should not be explored if there is any suspicion elevation may cause major haemorrhage or sinus thrombosis. Frontal skull fractures may be suggested by peri orbital ecimosis (Raccoons eyes), rhinorrhoea or anosmia. Petrous bone fractures may cause Battle's sign, discolouration over the mastoid bone, haemotympanum or otorrhoea or bleeding from the ear. Otorrhoea is usually self limiting. Leaking CSF should not be plugged. If otorrhoea persists then serial lumbar puncture may be performed. Prophylactic antibiotics have not been demonstrated to be beneficial. Hearing may be impaired following petrous bone fractures. Equally the facial nerve might be injured and when full paralysis occurs the nerve may be transected. Delayed facial nerve paresis usually indicates swelling with a better prognosis.

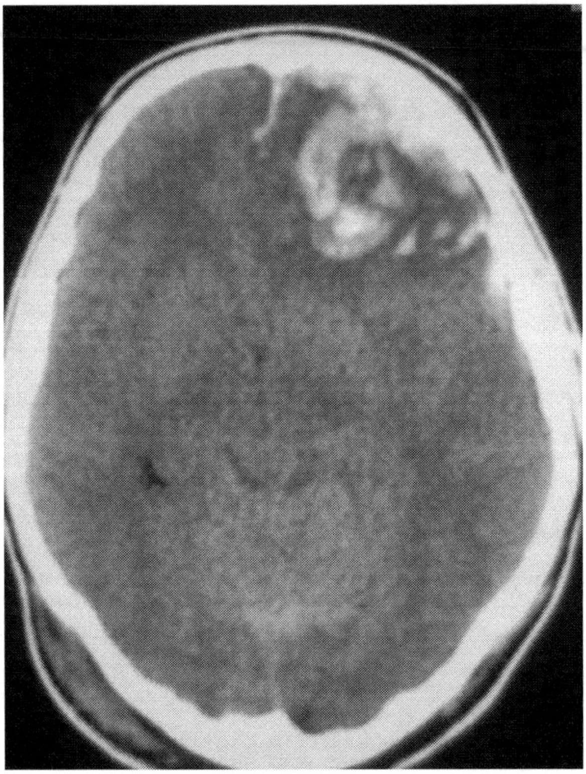

Figure 2.

3.3 Intra Cerebral Contusion (Figure 2)

Contusions of the brain occur following acceleration and deceleration movement of the frontal and temporal lobes which make contact with a rough inner skull. Contra coup injuries occur on the opposite side to the impact. The contusion produces symptoms reflecting the affected lobe of the brain. In general the management is not surgical but if intra cranial pressure is raised it may be necessary occasionally to resect a burst temporal or frontal lobe.

Repeat CT scanning is necessary in a patient who deteriorates with a known cerebral contusion. In circumstances where delayed haematoma formation is detected surgical decompression may be necessary. The incidence of delayed haemorrhage or haematoma may be as high as ten percent.

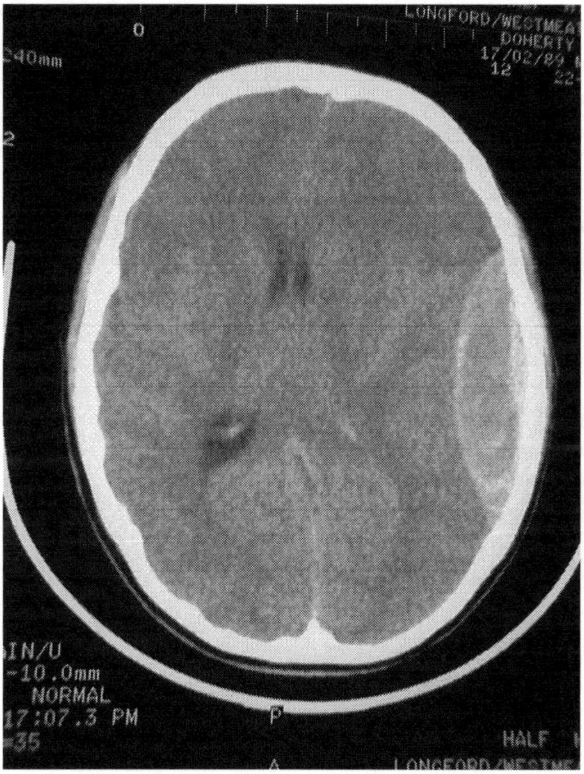

Figure 3.

3.4 Extra Dural Haematoma (Figure 3)

Extra dural haematomas occur most frequently in children and young adults. They are usually located in the temporal region following tearing of the middle meningeal artery associated with a fractured temporal fossa. Early extra dural haematomas may occur in the posterior fossa associated with a lacerated dural venous sinus. In sixty percent of patients an extra dural haematoma is associated with a blow to the head followed by a lucid interval and thereafter by headache, vomiting, loss of consciousness, hemiparesis and ipislateral pupillary dilatation. Death may then follow due to mid brain compression by the herniating temporal lobe over the edge of the tentorium. The patient with an extra dural haematoma should have the clot evacuated within two hours of early deterioration. A patient with an extra dural haematoma and an epislateral dilated pupil should have the clot evacuated within one hour or earlier.

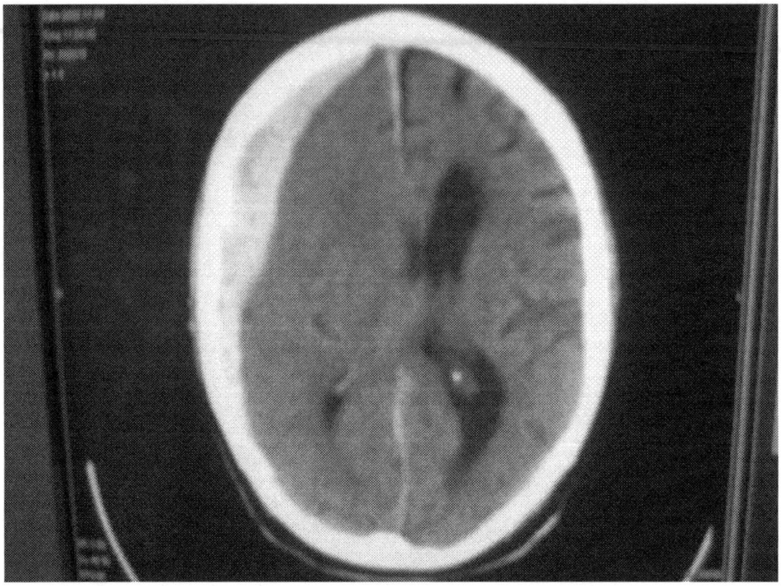

Figure 4.

3.5 Acute Subdural Haematomas (Figure 4)

Acute subdural haematoms are associated with lacerated cerebral arteries or veins and are often associated with high speed acceleration deceleration injuries. Frequently the cause of the major morbidity is associated contusion of the underlying brain. As for extra dural haematoma these lesions are diagnosed readily by CT scanning. Subdural haematomas should be evacuated early. If there is major contusion associated with a subdural collection morbidity and mortality may be high despite removal of the clot. Treatment which is delayed more than four to six hours following diagnosis of large subdural haematomas with midline shift may result in a ninety percent mortality.

3.6 Chronic Subdural Haematomas (Figure 5)

Chronic subdural haematomas may follow if acute subdural collections are managed medically. More commonly chronic subdural haematomas may occur after relatively minor head injuries especially in the elderly patients or those receiving Warfarin or Aspirin. The incidence of subdural haematoma

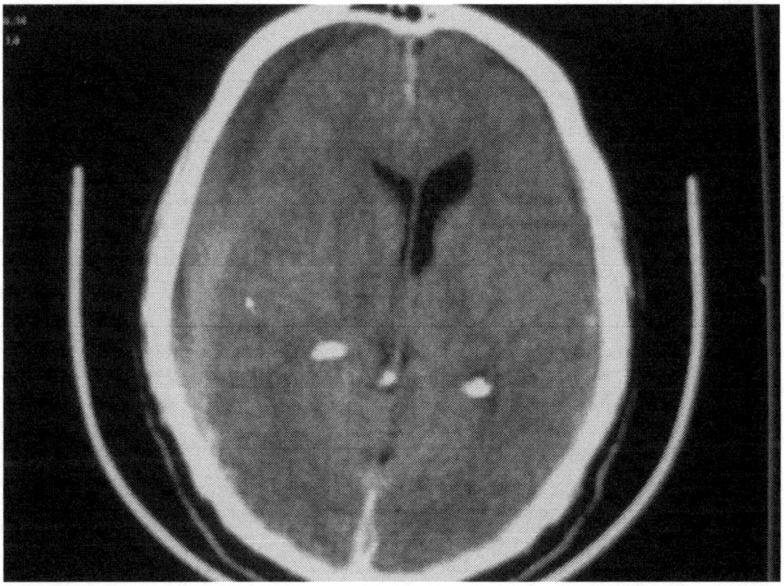

Figure 5.

in the latter group is rising alarmingly. A chronic subdural haematoma may remain undetected for several weeks and may present with progressive cognitive impairment or fluctuation in level of consciousness. These haematomas are not normally associated with major underlying brain injury. The treatment of subdural hygroma containing altered blood products may be multiple burr hole drainage. Recurrent chronic subdural haematomas may need craniotomy with excision of subdural osmotic membranes.

3.7 Posterior Fossa Haematomas (Figure 6)

Traumatic posterior fossa haematomas are uncommon. They may occur as a result of dural venous sinus tear over a fractured occipital bone. Large posterior fossa haematomas require surgical evacuation because of the risk of brain stem compression. Temporary frontal cerebro spinal fluid drainage may be indicated for traumatic obstructive hydrocephalus. Alert patients may be treated conservatively.

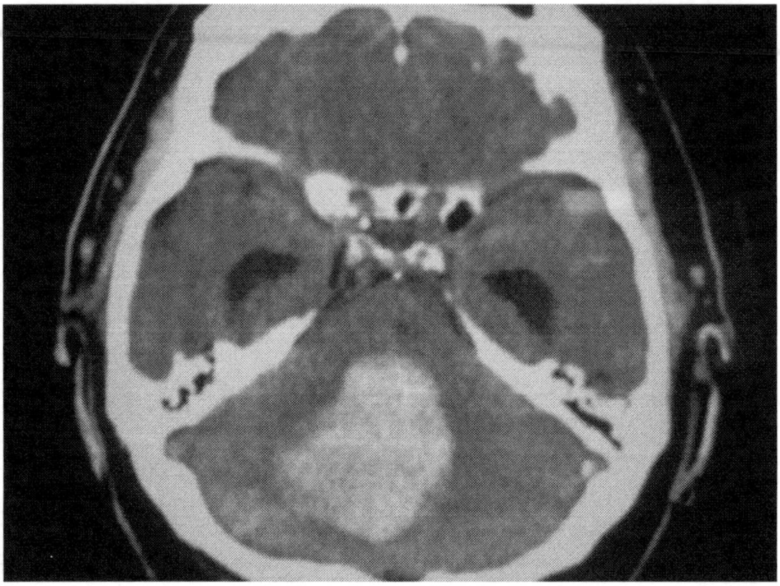

Figure 6.

4. DISCUSSION

This paper briefly outlines the pathological mechanisms and the surgical treatment of traumatic brain injuries.

In the Republic of Ireland a neurosurgical audit was carried out on head injured patients admitted to the National Neurosurgical Centre over a one year period [3]. That study confirmed findings which are recorded in other Western countries indicating that traumatic brain injury is a major medical problem disabling a young population with often unknown socio economic consequences. Road traffic accidents accounted for the largest proportions of injuries in our study (48%). Falls were responsible for 36% of head injuries. While the number of fatalities associated with road traffic accidents have decreased steadily over the last two decades non fatal injuries have increased over the same period. Death from head trauma has a trimodal distribution. Half of all trauma deaths occur within minutes of the injury and this is due to extensive cerebral brain stem, spinal, cardiac or major blood vessel injuries. The second peak fatality occurs within hours of the injury and is the result of intra cranial injuries that is to say haematoma and systemic haemorrhage. The final peak in trauma fatalities occurs days to weeks later and is as a result of sepsis and major organ failure. Medical treatment in the first peak will invariably be too late to be of any consequence. Early intensive medical management including surgery soon after injury will reduce the second and third peaks of the trauma fatality curve.

The initial treatment of severe brain injury is both labour intensive and involves high levels of sophisticated technology. The rehabilitatory phase of treatment is prolonged and the overall cost is difficult to calculate. The estimated cost to the United States is approximately four hundred million dollars per day. It is recognised that for road traffic accidents brain and lower extremity injuries account for the large proportion of medical costs and the cerebral and spinal cord patients are the most costly.

Neurosurgeons and neuroscientists continue to strive to understand the mechanisms of traumatic brain injury caused by falls and road traffic accidents. To date the pharmacological industry has failed to find suitable agents to counter the devastating hypoxic effects of diffuse axonal injury in patients for whom surgery has no role. There is room for fruitful co-operation with fellow professionals in the mechanical engineering and computer sciences with a special interest in traumatic brain injury.

REFERENCES

1. Trunkey D.D., Is a trauma care system necessary for Ireland?, *J. Irish Coll. Physicians Surg.*, 21, 1992: 261–270.
2. Miller, T.R., Pindus, N.M. and Douglass, J.B., Medically related motor vehicle injury costs by body region and severity. *J. Trauma, 34m* 1993:270–275.
3. O'Brien, D. and Phillips, J.P., Head injuries in the Republic of Ireland: A neurosurgical audit, *Irish Med. J., 89,* 1994: 216–218.

A SELECTIVE DEPOLARISATION-INDUCED INCREASE IN EXCITATORY AMINO ACID NEUROTRANSMITTER RELEASE IN RAT MEDIAL PREFRONTAL CORTEX USING A MICRODIALYSIS MODEL OF TRAUMATIC BRAIN INJURY

Aoife Smyth[1], Michael D. Gilchrist[2] and William T. O'Connor[1]
[1]Applied Neurotherapeutics Research Group, Conway Institute of Biomedical and Biomolecular Research, University College Dublin, Ireland; Tel.: +353 – 1 – 716 6759, Fax: +353 – 1 – 716 6290, E-mail: bill.oconnor@ucd.ie
[2]Department of Mechanical Engineering, University College Dublin, Ireland

Abstract. This study describes a microdialysis model that investigates the biochemical response of the brain to non-fatal impact trauma. A controlled cortical impact (mild and severe) was performed to the left medial prefrontal cortex (mPfc) in the isoflurane-anaesthesised rat. This was followed by intracerebral microdialysis whereby a microdialysis probe was implanted into the site of the injury. Changes in dialysate glutamate, aspartate and GABA levels were investigated immediately (*i.e.* 25 min) and 265 min following a local mild and severe impact to the brain. In addition, the effect of local perfusion with a depolarizing concentration of KCl (100 mM, 20 min) was also investigated 165 min after impact.

Dialysate levels measured 25 min after impact (n=14) showed an impact-dependent increase in glutamate (6 and 8-fold), aspartate (4 and 5-fold) and GABA (3 and 6-fold) following mild and severe impact respectively compared to non-impact controls. Dialysate levels measured 265 min after mild (n=12) and severe (n=13) impact had stabilized and continued to show a local 5-fold (mild) and 4-fold (severe) increase in local glutamate, a 6-fold (mild) and 3-fold (severe) increase in aspartate and a (3-fold (mild) and 5-fold (severe)) increase in GABA levels compared to control. Intra-mPfc KCl (n=14) increased local dialysate glutamate levels (4-fold following mild impact and 3-fold following severe impact) and aspartate levels (2-fold after both mild and

Michael D. Gilchrist (ed.), IUTAM Proceedings on Impact Biomechanics: From Fundamental Insights to Applications, 393–404.

severe impact) while GABA levels did not differ from non-impacted controls following either a mild or severe impact.

The present findings show that microdialysis in intact brain can be combined with the controlled cortical impact model to reveal selective impact-dependent and prolonged increases in local dialysate amino acid neurotransmitter levels. Furthermore, we reveal that 165 min following either a mild or severe impact to the left mPfc KCl-stimulated glutamate and aspartate release is abnormally increased while GABA release is not different compared to non-impacted controls. This finding may in part explain the excitotoxicity that contributes to diffuse posttraumatic lesions associated with secondary injury.

Key words: glutamate, GABA, KCl, HPLC, rat, basal, stimulated levels.

1. INTRODUCTION

Traumatic brain injury is the main cause of death in children and young adults living in the industrialized world. Current estimates suggest that in the U.S. between 2.5 million and 6.5 million individuals are living with the consequences of traumatic brain injury, much of it caused by motor vehicle accidents. Studies in human subjects have shown that brain trauma induces an immediate and profound increase in local extracellular concentrations of excitatory amino acids such as glutamate. However, major discrepancies have been noted with respect to the magnitude of the posttraumatic event. Baker et al. (1993) have reported that in humans, glutamate concentrations are elevated up to 28-times the normal CSF concentration up to 3 days post-injury whereas Palmer et al. (1994) have reported a 2 to 8-fold increase in CSF aspartate, glutamate and glycine up to four days post injury. This excess glutamate is believed to initiate a chain of events at pathological concentrations. An excitotoxic cascade results in critical concentrations of calcium in the cell and is one of the most important factors in the premature death of neurons. In man, the majority of brain trauma is directed towards the forehead – the medial prefrontal cortex (mPfc) in the rat. Here the main excitatory amino acid neurotransmitters glutamate and aspartate are derived from the dendrites and axon collateral terminals of corticofugal outputs (Carr and Sesack, 2000) while the inhibitory amino acid neurotransmitter GABA is derived from a separate population of local interneurons (Hendry et al., 1987).

Animal models of brain trauma have generated data that has helped gain an insight into the events that occur during and after injury. In particular, a number of microdialysis models of traumatic brain injury in rats have shown increased dialysate concentrations of the excitatory amino acid neurotransmitters following brain injury and these increases are generally related to the severity of injury. One of the most popular animal models of traumatic brain injury is the controlled cortical impact model (Dixon et al., 1991). This model utilises a pneumatic impactor that delivers an injury (generally through a craniotomy) to the intact dura. The head of the rat is restrained during the delivery of the impact. The controlled cortical impact model was chosen for investigation in the present study as the depth of penetration and velocity of impact can be accurately quantified and modified according to the experiment (Dixon et al., 1991) and for its ability to be combined with microdialysis in intact rat brain as reported by Tayag et al. (1996); Rose et al. (2002); Dixon et al. (1991); Stover et al. (2000) and Krishnappa et al. (1999).

The present study describes a model of brain injury that combines controlled cortical impact (Dixon et al., 1991, Gilchrist, 2004) with brain microdialysis (O'Connor, 2001). Towards this aim, we investigated the biochemical response of discrete sets of neurons within the left mPfc of an isoflurane anasethetised rat to non-fatal mild or severe impact.

2. MATERIALS AND METHODS

2.1 Animals

A total of 39 adult male Sprague Dawley rats (250-350 g) (Harlan Laboratories, United Kingdom) were used in this study. Prior to the experiment the animals were kept under regular lighting conditions (12 hr light/dark cycle) were given food and water ad libitum. Experimental protocols for ethical experiments on laboratory animals were approved by the local committee (license number B100/3366).

2.2 Surgery

On the day of the experiment rats are anaesthetised with isoflurane (4% - 2% in air delivered at 400 ml/min) in a vented anaesthesia chamber. Animals are placed in a stereotaxic frame (David Kopf Inc.) and the head adjusted until the skull between the bregma and lambda is level. An incisor bar hook and blunt ear bars are used to stabilise the rat's head. Anaesthesia is maintained throughout the experiment using a mask fitted over the nose of the animal, which allows the rat to breathe free. Body temperature is continuously monitored and maintained at 37.5 °C using a temperature controlled heating pad (CMA 150 Carnegie Medicin, Sweden). Using sterile techniques, a sagittal incision of the scalp is made along the midline from the level of the eyes to the occipital protuberance so that the frontal bones are exposed. A craniotomy 4 mm in diameter is drilled through the skull above the mPfc and the dura above it is exposed.

3. CONTROLLED CORTICAL IMPACT

Cortical contusion injury is performed at the left mPfc using a pneumatically driven vertical impactor (Figure 1). The device consists of a pneumatic cylinder mounted on an adjustable crossbar, which is positioned above the left mPfc to provide an impact by a 3.5 mm rounded impactor tip. Air pressure is set at 4 bar. The depth of penetration is determined by zeroing the piston to the cortical surface, withdrawing it and then lowering it to the required impact deformation. Animals are injured directly to the brain through a 4 mm craniotomy to the intact dura. To characterise the injury response, one of two injury levels is produced; mild (0.87 mm deformation) or severe (2.62 mm deformation). The velocity of impact (1.2 m/s) is verified using the storage oscilloscope.

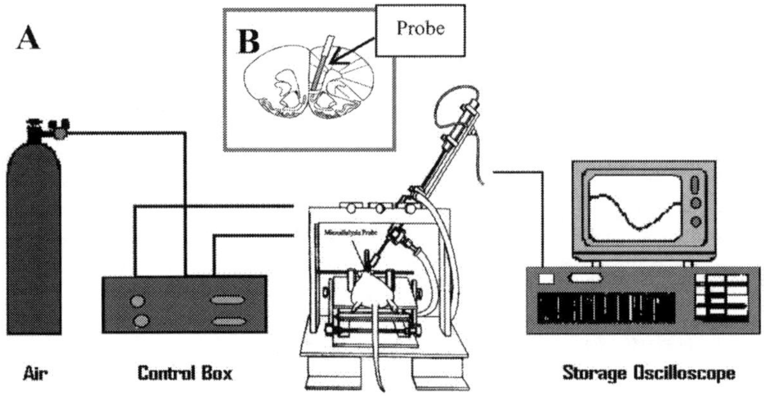

Figure 1. (**A**) Schematic representation of the experimental set-up employed in the present study. Pressurised air is linked to the pneumatic piston loaded on the controlled cortical impact rig with isoflurane-anaesthetized rat in situ (adapted from Cherian et al., 1994). An impactor tip is attached to the lower piston rod. Insert (**B**) Transverse section through the rat brain showing implantation of a 4 mm microdialysis probe in the left medial prefrontal cortex (mPfc). The left mPfc is the site of impact, local KCl perfusion and collection of dialysate samples.

A. Mild or Severe Impact

B. Non-Impacted Control

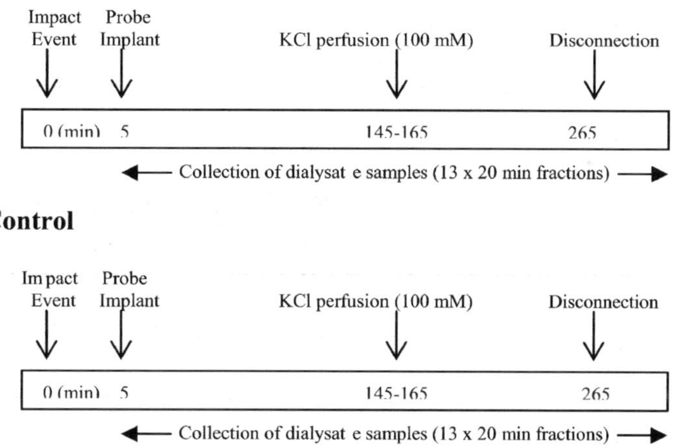

Figure 2. Schematic illustration of the experimental procedure employed in the present study. **A**. A mild or severe impact is applied to the left medial prefrontal cortex (mPfc). Within 5 min of injury a microdialysis probe is inserted into the mPfc and 13 x 20 min dialysate samples are collected for HPLC analysis. **B**. Animals in the control group are subjected to identical surgery and microdialysis probe implantation procedures except for the omission of a brain injury.

3.1 Microdialysis

As shown in Figure 2 immediately following impact to the left mPfc the piston is removed from the rig and a microdialysis probe (4 mm long, outer diameter 0.5 mm, CMA/12; CMA Microdialysis, Stockholm, Sweden) is carefully implanted into the left mPfc (co-ordinates (mm) from bregma: AP + 2.7, ML + 1.4, DV – 6.5 from bone; incisor bar: - 3.3). Samples are collected every 20 min, 265 min (4.3 hrs) post-injury with the samples stored on ice at –70°C until analysis by high pressure liquid chromatography (HPLC). The microdialysis probe is perfused with sterile Ringer solution (Baxter, UK; composition (mmol/L): Na^+ 147; K^+ 4; Ca^{2+} 2; Cl^- 156, pH 6) at a rate of 2 μl/min for 20 min prior to implantation. Dialysate sample collection is adjusted to allow for dead space in the probe and exit tubing. The flow rate of the perfusion medium through the probe is maintained at a constant flow by a microperfusion pump (CMA 100, Carnegie Medicin AB, Sweden) during implantation and for the duration of the microdialysis experiment.

3.2 Experimental Design

Rats were randomly divided into three groups namely those subjected to a (1) mild (n=12) or (2) severe impact (n=13) and (3) a control group (n=14). Animals in the control group underwent identical surgery and microdialysis probe implantation procedures except for the omission of a brain injury.

3.3 Neurotransmitter Analysis; HPLC

Each perfusate sample (40 μl) is divided into two aliquots with 10 μl used for glutamate/aspartate, 10 μl for GABA and 20 μl for back-ups. The dialysate samples are placed in separate refrigerated microsampler/injectors (CMA 200/240; Carnegie Medicin, Sweden) and are automatically injected onto high pressure liquid chromatography columns (HPLC) for detection of glutamate/aspartate and GABA.

3.4 Glutamate and Aspartate Analysis

Glutamate is determined by precolumn derivatisation with o-phthaldialdehyde/ mercaptoethanol reagent followed by separation by reverse phase HPLC and fluorimetric detection. A CMA 280 fluorescence detector (Carnegie Medicin, Sweden) with excitation wavelength set at 315 nm and emission cut off filter set at 395 nm is used. The limit of detection is 0.5 pmol/sample.

3.5 GABA Analysis

The GABA assay is based on precolumn derivatisation with *o*-phthaldialdehyde/ *t*-butylthiol reagent and separated by reverse phase HPLC on a Nucleosil 3,C18 column electrochemical detector (BAS LC-4C; Bioanalytical Systems, West Lafayette, IN, USA) under isocratic conditions. The limit of detection is 20 fmol/ sample.

3.6 Verification of Probe Placement

At the completion of the experiment the animals are euthanased with an overdose of isoflurane, the brain removed and placed in a 4% formaldehyde solution. Probe placement is verified by cryostat sectioning (10 μm) on a Leitz freezing microtome (Wetzlar, Germany) followed by microscopic examination to qualitatively confirm the morphologic extent of injury. Rats displaying improper probe implantation are omitted from the study.

3.7 Statistical Analysis

Data is calculated as the mean ± SEM. Investigated variables are compared for significant differences using one-way analysis of variance (ANOVA).

4. RESULTS

4.1 Impact Velocity

There was no difference in the mean impact velocity for mild and severe impact to the mPfc as verified using a storage oscilloscope. Thus the mean impact velocity (mean ± SEM) for mild impact was 1.13 ± 0.01 m/s (n=4) and for severe impact was 1.06 ± 0.01 m/s (n=6).

4.2 Dialysate Amino Acid Levels

Dialysate glutamate, aspartate and GABA levels in non-impacted control rats, decreased gradually over the duration of the experiment. Glutamate levels decreased by 75% and stabilised at a concentration of 0.50 ± 0.18 µM 120 min following probe implantation. Aspartate levels decreased by 76% and stabilised at a concentration of 0.16 ± 0.03 µM 60 min following probe implantation, while GABA levels decreased by 85% and stabilised at a concentration of 33.6 ± 10.5 nM 80 min following probe implantation.

Table 1. Local dialysate glutamate, aspartate and GABA levels measured in the left medial prefrontal cortex (mPfc) 25 min and 265 min after mild or severe impact. Non-impacted animals acted as controls. Each data point represents the mean + SEM raw dialysate level from 7-14 animals. (* = $p < 0.05$ v's Control, * * = $p < 0.01$ v's Control, * * * = $p < 0.001$ v's Control).

Group	Glutamate (µM)	Aspartate (µM)	GABA (nM)	Glutamate (µM)	Aspartate (µM)	GABA (nM)
	25 min after impact			265 min after impact		
Control	2.04 ±0.58	0.71 ±0.22	222.12 ±60.37	0.19 ±0.04	0.07 ±0.02	11.19 ±2.05
Mild	12.49 ±2.72 * *	2.9 ±0.54 * *	607.71 ±176.7 *	1.03 ±0.16	0.42 ±0.08 * *	30.69 ±12.69 * * *
Severe	16.88 ±1.86 * * *	3.60 ±0.42 * * *	1189.77 ±131.61 * * *	0.7 ±0.11 * *	0.22 ±0.04 *	54.36 ±8.89 * * *

4.3 Impact Levels

As shown in Table 1 a local impact-dependent increase in dialysate glutamate (6 (mild) and 8-fold (severe)), aspartate (4 and 5-fold) and GABA (3 and 6-fold) levels was observed 25 min after impact compared to non-impacted controls. This increase was still present 245-265 min after impact, glutamate (5 (mild) and 4-fold (severe)), aspartate (6 and 3-fold) and GABA (3 and 5-fold).

4.4 Stimulated Levels

As shown in Figure 3 local perfusion with KCl (100 mM, 20 min) increased dialysate glutamate (4-fold (mild) to 3-fold (severe) and aspartate levels (2-fold following both mild and severe impact). In contrast, GABA levels did not statistically differ from non-impacted controls following either a mild or severe impact.

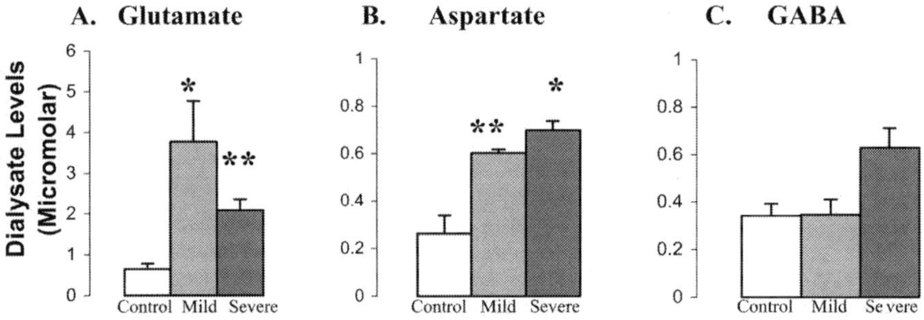

Figure 3. KCl (100 mM, 20 min) stimulated dialysate A. glutamate, B. aspartate and C. GABA levels measured 145-165 min following mild and severe impact to the left medial prefrontal cortex (mPfc). Non-impacted animals acted as controls. Each data point represents the mean ± SEM raw dialysate level from 6-13 animals. For glutamate (μM); control 0.65 ± 0.13; mild 3.77 ± 0.9; severe 2.08 ± 0.27. Aspartate (μM); control 0.26 ± 0.07; mild 0.6 ± 0.01 severe 0.7 ± 0.04. GABA (nM); control 340.84 ± 51.05; mild 346.31 ± 65.97; severe 628.31 ± 83.72. * = $p < 0.05$; * * = $p < 0.01$ (v's non-impacted control).

5. DISCUSSION

Head injury is a spontaneous, unpredictable event and while no single animal model is entirely successful in reproducing the complete spectrum of pathological changes observed after traumatic brain injury the controlled cortical impact model is superior to devices that are driven by gravity or a free falling guided weight because deformation parameters can be controlled with pneumatically driven devices (time, velocity, and depth of impact). In addition, the rigid impact model can be employed to mimic the whole spectrum of focal-type damage including contusions with haemorrhages and/or subdural haematoma formation. Furthermore, when combined with microdialysis the quantifiable nature of the single mechanical input used to produce the injury allows correlations to be made between the amount of deformation and the resultant pathology and functional changes (Lighthall et al., 1989).

Therapeutic innovations in the field of brain injury rely on advances in the understanding of the underlying pathophysiological cascade that causes neuronal damage. Indeed, the pathogenesis of traumatic brain injury is incompletely understood, in large part because patients often present with a variety of lesions of varying severity and distribution. However, it is widely accepted that there are two main stages in the development of brain damage after injury to the head; primary (mechanical) and secondary (delayed non-mechanical). The direct, primary event may be due to traumatic brain injury – blunt or penetrating, a neoplasm or a cerebrovascular accident – ischaemic stroke or haemorrhage and it is generally thought to be an irreversible event that warrants preventative measures. The concept of secondary injury was initially proposed at the end of the 19th century and was proven to be of significance in the 1960s and 1970s. It is produced by complicating processes that are initiated at the moment of injury, although the symptoms do not present clinically for a period of hours to days after injury. Ischaemia, axonal injury, focal haematomas, contusions and oedema all contribute to the progressive loss of brain cells involved in the process of delayed damage to the brain tissue. This delayed pathophysiological cascade is now believed to result from a combination of pathological factors including monumental rises in neurotransmitter release, notably the excitatory amino acid glutamate which causes excitotoxicity.

The findings in the present study show that immediately following mild and severe unilateral injury to the mPfc a local impact-dependent increase in dialysate glutamate (6 (mild) and 8-fold (severe)), aspartate (4 and 5-fold) and GABA (3 and 6-fold) levels is observed compared to non-impacted controls and this increase in still present 265 min after impact. Thus, we report an acute and prolonged increase in extracellular excitatory and inhibitory amino acid neurotransmitter levels in the mPfc after impact. Nilsson et al. (1990) combined microdialysis with the weight-drop model of traumatic brain injury in the rat and reported an increase in dialysate glutamate (8-fold (mild -1.5 mm deformation), 13-fold (severe - 2.5 mm deformation)) and aspartate levels (6-fold (mild), 17-fold (severe)) following injury to the parietal cortex. Maximal levels were observed in the first 10 min after impact and returned to stable basal levels within 20-30 min after injury. Using a separate model Faden et al, (1989) combined fluid percussion injury to the left parietal cortex of the rat with microdialysis in the hippocampus and reported an injury-related increase in dialysate glutamate – (3-fold (mild) to 9 fold (severe)) and aspartate (3-fold (mild) to 18-fold (severe)) compared to controls. Maximal levels were observed in the first 10 min after trauma with elevations sustained for more than 60 min in the severely injured group. However, injury related increases are not solely confined to excitatory amino acid neurotransmitters. GABA levels were reported to rise from a non-detectable basal level to 0.6–1.2 μM following injury to the parietal cortex in the rat (Nilsson et al., 1990). In humans, Palmer et al. (1994) reported CSF GABA levels to be 56 to 317-fold higher following injury compared to controls. Taken together, these findings strengthen the evidence that the extracellular levels of both excitatory and inhibitory amino acid neurotransmitters are increased following injury.

The mechanisms by which excitatory amino acids contribute to secondary tissue injury remain speculative, although increases in intracellular Na^+ and Ca^{2+} have been implicated in this delayed injury process (Choi, 1987). The release of glutamate from the contusion core is believed to be related to primary disruption of the cell membrane or vascular wall by the physical force of the impact, resulting in leakage from the metabolic pool in the cytosol or the blood stream (Maeda et al., 1998). In contrast, areas peripheral to the contusion may have a presynaptic mechanism of glutamate release, which consists in part, of a Ca^{2+}-dependent exocytotic release from depolarized nerve terminals. The glutamate released in the contusion core may diffuse towards a peripheral direction and act on the postsynaptic receptors, causing neuronal depolarization. Such a diffusion-reaction process would induce additional release of glutamate from the depolarised nerve terminals, inducing a cascade of excitotoxicity. The present findings of an abnormally high KCl-induced increase in both dialysate glutamate (4-fold (mild) and 3-fold (severe)) and aspartate (2-fold (mild and severe)) but not GABA levels after impact compared to non-impacted controls supports this hypothesis and may reflect the fact that glutamate and aspartate axon terminals are only lightly myelinated compared to the more heavily myelinated GABA interneuron. Thus, excessive depolarisation induced increase in excitatory amino acid release may underlie the excitotoxic component that actively destroys populations of neurons and contributes to diffuse posttraumatic lesions associated with the secondary injury linked to brain injury and may at least in part explain why glutamate receptor antagonists alone fail to successfully improve neurological outcome in severely brain injured patients in the clinical setting (Laurer and McIntosh, 2001). Furthermore, on the basis of the present findings we suggest that in some clinical head trauma situations surgical removal of contused brain tissue to prevent excessive depolarisation induced release may serve to protect the surrounding brain areas from glutamate-induced cytotoxicity.

In conclusion, we show that the controlled cortical impact model can be successfully combined with microdialysis in intact rat brain to reveal a selective impact-dependent and prolonged increase in local excitatory and inhibitory amino acid neurotransmitter levels. Furthermore, this microdialysis model differentiates the responsiveness of excitatory and inhibitory neuronal pools within the mPfc to reveal a selective increase in depolarisation-induced excitatory amino acid release following injury which may be useful for the investigation of new therapeutic strategies and pharmacological testing for an effective treatment for patients with head-injury.

ACKNOWLDEGEMENTS

The National Neuroscience Network (Ireland), Science Foundation Ireland (SFI) and Enterprise Ireland.

REFERENCES

Baker, A.J., Moulton, R.J., MacMillan, V.H. and Shedden, P.M. (1993) Excitatory amino acids in cerebrospinal fluid following traumatic brain injury in humans. *J. Neurosurg.* **79**(3):369–372.

Carr, D.B. and Sesack, S.R. (2000) Projections from the rat prefrontal cortex to the ventral tegmental area: Target specificity in the synaptic associations with mesoaccumbens and mesocortical neurons. *J. Neurosci.* **20**(10):3864–3873.

Cherian, L, Robertson, C.S., Contant, C.F. Jr. and Bryan, R.M. Jr. (1991) Lateral cortical impact injury in rats: Cerebrovascular effects of varying depth of cortical deformation and impact velocity. *J. Neurotrauma* **11**(5):573–585.

Choi, D.W. (1987) Ionic dependence of glutamate neurotoxicity. *J. Neurosci.* **7**:369.

Choi, D.W., Maulucci, G.M. and Kriegstein, A.R. (1987) Glutamate neurotoxicity in cortical cell culture. *J. Neurosci.* **7**:357–368.

Dixon, C.E., Clifton, G.L., Lighthall, J.W., Yaghmai, A.A. and Hayes R.L. (1991) A controlled cortical impact model of traumatic brain injury in the rat. *J. Neuroscience Methods*. **39**:253–262.

Faden, A.I., Demediuk, P., Panter, S. and Vink R. (1989) The role of excitatory amino acids and NMDA receptors in traumatic brain injury. *Science* **244**:798–799.

Gilchrist, M.D. (2004) Experimental device for simulating traumatic brain injury resulting from linear accelerations. *Strain* **40**:180–192.

Hendry, S.H., Schwark, H.D., Jones, E.G. and Yan, J. (1987) Numbers and proportions of GABA-immunoreactive neurons in different areas of monkey cerebral cortex. *J. Neurosci.* **7**:1503–1519.

Hong, Z., Xinding, Z., Tianlin, Z. and Liren C. (2001) Excitatory amino acids in cerebrospinal fluid of patients with acute head injuries. *Clin. Chem.* **47**(8):1458–1462.

Krishnappa, I.K., Contant, C.F. and Robertson, C.S. (1999) Regional changes in cerebral extracellular glucaose and lactate concentrations following severe cortical impact injury and secondary ischemia in rats. *J. Neurotrauma* **16**(3):213–224.

Laurer, H.L. and McIntosh, T.K. (2001) Pharmacologic therapy in traumatic brain injury: update on experimental treatment strategies. *Curr. Pharm. Des*. **7**:1505–1516.

Nilsson, P., Hillered, L., Ponten, U. and Ungerstedt, U. (1990) Changes in Cortical Extracellular Levels of Energy-Related Metabolites and amino acids following concussive brain injury in rats. *J. Cereb. Blood Flow Met*. **10**:631–637.

O'Connor, W.T. (2001) Functional neuroanatomy of the ventral striopallidal GABA pathway. New sites of intervention in the treatment of schizophrenia. *J Neurosci Meth*. **109**(1):31–39.

Maeda, T., Katayama, Y., Kawamata, T. and Yamamoto, T. (1998) Mechanisms of excitatory amino acid release in contused brain tissue: effects of hypothermia and *in situ* administration of Co^{2+} on extracellular levels of glutamate. *J. Neurotrauma* **15**(9):655–664.

Palmer, A.M., Marion, D.W., Botscheller, M.L. and Redd E.E. (1993) Therapeutic hypothermia is cytoprotective without attenuating the traumatic brain injury-induced elevations in interstitial concentrations of aspartate and glutamate. *J. Neurotrauma* **10**(4): 363–372.

Rose, M.E., Huerbin, M.B. Melick, J., Marion, D.W., Palmer, A.M., Schiding, J.K., Kochanek, P.M. and Graham, S.H. (2002) Regulation of interstitial excitatory amino acid concentrations after cortical contusion injury. *Brain Res*. **943**:15–22.

Stover, J.F., Schoning, B., Beyer, T.F., Woiciechowsky, C. and Unterberg (2000) Temporal profile of cerebrospinal fluid glutamate, interleukin-6 and tumour necrosis factor-α in relation to brain edema and contusion following controlled cortical impact injury in rats. *Neurosci Lett*. **288**:25–28.

Tayag, E.C., Nair, S.N., Wahhab, S., Katsetos, C.D., Lighthall, J.W. and Lehmann J.C. (1996) Cerebral uric acid increases following experimental traumatic brain injury in rat. *Brain Res*. **733**:287–291.

VICTIMS OF PENETRATING AND INCISED WOUNDS

A Review of 200 Cases

M.T. Cassidy and M. Curtis

Department of State Pathology, c/o Fire Brigade Training Centre, Malahide Road, Marino, Dublin 3, Ireland; Tel.: 01-8534872, Fax: 01-8534874, E-mail: mcassidy@statepathology.gov.ie, mcurtis@statepathology.gov.ie

Abstract. A review of 200 fatal victim cases of penetrating and incised wounds was conducted. The cases were entered into an electronic database. Analysis provided data concerning various aspects including:

- the age and sex of victims
- the number and type of wounds inflicted
- the wound sites
- the contour of wounds
- additional wound features such as hiltmarks, withdrawal scratches, notching, "fishtailing"
- the presence of defensive wounds
- associated blunt force trauma
- post-injury activity
- post-injury surgical intervention
- post-injury survival times
- the availability of victims' clothing for examination by the pathologist.

Key words: penetrating, incised wounds, stab wounds, defensive wounds.

1. MATERIAL AND METHODS

Post-mortem reports of 200 fatal victim cases of wounding with a sharp weapon, which were handled by one or other (or both) pathologists, were

Michael D. Gilchrist (ed.), IUTAM Proceedings on Impact Biomechanics: From Fundamental Insights to Applications, 405–414.

retrieved. The cases had been handled over a period of approximately 10 years in several cities, i.e. Glasgow, Dublin, Belfast and Sydney. Details from the postmortem reports were transferred onto a proforma which more readily allowed the computer analyst to compile an electronic database. From this, data were then extracted in response to specific questions posed by the pathologists. A 200-case database was considered to be of suitable size from which to extract meaningful information concerning some of the specific aspects and features of cases in which the pathologists were interested.

2. RESULTS

Of the 200 cases, 175 were males and 25 were females. A total of 1680 wounds were recorded, of which 882 were penetrating wounds and 798 were incised.

The number of wounds inflicted on the victims is displayed in Table 1. In 93 cases the victims had sustained penetrating wounds only, 6 cases had sustained incised wounds only, while a combination of both penetrating and incised wounds was seen in 101 cases.

Table 2 shows the anatomical sites of penetrating wounds.

Table 1. Number of Wounds.

(Penetrating and/or Incised)	
Wounds	**No. of Cases**
1	58
2-3	41
4-10	62
11-20	24
21-30	2
31-40	3
41-50	3
More than 50	7

Table 2. Sites of Penetrating Wounds.

Total 882 Wounds	
Head and Neck	183
Trunk	545
Upper Limbs	74
Lower Limbs	80

The anatomical sites of incised wounds for males and females are displayed in Table 3.

Table 3. Sites of Incised Wounds.

Total No. of Cases - 106		
	No. of Cases	Wound Sites
Females	6	Face and Neck (and Limbs)
(10 cases)	10	Limbs (1-3 wounds in 7 cases)
		(10 wounds in 3 cases)
Males	56	Face and Neck
(96 Cases)	57	Limbs (1-3 wounds in 51 cases)
	34	(10 wounds or more in 6 cases)
		Face, Neck and Limbs

Associated blunt force trauma was seen in 97 cases. Anatomically, this involved the head region in 83 cases, the trunk in 33 cases and the limbs in 43 cases.

A total of 437 defensive wounds, either penetrating or incised wounds, were seen in 83 cases (72 males, 11 females). Of these 437 wounds, 118 were present on the forearms, 285 were present on the hands and 34 were situated on the thighs.

The contours of the 882 penetrating wounds are shown in Table 4.

Table 4. Penetrating Wound Contours.

	No. of Wounds (882)	% of Penetrating Wounds
Indeterminate	426	48-30
Sharp/Blunt	384	43-54
Blunt/Blunt	39	4-42
Sharp/Sharp	33	3-74

In 80 cases, a mix of wound contours was present, as shown in Table 5. Only 17 cases (8.5% of the total) exhibited wounds with two sharp edges (sharp/sharp), of these, 2 cases had each sustained only a single penetrating wound.

Table 5. Cases with a mix of wound contours (total 80 cases).

Sharp/Blunt + Indeterminate	45
Sharp/Blunt + Blunt/Blunt + Indeterminate	16
Sharp/Sharp + Sharp/Blunt + Indeterminate	6
Sharp/Sharp + Sharp/Blunt	4
Sharp/Sharp + Indeterminate	3
Sharp/Blunt + Blunt/Blunt	3
Sharp/Sharp + Sharp/Blunt + Blunt/Blunt + Indeterminate	2
Blunt/Blunt + Indeterminate	1
Sharp/Sharp + Blunt/Blunt	0

Associated wound features, i.e. hiltmarks, "fishtailing", notching and withdrawal scratches are documented in Table 6.

Table 6. Associated Wound Features.

Feature	Total Wounds	Number of Cases
Hilt	43	28
Fishtailing	58	36
Notched	95	48
Withdrawal Scratch	115	62

In 82 cases, a single penetrating wound, with or without associated incised wounds, had been sustained. The age and sex distribution of these cases is seen in Table 7.

Table 7. Single Penetrating Wound +/- Incised Wounds.

(Age and Gender)

	Females	Males
Under 10 years	0	0
11 – 16	1	1
17 – 21	1	16
22 – 30	1	29
Over 31	5	28
Total	8	74

N.B. * 45 out of 82 received surgical intervention.
* Only one female had defensive wounds.

Surgical intervention had been carried out in 45 of these cases. Only one of the eight female cases with a single penetrating wound had defensive wounds. The single penetrating wounds were located on the front of the trunk in 64 cases, on the back of the trunk and in 10 cases, on the neck in 5 cases and on the limbs in 3 cases. In 70 of these cases, one or more vital structures had been injured, as displayed in Table 8.

Table 8. Single Fatal Wound Cases.

	No. of Cases	Organs Involved
Females	7	Heart and/or Great Vessel
(8 Cases)	1	Lung only
Males	63	Heart and/or Great Vessel
(74 Cases)	5	Lung
	2	Liver
	1	Spleen
	1	Bowel
	2	Soft tissue only

Of the total 200 cases, 88 had received surgical intervention, 99 had not received surgical intervention and in 13 cases the records did not reveal whether or not there had been surgical intervention. For the purposes of this study, a case was categorized as having received surgical intervention if a chest drain had been inserted or an operative procedure such as thoracotomy, laparotomy or surgical exploration of a wound had been performed. Routine resuscitation, including basic life support and the insertion of intravenous lines, was not categorized as surgical intervention. Of those who received surgical intervention, the number of penetrating wounds inflicted is shown in Table 9.

Table 9. Deaths Despite Surgical Intervention (Total 88 Cases).

No. of Wounds	No. of Cases
1	51
2	13
3	7
4 – 10	12
Over 10	5

In 51 of the 88 cases that received surgical intervention, only a single penetrating wound had been sustained. In the total 88 cases who had received surgical intervention, having sustained one or multiple penetrating wounds, the organ injuries are shown in Table 10.

Table 10. Deaths Despite Surgical Intervention (Total 88 Cases).

No. of Cases	Organs Injured
65	Heart and/or Great Vessels
11	Lung Wound Only
6	Liver Injury Only
1	Spleen Injury Only
1	Bowel Wound Only
4	Soft Tissue Injury Only

Post-injury activity was documented in 22 cases. Immediate collapse and immobility was documented in 35 cases. In the majority of cases (143 cases) it is unknown whether or not the victim had been capable of post-injury activity.

Survival for a period of 24 hours or more occurred in 15 cases. From this group, the number of wounds sustained and the organs injured are shown in Table 11.

Table 11. Survival More Than 24 Hours (Total 15 Cases).

No. of Wounds	No. of Cases
1	9
2 - 12	6

Of these: 6 cases had injury to heart or major vessel
5 cases had injury to lung
1 case had injury to liver
1 case had injury to soft tissue only
2 cases remain unknown

Clothing either on the body of the deceased or accompanying the body to the mortuary and available for examination by a pathologist, was noted in 108 cases, i.e. 88 of the males and 20 of the females.

3. DISCUSSION

Almost half of the cases (49%) had sustained three wounds or fewer, and the majority (80%) had sustained ten wounds or fewer (Table 1). As might have been expected, the majority of penetrating wounds were situated on the trunk, but a surprisingly large minority involved the head and neck (Table 2). Over half of the cases displayed a combination of penetrating and incised wounds.

Approximately half of all cases (48.5%) showed evidence of significant blunt force trauma. Not surprisingly, the head region was the predominant site.

Incised wounds were seen in approximately half of the male cases and fewer than half of the female cases. Of these, all the females had defense-type injuries to the limbs though such injuries were seen in fewer than half of the male cases. The defensive injuries were seen most commonly on the hands, the forearms being the next most frequent site. As a speculation, one might envisage female victims being more likely to be involved in a violent struggle with an assailant whereas in males a more rapid confrontational situation may develop between victim and assailant. No gender difference was observed in the frequency of incised wounds to the head and neck.

Clothing was only available for examination by the pathologist at the time of post-mortem examination in 108 cases (88 males, 20 females). In most other cases, the clothing had already been seized by the police for scientific examination.

The results of examination of the wounds themselves yielded some interesting results. Despite weapons with a single-edged blade being the commonest type used to inflict penetrating wounds, only half of the penetrating wound cases in this study showed the typical associated wound pattern, with one sharp wound edge and one blunt wound edge. In almost half of the cases, wounds of indeterminate contour were encountered.

Wounds displaying a two sharp-edged (sharp/sharp) contour were uncommon (%). In some, a knife with a double-edged blade was known to have been used to inflict them. Of those cases with wounds displaying a sharp/sharp contour, in only two cases was there a single penetrating wound. The other 15 cases had multiple wounds displaying a mix of contours. Therefore, in these 15 cases the sharp/sharp wounds could have been inflicted by weapons with single-edged blades. Even in those cases with wounds exclusively of sharp/sharp contour, the use of a single-edged blade cannot be excluded.

Special wound features, i.e. hiltmarks, notching, "fishtailing" and withdrawal scratches were seen uncommonly, even in those cases with multiple wounds.

Single penetrating wounds, most usually on the trunk, especially the front, were encountered in about half of the cases, both male and female. The heart or great vessels were the organs more frequently injured. Interestingly, about half of these cases survived long enough to receive surgical intervention.

Surgical intervention (more than simple resuscitation), such as the insertion of chest drains, thoracotomy and laparotomy was carried out on 44% of the cases. While the majority of these cases had sustained only one or 2 wounds, a surprising number (24 cases) had a multiplicity of wounds. Therefore, it cannot be assumed that multiple wounds necessarily lead to rapid death, even when there are injuries to the heart and great vessels, as seen in 9 of these cases.

A prolonged survival time, in excess of 24 hours, was a feature of 15 cases. Of these, 9 individuals had a single penetrating wound. Six individuals had suffered injury to the heart or great vessels, 4 of whom had a single penetrating wound. Six individuals had suffered injury to the heart or great vessels, four of whom had a single penetrating wound.

In the overwhelming majority of cases, it was unknown whether or not the victim had been capable of post-injury activity. This is not surprising, as the events surrounding most incidents of wounding are poorly recorded. Only in a few cases where events have been witnessed was there definite evidence concerning activity by the victim after injury was substantiated. Of the 57 unknown cases, 35 individuals apparently collapsed immediately at the scene.

The bio-mechanics of wound injuries have been well summarised by Knight [1]. Studies recording the force of penetration of the body, usually performed by means of a dynamometer or transducer being attached to the knife, have revealed that, apart from bone or calcified cartilage, the tissue most resistant to knife penetration is the skin. The most important factor in skin penetration is the sharpness of the tip of the knife blade. Stretched skin is easier to penetrate than slack skin. The thick skin of the palms and soles, although tougher than the skin of the rest of the body, offers little variation in resistance to the rest of the skin in terms of its vulnerability to penetration. Similarly, the age of the individual or the gender has little effect on the vulnerability to penetration. The speed of approach of the knife towards the body is particularly important in achieving penetration, a knife entering the body at high speed requiring less force to penetrate the skin than a knife held against the skin. When a knife point penetrates the skin, the skin dimples and resists penetration until the threshold is exceeded and the knife passes through under its own momentum unless impeded by bone or cartilage. Indeed, the suddenness of break through and the momentum of the blade renders it difficult or even impossible to prevent deep penetration of the body. These factors are particularly important, especially if the tip of the blade is sharp, in that a body can impale itself on a static knife even if the knife is not firmly supported. This is of particular importance when a defence of falling or running onto the knife is proposed.

Such studies have thus failed to produce an arithmetic scale of force concerning the penetration of a body by a knife. When pressed by Legal Counsel as to the degree of force required to produce a penetrating wound by a knife or similar implement, the pathologist can only employ a simple scale or force i.e. mild, moderate, considerable, severe, the latter being reserved for exceptional cases where the knife has impacted in dense bone, penetrated the skull, or left a marked bruised impression of the hilt guard on the skin.

The compressibility of soft tissues, and its effect on the apparent length of a wound track, is well-documented (i.e. due to compression, the wound length may appear longer than the length of the blade). This effect is most marked in the case of abdominal wounds but is also significant in the case of chest wounds.

4. SUMMARY

Post-mortem reports of 200 victims of penetrating and incised wounds were reviewed and the information entered into an electronic database.

Although this paper does not concentrate on the bio-mechanics of wounding, the main features of wounds have been recorded along with the patterns of injury, the visceral injuries sustained and certain clinical features such as post injury activity and survival times.

The size and contour of wounds is known to vary not only with the shape of the blade and its depth of penetration, but also due to relative movement i.e. rocking, when the blade is in the body, between the knife and the victim, or a combination of the two. Thus the leverage or angulation in the plane of the wound can extend the size of the surface wound. Typically, a single edged blade, the most common type of weapon used in wounding, produces a wound with a single sharp edge and a single blunt edge which may, due to splitting of the stretched skin, produce the classical 'fishtail' appearance. However, it is known that a knife with a single sharp edge and a blunt edge can produce a sharp split behind the blunt edge producing a wound with two sharp contours. Extreme caution must therefore be used in the assessment of any particular weapon causing such a wound. When a blade is withdrawn from the body, the sharp cutting edge may produce a withdrawal scratch on the skin surface. A knife plunged into its full extent may produce a hilt bruise or abrasion on the skin surrounding the wound. In the study, particular attention has been focused on the frequency with which such additional wound features are encountered.

Study of the data obtained reinforced several of the long-held impressions of various aspects of fatal wound victims, such as the age and gender of the victims, the prevalent wound sites and the associated visceral injuries [2-7]. However, a new perspective was perhaps gained regarding other factors such as the relative frequency of specific wound contours, special wound features (hiltmarks, etc.) and presence of defence wounds. When available the information regarding length of survival, post injury activity and surgical intervention was examined revealing no distinct correlation with the number of wounds or the organ injuries sustained.

REFERENCES

1. Knight's Forensic Pathology Third Edition Pekka Sauko Bernard Knight Arnold Publishers.
2. Miller, S.A. and Jones, M.D., Kinematics of four methods of stabbing: a preliminary study, Forensic Science Internation, Vol. 82, 1996, pp. 183-190.
3. Gresham, G.A., A Colour Atlas of Forensic Pathology, 1975.
4. Mason, J.K., The Pathology of Trauma, 2nd Edition, Arnold, 1993.
5. Knight, B., The dynamics of stab wounds, Forensic Science, Vol 6, 1975, pp. 249-255.
6. Green, M.A., Stab wound dynamics – A recording technique for use in Medico-Legal Investigations, Journal of Forensic Science Society, Vol. 18, 1978, pp. 161.
7. Hunt, A.C. and Cowling, R.J., Murder by stabbing, Forensic Science International, Vol. 52, 1991, pp. 107-112.

SESSION 8

COMPUTATIONAL HUMAN BODY MODELS

Jac Wismans[1], Riender Happee[2] and J.A.W. van Dommelen[3]
[1]TNO Automotive, Delft, The Netherlands, [2]TNO Automotive Safety Solutions, Delft, The Netherlands, [3]Eindhoven University of Technology, Eindhoven, The Netherlands

Abstract. Computational human body models are widely used for automotive crash-safety research and design and as such have significantly contributed to a reduction of traffic injuries and fatalities. Currently crash simulations are mainly performed using models based on crash-dummies. However crash dummies differ significantly from the real human body and moreover crash dummies are only available for a limited set of body sizes. Models of the real human body offer some promising advantages including the prediction of injury mechanisms and injury criteria. In this paper, a review will be given of a number of developments in the field of occupant crash simulations in the past 40 years. Topics presented include history of occupant crash simulation codes, human body geometry, human body material modelling and model quality rating. A discussion on foreseen future developments in this field will conclude this paper.

Key words: vehicle safety, injury biomechanics, simulation models, vehicle design.

1. INTRODUCTION AND HISTORICAL REVIEW

In the crash safety field, mathematical models can be applied in practically all areas of research and development including:

- design (CAD) of the crash response of vehicles, safety devices and roadside facilities;
- reconstruction of actual accidents;
- human impact biomechanics studies.

Dependent on the nature of the problem, several types of crash analysis programs have been developed each with their own, but often overlapping, area of applicability. Most of the models are of the deterministic type, that is,

Michael D. Gilchrist (ed.), IUTAM Proceedings on Impact Biomechanics: From Fundamental Insights to Applications, 417–429.

based upon measured or estimated parameter values, representing "average" characteristics of the human body, safety devices, the vehicle and its surroundings and using well established physical laws, the outcome of the crash event is predicted. Although the various deterministic models may differ in many aspects, they are all dynamic models. They account for inertial effects by somehow deriving equations of motions for all movable parts and solve these equations by a numerical method. The mathematical formulations used for these models can be subdivided into lumped mass models, multi-body models and finite element models. In a lumped mass model a system is represented by one or more rigid elements often connected by mass-less elements like springs and dampers. An example of a lumped-mass model in human body modelling is the well-known one-dimensional model of the human thorax developed by Lobdell in 1973 [**1**]. This model simulates the thorax response in case of a loading by an impactor. The model consists of 3 rigid bodies connected by springs and dampers. One mass represents the impactor mass and other masses the sternal and vertebral effective mass, respectively. Springs and dampers represent the skin and flesh between impactor and sternum and the connection between sternum and thoracic spine. The response of this model was shown to correlate well with post mortom human subject tests, further referred here to as PMHS tests.

1.1 Multi-Body Models

The most important difference between a lumped mass model and a multi-body model is that bodies in a multi-body formulation can be connected by various joint types, due to which the number of degrees of freedom between the elements can be constrained. A lumped mass model in fact can be considered as a special case of the more general multi-body model formulation. External forces generated by so-called force-interaction models cause the motion of the joint-connected bodies in a multi-body model. Examples of force-interaction models in a multi-body model for crash analyses are the model that accounts for an acceleration field, spring-damper elements, restraint system models and contact models. Another difference with lumped mass models is that in a multi-body formulation, instead of rigid bodies, also flexible bodies can be specified. Multi-body models in which the complete human body is simulated for the purpose of crash analyses are often referred to as Crash Victim Simulation (CVS) models, human body gross-motion simulators or whole body response models.

One of the early human CVS models, illustrated in Figure 1, was developed already more than 40 years ago by McHenry [**2**]. The model that represents the human body together with restraint system and vehicle is 2-dimensional and has 7 degrees of freedom. McHenry compared his model calculations with experimental data and was able to show quite good agreement for quantities like hip displacements, chest acceleration and belt loads.

The results of this model were so encouraging that since then several, more sophisticated, models have been developed. The most well known are the two-dimensional 8-body MVMA-2D [**3**], the three-dimensional CAL3D allowing up to 20 elements [**4**] and MADYMO 2D/3D allowing an arbitrary number of bodies, both rigid and flexible ones and a range of joint types [**5**]. For a review of the status of these models at the end of the eighties the reader is referred to Prasad and Chou [**6**].

MADYMO, which is developed and supported by TNO Automotive in the Netherlands, has since then gone through an extensive further development and validation program and includes among others a finite element part for crash analyses. The multi-body module of MADYMO calculates the contribution of the inertia of bodies to the equations of motion; the other modules calculate the contribution of specific force elements such as springs, dampers, muscles, interior contacts and restraint systems. Special models are available for vehicle dynamic applications including tyre models and a control module offers the possibility to apply loads to bodies based on information extracted from sensors.

Apart from MADYMO above models are hardly used anymore nowadays. One exception is a special version of the CAL3D program called the ATB (Articulated Total Body) program developed by the Air Force Aerospace Medical Research Laboratory in Dayton [7] for aircraft safety applications.

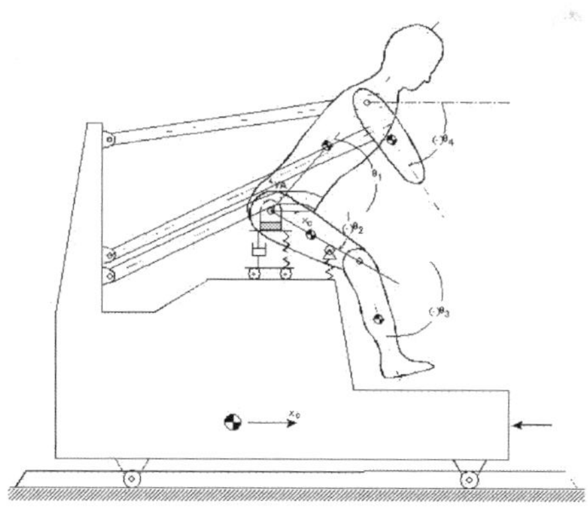

Figure 1. Example of a two-dimensional multi-body model by McHenry [**2**].

1.2 Finite Element Models

The finite element method is a numerical technique to solve equilibrium equations for a domain with arbitrary shape and constitutive properties. In a finite element model the system to be modelled is divided in a number of finite volumes, surfaces or lines. These elements are interconnected at a discrete number of points: the nodes to which degrees of freedom are associated. In the displacement-based finite element formulation, which is applied in practically all major finite element software packages, the motion of the points within each finite element is defined as a function of the motion of the nodes. The state of stress follows from the deformations and the constitutive properties of the material modelled. One of the earlier examples of using the finite element method for human body impact modelling is a model of the human head developed in the seventies by Shugar [**8**] and included a representation of the skull and brain. Linear elastic and linear viscoelastic material behaviour was assumed. The skull bone response and the brain response from the model were compared with experimental results of head impact tests with primates.

Three of the currently most often used software packages for crash simulation using the finite element method are LSDYNA 3D, RADIOSS and PAMCRASH. These packages are based on the public domain version of the DYNA explicit finite element program that was developed at the Lawrence Livermore Laboratories during the seventies [**9**]. For the spatial discretization, the available elements include shell elements, solid elements, beam elements and membrane elements. A large number of material models are available, among others, describing elastic material behaviour, elastic-plastic material behaviour with isotropic hardening and failure, crushable foam with failure, orthotropic material behaviour and strain rate dependent material behaviour.

1.3 Hybrid Modelling

The above finite element crash codes allow the inclusion of rigid bodies and are able to simulate some of the specific features of multi-body crash simulation programs. In the MADYMO crash simulation program that originally was developed as a multi-body code, most of the capabilities of the finite element based crash codes are provided in the integrated FE module. Moreover external interfaces between MADYMO and the FE based crash codes are available allowing integrated multi-body finite element simulations further referred to as hybrid simulations. An early example in which both the multi-body approach and the finite element approach are used in an integrated approach is the model of a car occupant interacting with a passenger airbag developed in the eighties by Bruijs [**10**]. The airbag (and airbag straps) was modelled in the PISCES 3D-ELK program (now MSC-DYTRAN) using almost 2000 triangular membrane elements. The finite element airbag model interacts with a multi-body model of the Hybrid III crash dummy modelled in MADYMO 3D.

2. HUMAN BODY MODELLING

2.1 Introduction

Human body models for crash analyses can be subdivided into models of crash dummies and models of the real human body. For a long time the focus has been on crash dummy modeling due to the need, particularly from the design departments in the automotive industry, for well-validated design tools which can reduce the number of regulatory tests with crash dummies in order to shorten and optimize the development process of a new car model and its safety features. Most model input data in case of crash dummy models can be measured relatively easily and moreover, results of experiments with crash dummies often are available for model validation and if not, such experiments, unlike tests with biological models, readily can be carried out in a well-equipped crash laboratory.

The need for well-validated databases of crash dummies has been recognised by many organisations in the past and has resulted in a number of (co-operative) research efforts to develop such databases. A detailed presentation of these efforts would be out of the scope of this paper, however worthwhile to mention here is the first effort of this kind in the mid-eighties concerning the Hybrid III crash dummy. Prasad [**11**] conducted in 1985 a series of frontal sled tests using a Hybrid III dummy on a rigid seat at 3 different severity levels. The test results were available for a SAE (Society of Automotive Engineers) subcommittee for the purpose of validation of dummy databases of the ATB and MADYMO programs. These validation efforts were presented at the 1988 SAE congress by several authors [see e.g. **12**], which resulted in a number of recommendations for further improvement of the quality of the Hybrid III dummy model database.

For most of the current crash dummies, nowadays, databases (often well-validated) for the various crash codes are available and continuous activities take place in various organisations and user groups to further improve such databases as well as to develop databases for new crash dummies.

This paper will focus further on models of the real human body. A model of the real human body is much more difficult to develop than a model of a physical crash dummy. This type of models offers improved biofidelity (human-likeness) compared to crash dummy models and allows the study of aspects like body size, body posture, muscular activity and post fracture response. Furthermore they potentially allow analysis of injury mechanisms on a material level. In the next sections, several important aspects of human body models will be discussed namely how to deal with the human body geometry (anthropometry), modelling of human body material and validation of human body models.

2.2 Anthropometry

One of the challenges in modelling the real human body compared to crash dummy modelling is how to deal with the large variations in human body sizes that exists. In case of a multi-body approximation several methods exist allowing the generation of an arbitrary sized human body model. Two of the most widely used methods will be mentioned here. The first one was developed in the early eighties and is available through the software package GEBOD [**13**]. This software generates geometric and inertia properties for a 15 segment ATB or MADYMO multi-body model. Computations for the geometrical parameters and mass distribution are based on a set of 32 body measurements to be specified by the user or generated by GEBOD using regression equations on the basis of body height and/or weight for both adult males and females. Also for children, regression equations are available. A major limitation of GEBOD is the approximation of body segments by simple geometrical volumes.

The second method is based on the use of software packages for ergonomic analyses like the RAMSIS software [**14**]. The RAMSIS model describes the human body as a set of rigid bodies connected by kinematic joints and the skin is described as a triangulated surface. RAMSIS provides a detailed geometric description of the body segments based on extensive anthropometric measurements on various civilian populations including automotive seated postures. In RAMSIS the body dimensions of each individual can be classified according to three dominant and independent features: body height, the amount of body fat, and body proportion, i.e. the ratio of the length of the limbs to the length of the trunk. A translator has been developed to convert RAMSIS models into MADYMO models [**15**]. The resulting database contains joint locations, joint ranges of motion, segment masses and centres of gravity and a triangulated skin connected to various body segments. This human model has been extended to a full dynamic human model suitable for simulation of impact and vibration loading [**15**].

The methodology available through RAMSIS can also be applied to scale crash dummies. Happee [**16**] created 30 different models by scaling male and female Hybrid III dummy models towards various RAMSIS anthropometries for the purpose of evaluation of restraint systems in a frontal collision mode. The models accounted for human variance with respect to length, corpulence and the proportion of seated height to standing height.

For detailed finite element based human body models the anthropometric information provided by RAMSIS is much too global. Detailed information is needed on the structures within the human body. One of the efforts to achieve this information for a specific subject is the work done in the European HUMOS-1 project [**17**]. For this purpose the method of physical slicing of a PMHS in driving position was chosen. The slices were photographed and afterwards digitised. The resulting MADYMO model from this HUMOS-1 project is illustrated in Figure 2.

The PMHS measured in HUMOS-1 approximated a 50th percentile human male. Alternative methods to achieve the detailed human body anthropometry are based on MRI techniques. A limitation of these models, although they are very detailed, is that they represent one unique human body size namely the one PMHS that was actually measured.

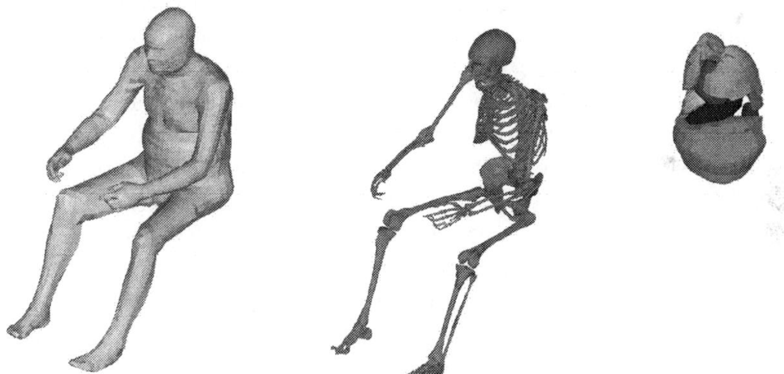

Figure 2. MADYMO Finite Element Occupant Model in automotive seating position. (soft tissues-left; skeleton-middle, organs-right)

2.3 Material Modelling

Simulations with human body finite element models require constitutive descriptions for the various materials that constitute the human body (such as bone, cartilage, ligament and tendon, muscular tissue, and various organs such as brain, heart, lung, etc.). Accurate representations of the mechanical behaviour of the various components are essential for reliable predictions of injury in impact situations.

In general, properties of biological tissues are viscoelastic (i.e. their response is rate-dependent and they show stress relaxation at constant strain level), non-linear, and anisotropic due to the specific microstructure (for example consisting of an arrangement of collagen fibres). A full characterisation of the constitutive behaviour considers the behaviour in various deformation modes (shear, uni-axial tension, compression, bi-axial deformation, etc.) and complex loading paths (e.g. reverse loading). Furthermore, the use of constitutive models for biological materials in impact biomechanics simulations requires a characterisation of these materials at high strain rates. The viscoelastic characteristics are typically determined in (small strain) oscillatory experiments, stress relaxation experiments, and constant strain rate tests at varying strain rates. Prior to the characterisation experiments, specimens are often preconditioned. The frequency range or stain rates that can be addressed is often limited by the capabilities of the experimental set-up. Characterisation experiments are often conducted *in-vitro*, on small specimens, either in compression/ extension or, more commonly, in shear [**18**]. *In-vivo* experiments are sometimes carried out by indentation of organs [**19**].

The mechanical properties of living tissues may depend on age and gender. Furthermore, properties vary largely between different subjects. Therefore, mechanical experiments on biological materials show a large scatter in obtained results. Additional scatter in data is due to varying test conditions (both physical and mechanical), handling and treatment, and post-mortem time (since most tests are conducted *in-vitro*). Furthermore, regional differences within an organ or within the body may exist. Tests are conducted either on human cadaveric material or on tissues from animal donors. The advantage of the latter is that material can often be tested at shorter post-mortem times or even in vivo. Consequently, the properties of biological materials as reported in the literature vary widely. For example, the mechanical properties reported for brain tissue vary over an order of magnitude. An overview of mechanical properties of various biological materials is given in [**17**].

The mechanical behaviour of biological tissue is the result of the properties of the individual microstructural components that constitute the material (for example collagen fibres and surrounding matrix material) and the interplay between these components. For soft tissues, often hyperelastic models are applied, which represent true elastic behaviour. Hyperelastic constitutive laws can be derived from a strain energy function. This energy function is given in terms of the deformation gradient tensor, often through invariants of the Cauchy-Green strain tensor. The strain energy function must be chosen such that the obtained constitutive relation matches experimental data. Some common choices for a hyperelastic model are for example the Mooney-Rivlin model or the Ogden model. Although isotropic models are successfully used for many applications, the incorporation of anisotropy in constitutive models can be essential. For soft tissues, a fiber-reinforced model may be used, where the strain energy function depends on the fiber stretch. Orthotropic elasticity is sometimes used for anisotropic materials such as bone [**20**].

A viscoelastic material model can be represented by a mechanical analogon consisting of a certain arrangement of springs and dashpots. By placing a number of so-called Maxwell elements in parallel, a general viscoelastic framework can be obtained. The linear viscoelastic theory considers the concepts of proportionality and superposition. In this theory, an arbitrary loading-history is assumed to be given by a Boltzmann integral over an infinite number of small steps. The so-called quasi linear viscoelastic (QLV) theory has been proposed by Fung [**21**]. This theory has become widely used in injury biomechanics and has been successfully applied for the constitutive or structural modelling of many soft biological tissues [**22**]. The QLV theory is a generalisation of the linear viscoelastic theory and is also formulated in terms of a convolution integral. In this integral representation, the elastic response is separated from the relaxation function. The quasi-linear viscoelastic theory assumes the time-dependent behaviour to be linear and a non-linear relation for the instantaneous elastic response can be used. This instantaneous elastic response is commonly derived from a strain energy function.

The properties of skeletal muscles can be separated in an active and a passive component. The active response may become important in low speed accidents such as, for example, rear-end collisions producing whiplash disorders [**28**]. One-dimensional phenomenological models are often used to describe the response of skeletal muscles. The Hill muscle model [**23**] is frequently used for bar elements that simulate the active and passive response of skeletal muscles.

Injury will develop if the mechanical response (e.g. strains, stresses, etc.) of the biological material attains a level at which either the structural integrity of the materials is affected or functionality is reduced. The latter may be the result of physiological processes that occur after the impact, at time scales that are much larger than the time scale of the loading conditions.

2.4 Validation

Validation is the process of assessing the reliability of a simulation model in comparison to one or more reference tests with human subjects. Very important in this process is that the reference tests, often also referred to in literature as biofidelity tests, are not the same tests as used for determination of model input data. If results from different tests are available, usually so-called biofidelity corridors are defined for instance consisting of envelopes of resulting time histories.

Human models have been validated for frontal, lateral, and rear impact as well as pedestrian loading [e.g. **15, 17, 28, 29**] using volunteer tests for low severity loading and PMHS tests for higher severity loading. Recently the validity range of the MADYMO multi-body human model has been extended to vertical vibration transmission [**27**].

FE human models have also been validated using bone segment testing [**17**] and some progress validating soft tissue responses has been made using marker and ultrasound techniques.

While human models have been validated extensively for kinematics, accelerations and for compliance (force-deflection) the next step is to demonstrate injury prediction capabilities of human body models which is an area still in its infancy. Some promising results in this respect have been achieved, among others, for long bones where initiation of bone failure was predicted based on yield stress and plastic strain [**26**].

3. DISCUSSION AND CONCLUSIONS

A general advantage of computer crash simulations over crash tests with mechanical human substitutes (crash dummies) is that the safety performance of design concepts and the effect of changes in the design can be studied efficiently, sometimes even without a prototype to be built (virtual testing). An important condition for virtual testing is that well-validated databases of the human body are available. Continuous efforts are needed to further improve the quality of models in order to allow their usage in even a wider range of applications and as assessment tools in crash regulations. Standards for validation procedures and performance criteria are needed for this purpose. In the past some attempts have taken place to develop such standards, for instance, the Validation Index developed in the early eighties by the "Analytical Human Simulation Task Force" of the SAE Human Biomechanics and Simulation Subcommittee (HBSS) [**24**]. A range of methods to generate objective rating methods for the quality of crash models has been evaluated within the European project VITES. As a results of this project the analysis tool called ADVISER has been developed that can assist the user/developers of crash models in assessing the quality of their models in an objective manner [**25**]. In addition the ADVISER stochastic tool enables prediction of the stochastic response of crash tests in relation to the scatter of the component responses in the system.

The earliest numerical models of the full human body have been based on multi-body techniques. More recently also finite element techniques have been used for this purpose. A major advantage of the multi-body approach is its capability of simulating, in an efficient way, spatial motions of mechanical systems with complex kinematic connections like they are present in the human body and in parts of the vehicle structure. The advantage of the finite element method is the capability of describing (local) structural deformations and stresses in a realistic way. But the creation of a finite element model is a time consuming job and the availability of realistic material data is still limited. Furthermore relatively large computer times are required to perform a finite element crash simulation, making the method less attractive for complex optimisation studies involving many design parameters or for stochastic type of simulations.

Models of the human body can be subdivided into models of crash dummies and models of the real human body. Many models of crash dummies have been developed and extensive series of validation studies have been conducted with rather impressive results. Also in the field of real human body models rather promising results have been achieved. An important advantage of real human body models is that they allow the study of the effect of body size, posture influence as well as muscular activity. The advantage of a design strategy based on real human body crash models over a design strategy based on crash tests with dummies (and crash dummy models) is the possibility to benefit without delay, in principle, from new scientific knowledge on injury mechanisms and injury criteria obtained through biomechanical research. In case of a crash test dummy based design strategy usually a long period elapses before new findings actually can be implemented in crash dummy hardware.

Several areas can be identified in the field of real human body models where further developments should take place. This includes further improvements in the description of the non-linear dynamic behaviour of muscles (incl. neuro-muscular control), the modelling of complex human joints and the study of constitutive equations and parameters for biological materials (e.g. brain, skin). The development of constitutive models that consider the behaviour in various deformation modes and complex loading paths remains a challenge.

Other areas of developments include detailed finite element models that fully take into account anthropometry and age variations over the population, like already is possible now to some extent in case of global multi-body based human body models. Also there is a clear need for models that account for the pre-crash response of the human body in view of development of pre-crash sensing based restraint systems.

REFERENCES

1. Lobdell T.E., Impact response of the human thorax, in: *Human Impact Response: Measurement and Simulation*, Plenum Press, New York, 1973, pp. 201-245.
2. McHenry, R.R., Analysis of the dynamics of automobile passenger restraint systems, *Proc. 7th Stapp Car Crash Conference*, 1963, pp. 207-249.
3. Robbins, D.H., Bowman, B.M. and Bennett, R.O., The MVMA two-dimensional crash victim simulation, *Proc. 18th Stapp Car Crash Conference*, 1974, pp. 657-678.
4. Fleck, J.T., Butler, F.E. and Vogel, S.L., An improved three-dimensional computer simulation of motor vehicle crash victims, *Final Technical Report No. ZQ-5180-L-1*, Calspan Corp. (4 Vols.), 1974.
5. *MADYMO Theory Manual*, Version 6.2.1, TNO Automotive, Delft, the Netherlands, December 2004.
6. Prasad, P. and Chou, C.C., A review of mathematical occupant simulation models, Crashworthiness and occupant protection in transportation systems*, Proceedings AMD-Vol. 106, BED-Vol. 13 of the Winter Annual Meeting of ASME*, December 1989.
7. Wismans, J. and Obergefell, L., Data bases and analytical modelling, Chapter 8 in *AGARD Advisary Report 330 Anthropomorphic Dummies for Crash and Escape System Testing,* AGARD/AMP/WG21, 1996.
8. Shugar, T.A., A finite element head injury model, *Final Report Vol 1 Contract DOT HS 289-3-550-IA,* NHTSA , Washington DC, July 1977.
9. Hallquist, J.O., A procedure for the solution of finite deformation contact-impact problems by the finite element method, *Report UCRL 52066*, Lawrence Livermore Laboratory, Livermore, CA, 1976.
10. Bruijs, W.E.M., Subcycling in transient finite element analysis, *PhD Thesis, Department of Mechanical Engineering, Eindhoven University of Technology*, Eindhoven, the Netherlands, ISBN 90-9003684-9, 1990.
11. Prasad P., Comparative evaluation of the dynamic response of the Hybrid II and Hybrid III dummies, SAE Paper No. 902318, *Proc. of the 34th Stapp Conference*, 1990.
12. J. Wismans and Hermans, J.H.A., MADYMO 3D simulations of Hybrid III dummy sled tests', SAE Paper No. 880645, SAE PT-44, *SAE Int. Congress and Exposition,* Detroit, Society of Automotive Engineers Inc., 1988.
13. Baughman, L.D., Development of an interactive computer program to produce body description data, *Report No. AFAMRL-TR-83-058*, NTIS Doc. No. AD-A 133 720, University of Dayton Research Institute, Ohio, USA, 1983.
14. Geuß H., Entwicklung eines anthropometrischen Mebsystems für das CAD-Menschmodel Ramsis. *PhD Thesis, München University*, 1994.
15. Happee, R., Hoofman, M., van den Kroonenberg, A.J., Morsink. P. and Wismans, J., A mathematical human body model for frontal and rearward seated automotive impact loading", in: *Proceedings of the 42nd Stapp Car Crash Conference*, Tempe, USA, November 2-4, 1998.
16. Happee R., Haaster R. van, Michaelsen L. and Hoffmann R., Optimisation of vehicle passive safety for occupants with varying anthropometry, *ESV Conference 1998*, Paper 98-S9-O-03, 1998.

17. Haug, E., Choi, H.-Y., Robin, S., and Beaugonin, M., Human models for crash and impact simulation, in: *Handbokof Numerical Analysis, Vol. XII,* P.G. Ciarlet (ed.), Elsevier, 2004.
18. Brands, D.W.A., Bovendeerd, P.H.M., Peters, G.W.M. and Wismans, J.S.H.M.. The large shear strain dynamic behaviour of in-vitro porcine brain tissue and a sillicone gel model material, *Proceedings of the 44th Stapp Car Crash Conference*, 2000.
19. Gefen, A. and Margulies, S.S., Are in vivo and in situ brain tissues mechanically similar? *Journal of Biomechanics* **37**, 2004, 1339-1352.
20. Ashman, R.B., Cowin, S.C., Buskirk, W.C. van and Rice, J.C., A continuous wave technique for the measurement of the elastic properties of cortical bone, *Journal of Biomechanics* **17**, 1984, 349-361.
21. Fung, Y.C., *Biomechanics: Mechanical Properties of Living Tissues,* Springer-Verlag, New York, 1981.
22. Woo, S.L.-Y., Hollis, J.M., Adams, D.J., Lyon, R.M., and Takai, S., Tensile properties of the human femur-anterior cruciate ligament-tibia complex. The effects of specimen age and orientation, *American Journal of Sports Medicine* **19**, 1991, 217-225.
23. Hill, A.V., *First and last experiments in Muscle Mechanics,* Cambridge, 1970.
24. Prasad, P., Occupant simulation models: experiment and practice, in: *Crashworthiness of Transportation Systems: Structural Impact and Occupant Protection*, J.A.C. Ambrosio et. al. (eds), Kluwer Academic Publishers, 1997, pp. 209-219.
25. Hoof J van, Puppini R., Baldauf H., Oakley C. and Kayvantash K., Adviser: A software tool for evaluation and rating of numerical models in crash safety analyses, *ESV Conference 2003*, Paper No. 483, 2003.
26. Rooij, L. van, Bours, R., Hoof, J. van, Mihm, J.J., Ridella, S.A., Bass, C.R. and Crandall, J.R., The development, validation and application of a finite element upper extremity model subjected to air bag loading, *Stapp Car Crash Journal, Vol. 47*, 2003.
27. Verver M.M., Hoof J. van, Oomens C.W.J., Wouw N. van de and Wismans J.S.H.M., Estimation of spinal loading in vertical vibrations by numerical simulation, *Clinical Biomechanics* **18(**9), 2003, 800-811.
28. Horst, van der M., Human head neck response in frontal, lateral and rear end impact loading – modelling and validation, *PhD Thesis, Eindhoven University of Technology*, ISBN 90-386-2843-9, 2002.
29. Happee R., Ridella S., Nayef A., Morsink P., de Lange R., Bours R. and van Hoof J., Mathematical human body models representing a mid size male and a small female for frontal, lateral and rearward impact loading, *IRCOBI Conference,* 2000.

RECONSTRUCTION OF HEAD INJURY CASES ARISING FROM FALLS USING THE UCD BRAIN TRAUMA MODEL

M.C. Doorly, T.J. Horgan and M.D. Gilchrist
Department of Mechanical Engineering, UCD, Belfield, Dublin 4, Ireland

Abstract. While Road Traffic Accidents continue to be the largest contributor to head injury, falls are usually second in prevalence. This paper looks at numerical modelling techniques, namely multibody body dynamics and finite element methods, in order to reconstruct two real-life accident cases arising from falls. Various modelling strategies are explored, and the results are compared with existing published brain injury tolerance levels.

Key words: impact biomechanics, falls, multibody dynamics, accident reconstruction, head injury, Finite Element Method (FEM).

1. INTRODUCTION

Traumatic head impact injuries occur when the human skull and brain are rapidly subjected to intolerable levels of energy. There exist many causes of neurotrauma: accidents, falls, assaults and injuries occurring during occupational, recreational and sporting activities. While Road Traffic Accidents tend to be the leading cause of injury related death, falls tend to be the leading cause of non-fatal hospitalisation [1, 2]. In Ireland, falls are the single greatest cause of hospital admissions for both males and females across most age groups, with head injuries occurring in approximately a quarter of all fall admissions [2].

The aim of this research is to reconstruct a number of real life falling accidents using numerical techniques in order to establish injury criteria for specific types of brain lesion. Firstly, the accidents are modeled using

Michael D. Gilchrist (ed.), IUTAM Proceedings on Impact Biomechanics: From Fundamental Insights to Applications, 431–440.

multibody dynamics software to recreate the overall movement of the body during the accident. Next, the output from the multibody dynamics simulation is then used as input for the University College Dublin Brain Trauma Model (UCDBTM) [3]. The UCDBTM is able to simulate the effect of the overall head movement on the cranial contents, so the local deformation parameters within the brain tissue can be examined and compared to the clinical results observed.

2. METHODOLOGY

The head injury research group at University College Dublin have been working in conjunction with the National Department of Neurosurgery at Beaumont Hospital, Dublin, obtaining head injury cases arising from simple falls. Using information from witness accounts and neurosurgery reports, two such cases were reconstructed using numerical methods [4].

Case 1: Boy fainting at a fountain.

Suffering from heat exhaustion, an 11 year old boy (height 152 cm, weight 37 kg) fainted after straightening up from drinking at a water fountain on a city street. According to witnesses he fell straight backwards and his head rebounded off the ground. The ground was reported to be level and concrete. The patient experienced a brief loss of consciousness. Upon revival his Glasgow Coma Scale (GCS) score was 14/15, indicating mild confusion. Detailed clinical examination of the patient revealed that the fall resulted in impact just over the occiput, mostly central, and slightly right-sided, as evidenced by subcutaneous bruising and swelling of the scalp overlying this region. There was no apparent skull fracture and no other injuries were noted on full clinical trauma survey of this patient (i.e. neck, chest, abdomen, pelvis, and extremities). CT imaging on presentation to hospital revealed a right lateral frontal intracerebral haemorrhagic contusion. There was also evident blood in the right Sylvian fissure seen on CT imaging, representing traumatic subarachnoid haemorrhage.

Case 2: Lady falling from step.

The second case involved a 76 year old lady (height 160cm, weight 60kg) who lost her balance while standing on the back step of her house, facing the door. She fell straight backwards, with the layout of the environment suggesting that she hit her head off a vertical concrete/cement wall. This lady did not lose consciousness and presented to hospital with a GCS score of 14/15 (representing a mildly confused state). Detailed clinical examination of the patient revealed that the fall resulted in impact to the occipital bone, slightly to the left of midline, and this was evident by a 3cm laceration on the scalp overlying this region. There was no apparent skull fracture. A graze on the right elbow was also sustained, but no other external injuries were noted on full clinical trauma survey of this patient. CT scans revealed a large parenchymal haemorrhage of the right temporal lobe, and a small focal bleed on the cortical surface of the left frontal lobe.

MADYMO [5] multibody dynamics software was used to reconstruct the falls. MADYMO contains a suite of pedestrian models and these cases were modeled using the 5th percentile female model, since this was closest in height and weight to both patients. The features of the environment that may have affected the fall were modeled using planes and cylinders. For both of these simulations, the model was positioned in the inertial space leaning slightly backwards (backward rotation of 0.1 radian) and allowed to fall with gravity without any initial velocity being applied. Contact was specified between the environment and the ellipsoids of the body relevant for each fall. The force-penetration characteristics of the head were altered to represent more realistic contact. An analysis of the sensitivity of the results to changes in initial conditions was carried out and the range of results was sufficiently small to not require multiple FE simulations to be run [4].

The baseline UCDBTM, which was refined to differentiate between the grey, white and ventricular matter, was used for this analysis. The scalp elements (which completely encased the head apart from the foramen region where the brain elements are unconstrained) were redefined to be rigid and the full time histories of the six velocity components of the centre of the MADYMO head were applied to a reference node located at the COG of the head. The FE head model was also scaled so as to represent the same weight head, which was used in the multibody simulations, since this affects the intracranial pressure, Von-Mises stress and shear stress responses [3].

Applying the full kinematic time-histories to the model required excessive CPU time with the current hardware resources (up to 17 days to solve) and consequently two other modeling strategies were explored in an effort to reduce the time required to simulate the impact. For studying brain injury mechanics it is the time during the impact event that is most important. Two methods were devised by which the simulation time could be reduced. In both cases the analysis was run from a time closer to when the impact occurred. The first method, referred to as "abrupt run", involved applying the velocity profiles at a short time before impact directly to the head at the start of the analysis. The head was orientated at the appropriate angle relative to the ground before the simulation began. This is a very abrupt method of starting the simulation, however it was established that there was enough time before the impact occurred for the starting effects to attenuate (by examining the evolving stresses in various elements). In the second approach, which is referred to as the "adjusted run", the velocities before impact were taken as the datum, and the velocities relative to these were applied to the rotated head. This meant that the analysis effectively started from stationary and the movement was applied more gently to the model. The pressure, Von-Mises response, shear stress response and logarithmic strain values were compared for all three analysis strategies in the frontal, parietal, occipital and midbrain regions. As seen in Figure 1 the shortened simulation strategies were identical to the results of the full simulation and thus the "adjusted run" method was chosen as the strategy for modelling further accident reconstructions: this method required the least computation time (typically only three days).

Having chosen this method, the FE model was tested for both cases, comparing the coupled node brain/CSF approach with a sliding brain/skull boundary approach [6]. In this second approach the solid elements previously defined as CSF material were redefined to correspond to the material of the dura, falx and tentorium (the shell elements which had previously been assigned to these regions were removed from the model). The simulation therefore modelled the brain in direct contact (sliding with friction) with the intracranial membranes and the inner surface of the skull.

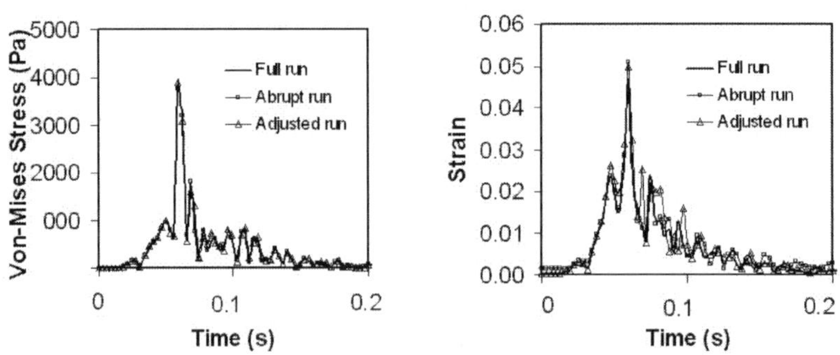

Figure 1. Comparison of methods for reconstructing case studies. These graphs are representative of the curves observed for all regions analysed. It is evident that the faster abrupt and adjusted modeling strategies work sufficiently.

The results of the multibody dynamics simulations are presented in Table 1 along with a summary of the relevant literature. A similar table for the finite element simulations is presented in Table 2, which reports the levels predicted in the area where the actual injury occurred as well as the maximum values predicted for the finite element brain.

Table 1. Table of outputs from MADYMO simulation for both cases.

Output Parameter	Simulation Value		Comments
	Boy	Lady	
Force Duration	12.1 kN < 2 ms	5.134 kN < 1.5 ms	8.8 – 14.1 kN → Fracture [7] 5.8 – 17 kN⟶ Fracture [8]
Resultant Linear Velocity (m/s)	6.88	3.77	
Resultant Angular Velocity (rad/s)	30.21	25	Below gliding contusion threshold [9]. Inside bounds for DAI and concussion to be absent [10].
Resultant Linear Acceleration (g)	584	308	Above lower bound for contrecoup contusion, SAH and SDH [11]. Above 100% concussion probability [12].
Resultant Angular Acceleration (rad/s^2)	20700	44400	Below tolerance curve for SDH [13]. Above sustainable levels observed [14]. Above 100% injury probability of MTBI [12].
HIC	8690	1150	Boy: Level indicates fatal injury sustained. Lady: Above 50% injury probability [15]
GAMBIT [16]	3.03	5.36	>1 is injurious
HIP (kW) [12] 50th %ile values: Scaled FE head*:	 92.3 66.7	 21.5 17.5	Head Impact Power - 50% injury probability at 12.8kW.
PI (kW) [12, 17] 50th %ile values: Scaled FE head*:	 118 87.5	 27 23	Threshold for power index subdural haematoma for posterior anterior rotation of 50kW.

*Mass: 3.136, Ixx: 0.01131, Iyy: 0.01102, Izz: 0.007714

Table 2. FE simulation results for both cases. Methods 1 and 2 refer to models using the tied CSF definition and the sliding brain dura boundary approach [6] respectively.

Parameter	Method	Simulated Value				Literature Value
		Injury area		Highest Value *		
		Boy	Lady	Boy	Lady	
Von Mises Stress (kPa)	1: 2:	3.2 3.26	6 5.09	12.67 9.64	18.8 15.17	15-20 → concussion [18,19] 30 → AIS 2+ [20] 7kPa for approach A, 8.6kPa for approach B → contusion [6]**
Strain	1: 2:	0.07 0.09	0.2 0.17	0.3 0.27	0.28 0.43	>0.25 → structural failure [21] >0.20 → functional deficit [21] <0.10 → reversible injury [21] 0.188 → BBB injury [22] 9% for approach A, 7.6% for approach B → contusion [6]
Strain Rate (s^{-1})	1: 2:	11.8 9.36	30 26.1	71.63 38.2	57.7 50	23-140 → injury [23] 11-67 → non injury [23]
Strain x Strain Rate (s^{-1})	1: 2:	0.48 0.14	2.4 0.62	19.5 2.4	20 4.44	10-80 → injury (concussion) [23] 0-20 → no injury [23]

* Maximum values achieved at different locations for both methods.

** Approach A is Miller *et al* version of Method 1, Approach B is similar to Method 2.

Figure 2 shows a cross section of the FE brain showing the state of maximum strain for Case1. Figure 3 shows a cross section of the FE brain showing the state of maximum Von-Mises stress for Case 2. For both simulations the differences between the two simulation approaches are evident. In both cases the highest value reported by the sliding model is in closer proximity to the location of the actual injury. There may be an unrealistic constraint put on the motion of the brain under the current configuration used for the CSF in the coupled node model.

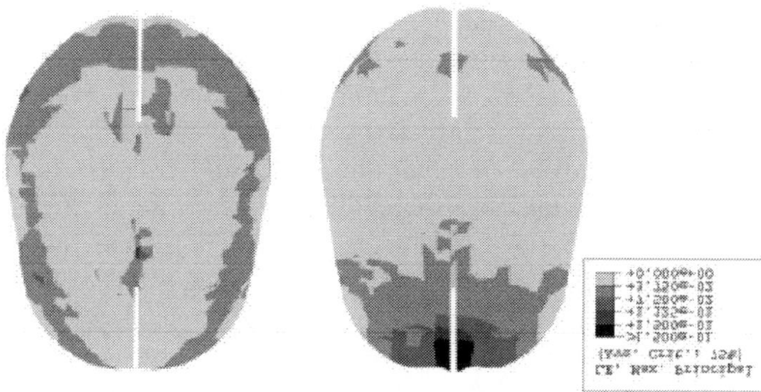

Figure 2. Comparison of strain results using the two modelling methods for Case 1. On the left is the fixed node definition and the right the sliding definition. An obvious higher strain exists in the frontal lobe area of the sliding model. This is also where contusion occurred.

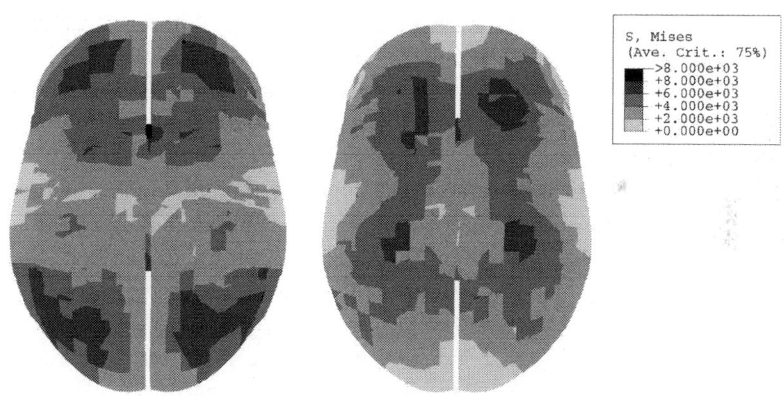

Figure 3. Off centre sagittal view of the Von-Mises stress in the FE head during Case 2. The left image shows the prediction of the tied boundary model, while the right is that of the sliding boundary model. The sliding boundary model shows stresses set deeper into the brain, with the coupled node model experiencing peaks nearer the cerebrum's periphery. Again the sliding model correlates better with injured regions.

3. DISCUSSION

The results from the two cases are considerably different, especially when looking at the results of the MADYMO simulations. Some of the results observed for the fountain case fall well above the various threshold values cited in literature, indicating that the boy should have sustained much more serious injury than that observed clinically. The case of the lady on the step gives much more realistic results in terms of HIC and linear acceleration, however angular acceleration values are very high here. The oblique nature of the impact is more likely to result in higher angular accelerations than linear impact would.

The MADYMO model used, while being the 5th percentile, was still heavier than the actual boy. Mohan *et al.* [24] suggest that at age 13, skull stiffness is approximately 90% of adult skull stiffness. This boy was 11, and it is possible that his skull is not fully calcified compared to an adult, leading to lower head stiffness. The joint stiffnesses of the model were not analysed for the purpose of this study. Tolerance limits are also scarce for children. If these factors could be represented more accurately, the linear and angular accelerations would be more accurate, and hence more insight into injury of children may be found from the FE model.

In the case of falls onto rigid surfaces the impact duration is usually quite short with an associated high peak value. Even though very high values for acceleration are observed in these cases, the duration for which these peaks occur is very short. This factor is not taken into account by injury measures such as GAMBIT.

The results of the finite element analysis have shown that the UCDBTM demonstrates stress and strain levels in agreement with those of the literature, though availability of tolerance limits for fall cases are fewer than for those obtained from tests conducted for investigating RTAs. For the study of contusion, the case studies suggest that a sliding boundary model has better prediction capabilities, with the sliding model showing higher strain and Von-Mises stress levels in closer proximities to the region that was injured. For the case studies analysed, the tied brain/CSF model always experienced maximum strain and stress in the midbrain region. The tied model also experienced higher levels of stress and strain in the occipital lobe than the sliding model, but no injury was observed in that location.

4. CONCLUSIONS

Due to the large amount of CPU time required to run a finite element analysis of a head impact resulting from a simple fall, other modelling strategies were explored in an effort to reduce this time. It was found that two alternative strategies produced very similar results, and so it was decided that the "adjusted run" method be used for future analyses since this required least CPU time.

In order for the prediction capabilities of the reconstructions to become more reliable improvements would be required for the UCDBTM. These include increasing the mesh density of elements next to the fluid elements and an improved representation of the boundary region of the ventricles. The MADYMO reconstruction of the case studies presented here suffered high levels of angular and linear acceleration over very short durations, with their values usually exceeding what is generally accepted as injurious.

The results observed for these two cases suggest that this method will be useful for investigations into tolerance limits acceptable to the human for different types of injury, provided correct modelling methods are used. Future work involves the analysis of more accident cases in the hope of finding tolerance levels of brain deformation which result in specific brain injuries.

REFERENCES

1. Watson, W.L. and Ozanne-Smith, J., Injury surveillance in Victoria, Australia: Developing comprehensive injury estimates, *Accid. Anal. Prev.*, **32,** 2000, 277-286.
2. Scallan, E., Staines, A., Fitzpatrick, P., Laffoy, M. and Kelly, A., Injury in Ireland, *Report of the Department of Public Health Medicine and Epidemiology*, University College Dublin, 2001.
3. Horgan, T. and Gilchrist, M., The creation of three-dimensional finite element models for simulating head impact biomechanics, *Int. J.Crashworthiness,* **8**(4), 2003, 353-366.
4. Doorly M.C., Phillips, J.P. and Gilchrist, M.D., Reconstructing real life accidents towards establishing criteria for traumatic head impact injuries, in *IUTAM Conference Proceedings on Impact Biomechanics,* Dublin, 2005 (this publication).
5. MADYMO, Version 6.0, TNO Automotive, The Hague, NL, 1999.
6. Miller, R., Margulies, S., Leoni, M., Nonaka, M., Chen, X., Smith, D. and Meaney, D. (1998), Finite element modeling approaches for predicting injury in an experimental model of severe diffuse axonal injury, in *42th Stapp Car Crash Conf.,* SAE Paper No. 983154 , 155-166.
7. Yoganandan, N., Pintar, F.A., Sances Jr., A., Walsh, P.R., Ewing, C.L., Thomas, D.J. and Snyder, R.G., Biomechanics of skull fractures, *J. Neurotrauma*, **12,** 1995, 659-668.
8. Allsop, D.L., Perl, T.R. and Warner, C.Y., Force/deflection and fracture characteristics of the temporo-parietal region of the human head, in *Proc. 35th Stapp Car Crash Conf.*, 1991, 269-278.
9. Löwenhielm, P., Mathematical simulation of gliding contusions, *J. Biomech.*, **8**, 1975, 351-356.
10. Margulies, S. and Thibault, L., A proposed tolerance criterion for diffuse axonal injury in man, *J. Biomech.*, **25**(8), 1992, 917-923.
11. Auer, C., Schonpflug, M., Beier, G. and Eisenmenger, W., An analysis of brain injuries in real world pedestrian traffic accidents by computer simulation reconstruction, in *Proc. Int. Soc. Biomechanics XVIIIth Congress*, Zurich, 2001.

12. Newman, J., Shewchenko, N. and Welbourne, E., A proposed new biomechanical head injury assessment function – the maximum power index, in *44th Stapp Car Crash Conf.,* SAE Paper No. 2000-01-SC16, 2000.
13. Gennarelli, T.A. and Thibault, L.E., 'Biomechanics of acute subdural haematoma', *J. Trauma*, **22,** 1982, 680-686.
14. Pincemaille, Y., Troisseille, P., Mack, P., Tarriere, C., Breton, F. and Renault, B., Some new data related to human tolerance obtained from volunteer boxers, in *Proc. of the 33rd Stapp Car Crash Conf.,* SAE Paper No. 892435, 1979.
15. Prasad, P. and Mertz, H., The position of the United States delegation to the ISO working group 6 on the use of HIC in the automotive environment, *Technical Report,* SAE Paper No. 851246, 1985.
16. Newman, J., A generalized acceleration model for brain injury threshold (GAMBIT), in *Proc. International Conf. on the Biomechanics of Impact (IRCOBI)*, 1986, 121-131.
17. Kleiven, S., Influence of impact direction to the human head in prediction of subdural haematoma, *J. Neurotrauma*, **20**(4), 2003, 365-379.
18. Willinger, R., Baumgartner, D., Chinn, B. and Neale, M., Head tolerance limits derived from numerical replication of real world accidents, in *IRCOBI Conf.,* Montpellier, 2000, 209-221.
19. Baumgartner, D., Willinger, R., Shewchenko, N. and Beusenberg, M., Tolerance limits for mild traumatic brain injury derived from numerical head impact replication, in *Proc. of the 2001 International IRCOBI Conf. on the Biomechanics of Impacts,* Isle of Man, 2001, 353-355.
20. Anderson, R., Brown, C., Blumbergs, P., Scott, G., Finney, J., Jones, N. and McLean, A., Mechanisms of axonal injury: an experimental and numerical study of a sheep model of head impact, in *IRCOBI Conf.,* Sitges, 1999, 107-120.
21. Galbraith, J., Thibault, L. and Matteson, D., Mechanical and electrical responses of the squid giant axon to simple elongation, *J. Biomech. Eng.*, **115**, 1993, 13-22.
22. Shreiber, D. I., Bain, A. C. and Meaney, D.F., In vivo thresholds for mechanical injury to the blood-brain barrier, in *Stapp Car Crash Conference*, SAE Paper No. 973335, 1997, 277-291.
23. King, A.I., Yang, K., Zhang, L. and Hardy, W., Is head injury caused by linear or angular acceleration?, in *"Bertil Aldman Award" Lecture, Proc. IRCOBI Conf.,* Lisbon, 2003, 1-12.
24. Mohan, D., Bowman, B.M., Snyder, R.G. and Foust, D.R., A biomechanical analysis of head impact injuries to children, *J. Biomech. Eng.,* **101,** 1979, 250-260.

NUMERICAL HUMAN MODEL TO PREDICT SIDE IMPACT THORACIC TRAUMA

P.A. Forbes[1], D.S. Cronin[1], Y.C. Deng[2] and M. Boismenu[3]

[1]*Department of Mechanical Engineering, University of Waterloo, 200 University Ave. W., Waterloo, ON, Canada, N2L 3G1, 519-888-4567 ext. 6583, paforbes@uwaterloo.ca;* [2]*General Motors Corporation, 6440 E.12 Mile Road, Warren, MI, USA, 48090-9000;* [3]*General Motors of Canada Limited, 1908 Colonel Sam Drive, Oshawa, ON, Canada, L1H 8P7*

Abstract. A previously developed thoracic numerical model was further enhanced for use in predicting thoracic trauma in auto crash simulations. The model consists of a detailed thoracic cage and organs developed from digital images. The first iteration of the detailed thoracic model incorporated three-dimensional finite element representations of the spine, ribs, heart, lungs and major blood vessels. The second iteration of the model was expanded to include rib cage surface muscles and upper limbs, with improvements to several material models. This detailed thoracic model correlated well with existing front and side pendulum impact experiment data. As this model will be integrated with numerical vehicle models it was necessary to develop a simplified head, pelvis, abdomen and legs.

The current version of the model was developed and evaluated in a stepwise fashion using existing experimental data including frontal and side thoracic pendulum impact tests. Side abdominal and pelvic pendulum impact tests were used to design simplified representations of the associated components to complete the human body model.

Complex loading via side sled impact tests was then investigated, where the body is loaded unbelted using a representative NHTSA-type and WSU-type sled test system. The predicted model response shows good agreement with the experimental data.

Key words: numerical thoracic model, Finite Element Method (FEM), side impact, trauma.

Michael D. Gilchrist (ed.), IUTAM Proceedings on Impact Biomechanics: From Fundamental Insights to Applications, 441–450.

1. INTRODUCTION

The need for virtual vehicle evaluation in the automotive industry has driven the development of virtual human models for the prediction of trauma and human body response in auto crash. This paper presents a continuation of an effort to develop human body models to predict thoracic trauma particularly from side impact thoracic loading, a significant source of automotive fatalities. In the 2001 Fatality Analysis Reporting System (FARS), 26.8% of fatal injuries were caused by side impact collisions [1]. A study of the same data by NHTSA reported that thoracic injuries account for 38% of fatalities and 59% of non-fatal AIS 3-5 injuries [2].

Virtual human models present unique challenges due to the complex geometry, complex material properties, and corresponding non-linear response to applied loads. The potential to improve the understanding of impact physics and predict injury at a local level, as with the detailed model presented in this paper, justifies the significant amount of effort devoted to these models.

In general, a highly detailed model of the entire human body is desired for predictive evaluation of injury under a variety of loading conditions. Although computing power has increased significantly over the past several years, some simplifications are still required to minimize the size of the calculation. Having identified the key structures of interest in a particular situation (i.e. thoracic injury for side impact) a very detailed model of these components can be completed with simplified, but representative models of the remaining components. This approach was used to develop the third iteration of a detailed human thorax model.

2. MODEL DEVELOPMENT

The current version of the model builds on developments by Deng et al. [3] who developed the detailed thoracic cage and Chang [4] who developed the shoulder complex, arms and surface muscles. Each development was based on data from a 50th percentile adult male. Figure 1 shows front and rear views highlighting the musculature and bony upper limb structure.

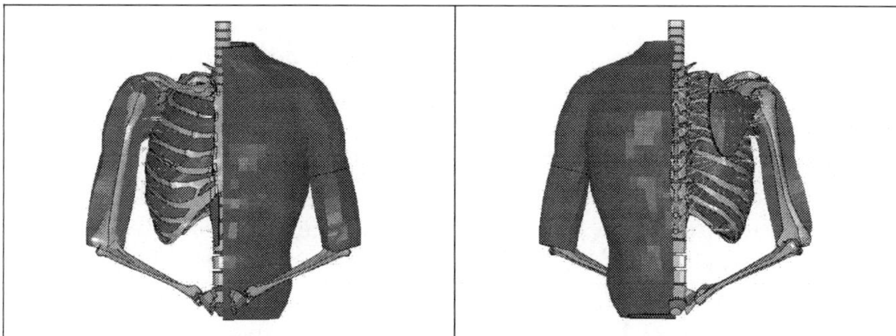

(a)	(b)

Figure 1. Detailed Thoracic Model (a) Front View (b) Rear View.

The model has been enhanced through the addition of a simplified abdomen and pelvis, lower limbs, a head and improved muscle properties. The model has been subjected to pendulum impacts and more complex loading such as side sled impacts. The results are compared against various post mortem human subject (PMHS) responses.

2.1 Full Body Model Completion

The principal enhancements to the detailed thoracic model included the addition of a simplified abdomen and pelvis, legs, head, and improvement of the material constitutive models. The model as implemented with the new components can be seen in Figure 2, with a description of each below.

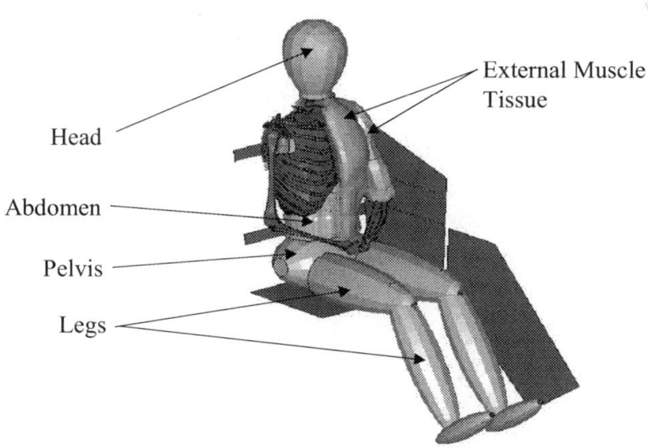

Figure 2. NHTSA Sled Test.

Initial side impact studies emphasized the need for a representation of the abdomen to model body response to side impact. A simplified abdomen was introduced to provide continuity between the diaphragm and the pelvis, which possesses appropriate deformable characteristics and mass approximations to better represent the global response of the human body during various impacts. Material properties were based on in vivo testing of Rhesus monkey kidneys and livers loaded in direct impacts by Melvin et al. [5]. The abdomen has been attached to the external muscle tissue using beam elements that act only in tension and to the pelvis through nodal constraints.

A simplified pelvis was implemented as a deformable component with geometry obtained from anthropometric databases. Attachment of the pelvis to the lower spine in the human body involves a series of complex joints and numerous ligaments. The simplified pelvis in the model has been rigidly attached to the existing sacral vertebrae, deemed appropriate since the sacrum is able to flex with the remainder of the spine. The pelvis material properties were designed to represent the global deformation observed during impact. Viano et al. [6] conducted pelvic pendulum impact testing at velocities of 4.5, 6.7 and 9.4 m/s, and measured the force-compression response throughout the impact. This data was converted into stress-strain curves based on the anthropometric measurements of each PMHS test and a test strain rate. The data was then implemented in a rate dependent material model.

The legs were modeled using anthropometric geometry from the simplified human body model Generator of Body Data (GEBOD) [7] and University of Michigan Transportation Research Institute anthropometric data [8]. The legs were modeled as ellipsoids similar to the GEBOD model with simplified internal bone structures and associated material properties. The surrounding soft tissue was assigned the same material model as that for the thoracic tissue (described below) with densities approximated to represent the mass of each component based on data from Robbins [8] and McConville et al. [9]. Connectivity was achieved through spherical joints, which were given moment-rotation curves as suggested by Cheng et al. [7].

The current model included only a simplified representation of the head, modeled as a rigid part connected to the top cervical vertebrae with moment/rotation curves as defined by Cheng et al. [7]. Future evolutions will incorporate a more detailed head model.

Modifications to the constitutive material models primarily involved the thoracic muscle tissue, which was identified as a key contributor to the loads transferred to the internal organs. An improved material model [10] was implemented based on high rate compressive properties of bovine muscle tissue from Van Sligtenhorst [11], low rate compressive properties of bovine muscles tissue from McHelaney [12] and tensile properties of human pectoralis muscles tissue from Yamada [13].

3. MODEL EVALUATION

The human body model response was evaluated based on progressively more complex loading cases including pendulum impact and side sled tests. Impact force and chest compression as measured during the PMHS experiments and obtained from the NHTSA Biomechanics Test Database [14] were used to evaluate the model. All force data was filtered using SAE 180 as recommended by Kuppa et al. [15] and normalized using the procedure outlined by Eppinger et al. [16].

The following four pendulum impact PMHS tests configurations were considered for the evaluation of the FE-model: pelvic, side abdominal, frontal thoracic and side thoracic. These tests were used to develop the individual components in terms of geometry and material properties under representative loading rates.

3.1 Pelvic and Abdominal Pendulum Impact Tests

Viano et al. conducted side impact tests on PMHS by impacting the abdomen and pelvis with a 150 mm diameter, 23.4 kg pendulum at velocities of 4.5, 6.7 and 9.4 m/s [6]. The human body model was subjected to abdominal impacts at three velocities and pelvic impacts at two velocities. The abdominal impact results were compared to thirteen PMHS impacts and the pelvic impact results to twelve PMHS impacts.

The general shape of the force and compression curves for the abdominal impacts was in agreement with the experimental data. However, the predicted peak force was generally higher and peak compression lower compared to the experiments. This is explained by the exclusion of the hollow organs in the simplified abdomen, but was considered acceptable for the goals of this model.

The force and compression curves of the pelvic pendulum impact simulations showed matching peak values and durations when compared to the experimental data, with minor differences related to the exclusion of the soft tissue surrounding the pelvis.

3.2 Thoracic Pendulum Impact Tests

Kroell et al. conducted frontal impact tests on PMHS by impacting the sternum with a 152.4 mm diameter, 23.4kg pendulum at velocities of 4.5 m/s and 6.7 m/s [17]. In this study, the model was impacted at the more aggressive velocity of 6.7 m/s and the results compared to five PMHS impacts. The impact force results from this simulation show good short-term correlation to the experimental data, however poor long term response. Two distinct force characteristics were noted in the PMHS data, an initial peak created by the impact followed by a plateau force associated with deformation of the chest [17]. In the simulation, the first peak correlates well, while the plateau force was higher. As well, the predicted peak compression was 37% below the experimental results. As the original model [4] correlated well with pendulum impact data, the current differences are attributed to the elevated stiffness of the simplified abdomen. The detailed thoracic model is capable of predicting rib fractures, and correlates very well to the experimental data. The simulation predicted 18 rib fractures while the experimentation recorded a maximum number of 17 rib fractures.

Chung et al. [18] conducted side impact pendulum tests on PMHS with a 152 mm diameter disk at a velocity of 5.6 m/s. The force and compression predicted by the simulation showed good correlation with the experimental data based on load cell and chestband results. The peak values of both predictions appear slightly elevated compared to the experimental average. However, when the model is compared to a PMHS test that closely matched the model in anthropometry, the results were similar. That single experiment also produced 15 rib fractures, the same number observed in the simulation.

3.3 NHTSA Side Sled Tests

Pintar et al. conducted simulated side sled impacts with PMHS conducted at the Medical College of Wisconsin through a cooperative agreement with the National Highway Traffic Safety Administration (NHTSA) [19]. The PMHS were loaded in a NHTSA side sled seat, which impacted the PMHS into a rigid steel load wall while the impact forces were measured through four different impact plates. The plates were oriented from top to bottom at the height of the thorax, abdomen, pelvis and legs, (Figure 2) chosen to represent the average windowsill height of automobiles [19]. PMHS were impacted in a seated position at velocities of 6.67 m/s and 9.04 m/s. For this study, the model was compared to three PMHS impacts at 6.67 m/s and five PMHS impacts at 9.04 m/s.

Figure 3 and Figure 4 show the force predicted and experimental results of the thoracic and abdominal plates for the 9.04 m/s. In general, there was good correlation between the model and experiment as the forces remained within one standard deviation of the experimental results at both impact velocities.

The forces measured at the pelvic and leg impact plates showed peak values higher than those observed in the experiments. Further investigation revealed that the majority of the impact force was arising from the impact of the leg. Since the leg is modeled as an ellipsoid, the point of initial impact occurs midway up the thigh and acts as a concentration point where the majority of the impact force is absorbed. These values are however noted to be within acceptable limits.

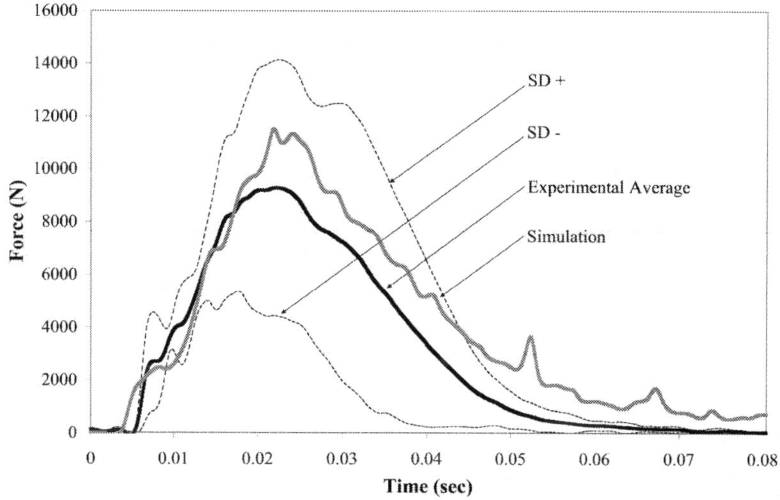

Figure 3. NHTSA 9.04 m/s Thoracic Impact Force.

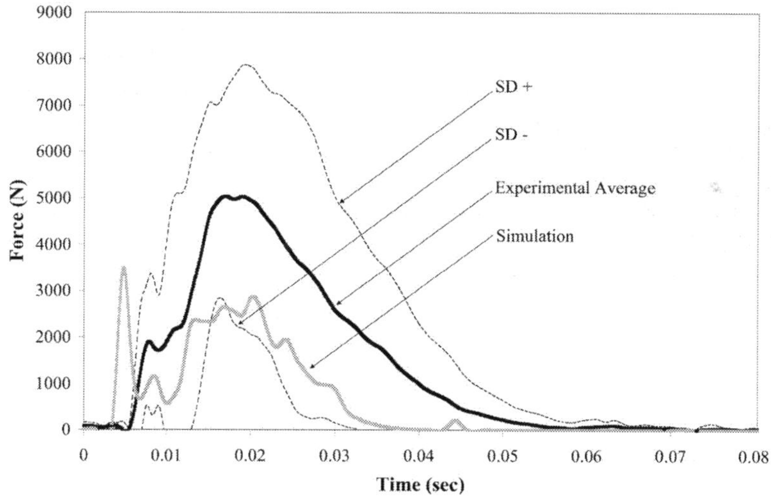

Figure 4. NHTSA 9.04 m/s Abdominal Impact Force.

The compression results of the top and bottom chest bands as measured during both velocities compared well with the experiments. In contrast, high levels of compression were predicted by the simulation at the middle band, and are attributed the initial position of the body. Figure 5 shows screen shots as taken from high-speed film for comparison to similar times in the simulation.

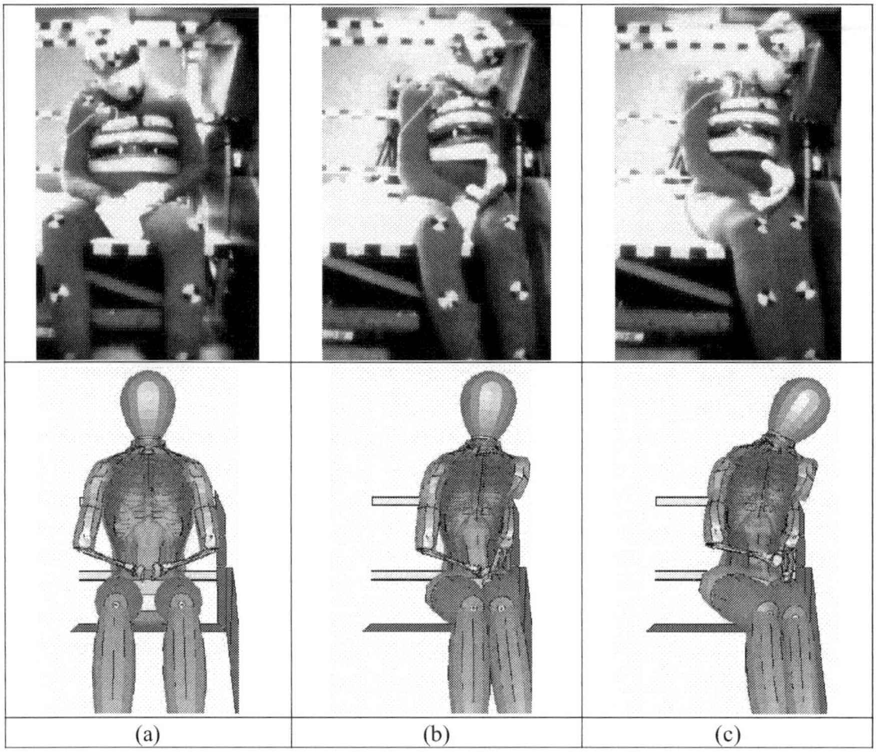

Figure 5. Experimental vs. Simulation (a) t=0 msec (b) t=27.6 msec (c) t=45.6 msec.

3.4 WSU Side Sled Tests

Cavanaugh et al. conducted similar side sled impacts with PMHS in a Wayne State University (WSU) side sled seat [20] but used a different configuration for the rigid steel impact plates. In particular this test included an impact plate at shoulder height. PMHS were impacted at velocities of 6.67 m/s and 9.04 m/s. The predicted peak forces at 6.67 m/s were in good agreement with the experimental data with higher predictions at the pelvic plate as noted above. The half width chest compression correlated well with the experiments.

The 9.04 m/s impacts have not been investigated due to severe rotations of the scapula during the impact, attributed to the lack of the coracoacromial, conoid and trapezoid ligaments. These ligaments are currently being implemented in the model.

4. CONCLUSIONS

A detailed thoracic human body model has been completed with a simplified abdomen and pelvis, as well as leg and head representations for use in thoracic injury studies. The goal of this model is to predict thoracic trauma in side impact auto crash, and required the addition of these components to model the global human body response.

The response of the model to various pendulum tests agrees with experimental data, although the abdomen appears to be overly stiff. Importantly, side impact sled test simulations are in good agreement with PMHS responses. Impact force at the shoulder, thorax and abdomen match well with the experiments, while impact force at the pelvic and leg plates remain slightly elevated but within acceptable limits. Chest compression of both simulations compared well with the experiments with slight differences attributed to the initial impact position.

Further development is currently underway to include important shoulder ligaments and improve the simplified abdomen model to accurately model the compressive response of the thorax.

REFERENCES

1. National Highway Traffic Safety Administration, A Compilation of Motor Vehicle Crash Data from the Fatality Analysis Reporting System and the General Estimates System, Traffic Safety Facts 2001, DOT HS 809 484
2. National Highway Traffic Safety Administration, Federal Motor Vehicle Safety Standards; Side Impact Protection; Side Impact Phase-In Reporting Requirements; Proposed Rule, Federal Register, Part IV, Department of Transportation 49 CFR Parts 571 and 598, Docket No. NHTSA-2004-17694, 2004
3. Deng, Y.C., Kong, W. and Ho, H., Development of a Finite Element Human Thorax Model for Impact Injury Studies, *SAE International Congress and Exposition*, Detroit Michigan, SAE Paper 1999-01-0715, 1999
4. Chang, F., The Development and Validation of a Finite Element Human Thorax Model for Automotive Impact Injury Studies, *Proceedings of the 2001 ASME International Mechanical Engineering Congress and Exposition*, AMD-Vol. 251
5. Melvin, J.W., Stalnaker, R.L., Roberts, V.L. and Trollope, M.L., Impact Injury Mechanisms in Abdominal Organs, *The 17th Stapp Car Crash Conference*, 730968, 1973
6. Viano, D.C., Lau, I.V., Asbury, C., King, A.I. and Begeman, P., Biomechanics of the Human Chest, Abdomen, and Pelvis in Lateral Impact, *Accident Analysis and Prevention*, **v. 21**, **n. 6**, 1989
7. Cheng, H., Obergefell, L. and River, A., Generator of Body (GEBOD) Manual, Wright-Patterson Air Force Base, Ohio, AL/CF-TR-1994-0051, 1994
8. Robbins, D.H., Anthropometric Specifications for Mid-Sized Male Dummy, Volume 2, The University of Michigan Transportation Research Institute, DTNH22-80-C-07502, 1983

9. McConville, J.T., Clauser, C.E., Churchill, Cuzzie, J. and Kaleps, I., Anthropometric Relationships of Body and Body Segment Moments of Inertia, Aerospace Medical Research Laboratory; Wright-Patterson AFB AFAMRL-TR-80-119, 1980
10. Du Bois, P., A Simplified Approach to the Simulation of Rubber-Like Materials Under Dynamic Loading, *4th European LSDYNA Users Conference*, Chapter D, Material, D-I-31, 2003
11. Van Sligtenhorst, C.R., Cronin, D.S. and Brodland, G.W., High Strain Rate Compressive Properties of Soft Tissue, *Proceedings of the 2003 ASME International Mechanical Engineering Congress and Exposition*, IMECE2003-41258
12. McElhaney, J.H., Dynamic Response of Bone and Muscle Tissue, *Journal of Applied Physiology*, **v. 21**, 1966
13. Yamada, H., *Strength of Biological Materials*, G. Evans (eds), The Williams and Wilkins Company, Baltimore, 1970
14. National Highway Traffic Safety Administration, Biomechanics Test Database, National Transportation Biomechanics Research Center, http://www-nrd.nhtsa.dot.gov/, 2004
15. Kuppa, S., Eppinger, R., Maltese, M., Naik, R., Pintar, F., Yoganandan, N., Saul, R. and McFadden, J., Assessment of Thoracic Injury Criteria for Side Impact, *IRCOBI Conference*, 2000
16. Eppinger, R.H., Marcus, J. and Morgan, R., Development of Dummy and Injury Index for NHTSA's Thoracic Side Impact Protection Research Program, SAE Paper No. 840885, 1984
17. Kroell, C.K., Schneider, D.C. and Nahum, A.M., Impact Tolerance and Response of the Human Thorax, *The 15th Stapp Car Crash Conference*, 710851, 1971
18. Chung, J., Cavanaugh, J.M., King, A.I., Koh, S.W. and Deng, Y.C., Thoracic Injury Mechanisms and Biomechanical Responses in Lateral Velocity Pulse Impacts, *The 43rd Stapp Car Crash Conference*, 99SC04, 1999
19. Pintar, F.A., Yoganandan, N., Hines, M.H., Maltese, M., McFadden, J., Saul, R., Eppinger, R., Khaewpong, N. and Kleinberger, M., Chestband Analysis of Human Tolerance to Side Impact, *The 41st Stapp Car Crash Conference*, 973320, 1997
20. Cavanaugh, J.M., Walilko, T.J., Malhotra, A., Zhu, Y. and King, A.I., Biomechanical Response and Injury Tolerance of the Thorax in Twelve Sled Side Impacts, *The 34th Stapp Car Crash Conference*, 902307, 1990

NUMERICAL SIMULATION OF SHOULDER RESPONSE TO A LATERAL IMPACT WITH THE HUMOS MODEL

Preliminary Results

S. Duprey, K. Bruyere and J.-P. Verriest
INRETS – UCBL. Laboratoire de Biomécanique et de Modélisation Humaine, 25, av. F. Mitterrand, 69675 Bron Cedex, France; E-mails: sonia.duprey@inrets.fr, karine.bruyere@inrets.fr, jean-pierre.verriest@inrets.fr, Phone: +33 4 72 14 23 93

Abstract. The HUMOS model, developed in the context of a European collaboration [**1, 2**], has been upgraded. The aim is to enhance the shoulder model biofidelity under lateral impact. First, modifications such as arms positioning and articulation modeling have been realised with the Radioss® (Mecalog) software. Secondly, the model was assessed using experimental data obtained at the LBMH [**3, 4**]. A comparison study of data provided by numerical simulations and experimental impacts was carried out. Main data analysed were deflections and impact contact force-time histories.

Key words: shoulder, lateral impact, biofidelity, Finite Element Method (FEM), injury prediction.

1. INTRODUCTION

According to Frampton [**5**], shoulder injury risk is becoming relatively important. Furthermore, this joint is poorly protected in the case of lateral impact and shoulder injuries have long term functional repercussions. Thus, there is a need to enhance shoulder protection. To assess user's protection, human substitutes, such as numerical models, are required. Numerical models offer numerous advantages compared to experimental approaches.

Several finite element (FE) models of the human body, including the shoulder, have already been achieved. For example, Lizee et al. [**6**] created, in 1998, a whole body model, partially validated with data from the "Laboratoire d'Accidentologie et de Biomécanique" (LAB). Nevertheless,

Michael D. Gilchrist (ed.), IUTAM Proceedings on Impact Biomechanics: From Fundamental Insights to Applications, 451–458.

the shoulder part was simplified (scapula and humeral head defined as rigid bodies) and the validation considered only the general motion of the joint. Only a few FE models focus on the shoulder. In 2000, Iwamoto et al. [**7**] developed a detailed shoulder model validated against three different types of tests (sled tests, pendulum impact tests and side airbag out-of-position deployment tests). In general, simulation results showed good correlations with test data, however no failure was observed for clavicle and scapula. In fact, the focus was on the interaction between shoulder and thorax rather than on shoulder injury prediction. Thollon et al. [**8, 9**] developed an upper limb model whose geometry was the same as the HUMOS model. The arm model was validated against bending tests of the humerus and of the whole arm. The thoracic member (arm and shoulder) was then linked to the HUMOS thorax and compared to experimental tests in which either the shoulder or the humerus was impacted. The validation was only partial because the humerus was submitted to flexion and rupture during the numerical simulations whereas no humerus injuries were found during the experimental tests. Thollon's model was based on the same HUMOS geometry as the model presented here. Concerning the HUMOS model, the shoulder part has not been validated yet. Its modelling is simplified: the scapula is defined as a rigid body, and numerical joints replace ligaments. Thus, the aim of this study is to enhance the shoulder biofidelity and to provide a model capable of predicting injuries.

2. METHODS

2.1 Validation Test Database

Several experimental studies have been carried out to better understand shoulder response and mechanisms under lateral impact. Compigne et al. [**3—5**] realised non-injurious and injurious impactor tests on the shoulder of seven Post Mortem Human Subjects (PMHS). PMHS gleno-humeral head was struck laterally by a 23.4 kg rectangular-shape impactor. Three low velocity impacts (1.5 m/s) were performed on the right shoulder with three different impact directions: 0°, –15° and +15° (Figure 1).

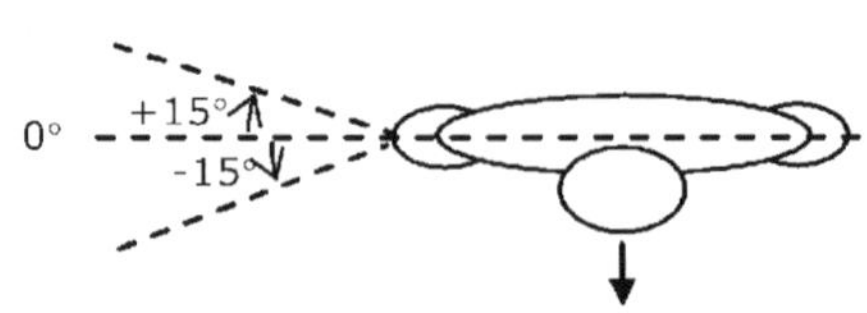

Figure 1. Impact directions with respect to PMHS in Compigne experiments [**3, 4**]: –15°, 0° and +15° (top view).

Tests were checked for repeatability and it was considered that low velocity impacts didn't cause any injury to the shoulder. Left shoulders were submitted to a high velocity impact (from 3 to 6 m/s) in order to assess shoulder injury threshold and to observe shoulder injury caused by a high level impact. Impact contact force-time histories, accelerations at different bony locations (acromions, internal and external clavicle, sternum, humerus, first thoracic vertebra) were recorded and filtered with low pass band Butterworth filter (CFC 180Hz). Three-dimensional shoulder deflections were measured through multiple video analyses of movements. Autopsies were performed after tests: injury threshold appears to be at a velocity of about 4 m/s. Lesions found were clavicle fractures (on the external part) and scapula fractures (acromion, coracoid process and glenoid cavity). For low level impacts, differences were observed between 0° and –15° impacts, for which load was transferred to the scapula, and +15° impact, for which load was directly transmitted to the clavicle. In the first case (0° and –15°), the scapula moved inward by translating along the ribcage, whereas in the second case there is no scapula translation. This difference implied a less important shoulder deflection and a higher amplitude for the impact force, as the clavicle imposed more resistance at +15°.

Experimental data based on these tests are used for the assessment of the numerical model.

2.2 Shoulder Finite Element Model

2.2.1 Mesh and mechanical properties

HUMOS model is the outcome of a European collaboration between research laboratories, automotive manufacturers and suppliers [**1**, **2**]. This FE model represents a 50th percentile male, composed of bones, flesh, major muscles and ligaments but also organs. Its geometry comes from slices of a frozen PMHS seated as a car driver. Modelling of scapula and main ligaments was improved. Constitutive elements and mechanical behaviour of the final model are reported in the following table. More particularly, the shoulder part, composed of scapula, humerus and clavicle bones, has the following features. The scapula is only represented by the cortical part, and its mechanical behaviour data come from Thollon [**8**]. Spongious and cortical parts of the clavicle are both represented. Ligaments' mechanical behaviour comes from [**10**]. Muscles are 3D, 2D or 1D passive elements. Interfaces, based on the penalty method (mostly type 7 of the RADIOSS® software), are used to model contacts.

Table 1. Model's constitutive elements and mechanical behaviour.

Parts			Elements	Number	Mechanical behaviour
Bones	Clavicle	Cortical	Shells	90	Elasto-plastic
		Spongious	Bricks	42	Elasto-plastic
	Humerus	Cortical	Shells	397	Elasto-plastic
		Spongious	Bricks	88	Elasto-plastic
	Scapula	Cortical	Shells	599	Elasto-plastic
Muscles		3D	Bricks		Elasto-plastic
		2D	Shells		Elasto-plastic
		1D	Springs		Non-linear elastic
Ligaments	Initial		Springs		Non-linear elastic
	Added		Shells		Linear elastic
Skin			Shells		Elasto-plastic

2.2.2 Loading and boundary conditions

Numerical simulation has to exactly reproduce experimental tests. The numerical impactor has the same dimensions as the experimental one, and is defined as a rigid body. To decrease the computation duration the model was reduced to torso, arms and head, as impacts are applied to the shoulder. A comparison was done between simulation with the whole model seated (with imposed gravity) and simulation with torso only. A mass was added at the bottom of the torso to represent the lower part of the body (abdomen, pelvis and legs) and in order to retrieve similar inertia. It appears that the difference was not significant, the same results were obtained for impact force time histories but also for accelerations and bone displacements. Secondly, to respect boundary conditions, the model had to be positioned similar to the PMHS. During tests, PMHS had their arms beside the body, whereas the model was in the driving position. The arms rotation realised was a kinematics simulation: a low velocity rotation was imposed to the hand and forearm, with shoulder articulation centre taken as rotation centre. No important strains were generated during the rotation and bone relative positions obtained were close to those of PMHS. The conclusion drawn was that the bone relative positions seemed anatomically correct.

2.2.3 Geometry personalisation

The PMHS tested were mostly aged and small while the HUMOS model is a 50th percentile. Hence, the model geometry was scaled in order to match the subject's anthropometry as close as possible. The scaling coefficient, in vertical plane, was sizes ratio. In frontal plane, the coefficient was shoulder widths ratio, and the third scaling coefficient was a mean of the first two coefficients.

2.2.4 Assessment method

Numerical model's instrumentation is similar to the experimental one and data from accelerometers are filtered with low pass band Butterworth filter too. Thus, simulation data can be compared with data found during experimental tests. Data compared are those provided by experimental tests on subject 1 (for 6 m/s and 1.5 m/s impacts) and subject 7 (for 4 m/s impacts) [**3**, **4**] and by simulations with the numerical model scaled in accordance with their anthropometry. These subjects were chosen, because their anthropometry was close to that of HUMOS, thus scaling processes did not imply such important variations.

Impact contact force-time histories, bones accelerations and three-dimensional shoulder deflections have been compared.

3. SIMULATIONS AND RESULTS

3.1 High Velocity Impacts

Force impact curves are illustrated below.

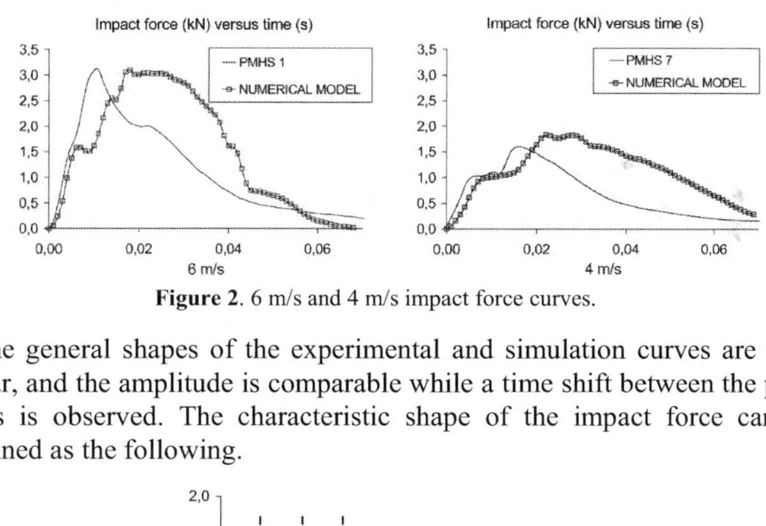

Figure 2. 6 m/s and 4 m/s impact force curves.

The general shapes of the experimental and simulation curves are very similar, and the amplitude is comparable while a time shift between the peak values is observed. The characteristic shape of the impact force can be explained as the following.

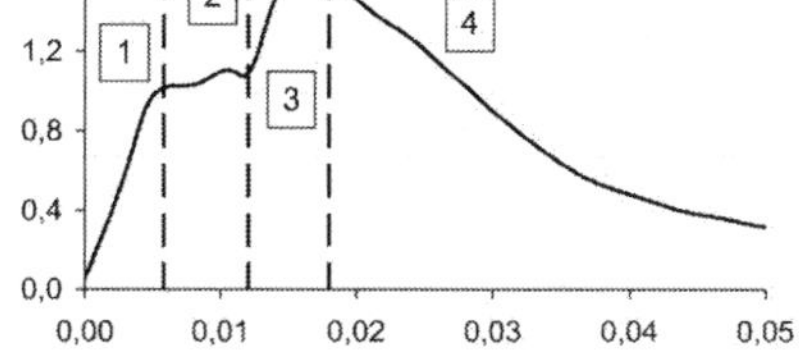

Figure 3. 4 m/s experiment impact force curve. Description.

The first step corresponds to the crush of soft tissues. Then the flat stage is a period during which the scapula translates, pushed inward by the humeral head, up to the ribcage. At the 3rd step, the scapula is against the ribcage and the whole body is carried along by the impactor. The 4th step corresponds to the impactor's velocity decrease. There are small amplitude discrepancies between simulated and experimental curves for the impact force time histories: 0.8% at 6 m/s and 15% at 4 m/s. However there are time-shifts: 7.5 ms at 6 m/s and 6 ms at 4 m/s. Concerning the kinematics, the analysis demonstrates that shoulder deflection (acromion-acromion) is too low whereas impactor to T1 deflection is correct (Figure 4 for a 4 m/s impact). The same discrepancies can be observed at 6 m/s.

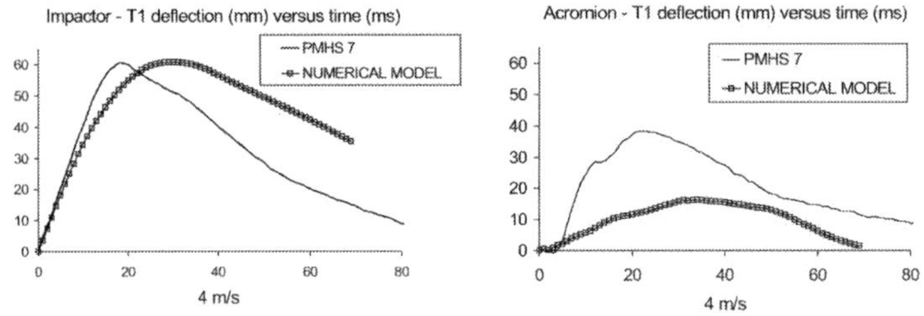

Figure 4. 4 m/s deflection curves.

This can be explained this way: during the impact, first, the impactor strikes the humeral head and this pushes the scapula via the glenoid cavity. Then the acromion should be carried along, but it drops behind, stopped by the clavicle. This is either due to a large stiffness within the clavicle, or to an insufficiently stiff scapula spine.

At 6 m/s, experimental tests induced injuries, such as clavicle fractures and coracoid process fractures. But during simulations, no injury was found. Furthermore distribution of the simulated maximum stresses is also wrong since the clavicle is poorly loaded during the impact. Thus there is a need to stiffen the scapula, by means of meshing its spongious part for example, in order to make a better load transfer to the clavicle possible.

3.2 Low Velocity Impacts: 0°, –15° and +15° Angles

Force impact curves at 0°, +15° and –15° are illustrated below.

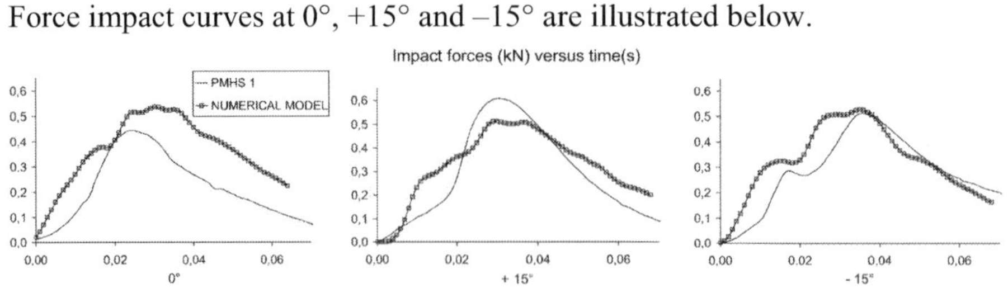

Figure 5. 1.5 m/s impact force curves.

At 0°, +15° and –15°, force amplitude discrepancies are respectively 20%, 15% and 3% and time shifts are respectively 6, 0 and 3.5 ms. These disparities are judged acceptable. Furthermore, results appear qualitatively correct since for the –15° impact force, the same flat level (between 15 and 22 ms) is found (Figure 4). This stage represents the period during which the scapula translates up to the ribcage, and as the scapula is solicited more during an anterior-posterior impact, it seemed logical to obtain a more pronounced translation at –15°.

Simulated bone relative displacements are still too low, for 0° and –15° impacts (Figure 5). Nevertheless, for +15° impact, amplitude deflection is correct.

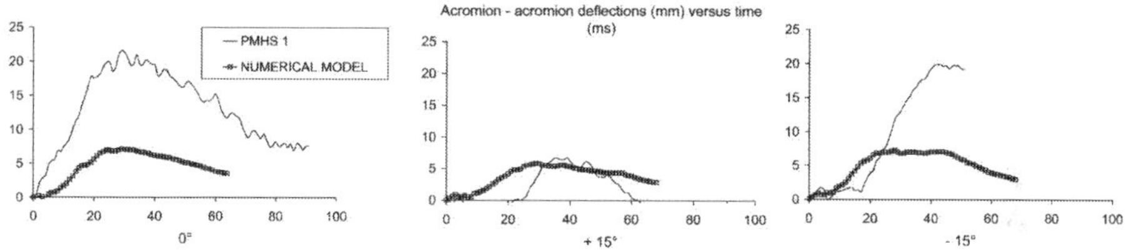

Figure 6. 1.5 m/s acromion- acromion deflection curves.

At this low velocity, bones don't sustain important deformation. Furthermore for +15°, load is transferred to the clavicle whereas at –15° and 0° load is transfer to the scapula. This points out that scapula kinematics couldn't be anatomically correct.

On the whole, there are still important discrepancies between experimental and simulation curves. That is why experimental tests will be realised on isolated bones in order to optimise the mechanical behaviour data and the mesh of the bones.

4. CONCLUSION

Changes have been brought to the initial HUMOS model in order to obtain a model which could be used for shoulder impact simulation. These simulations reproduced the experimental tests carried out at the LBMH. Simulated shoulder or acromion to T1 deflections appeared to be too low. This proves that the model has to be enhanced in order to improve the load transfer between acromion and clavicle and to improve the scapula's kinematics. However the model reproduces the same general motion as PMHS, since the same characteristic shapes of impact force time histories, linked with anatomical motions, are found. Thus, at this point, the shoulder behaviour of this numerical model is promising. The perspective of this work is a model personalisation with a better integration of individualised data such as geometries and mechanical behaviours of biological tissues.

ACKNOWLEDGEMENTS

The authors want to thank Kambiz Kayvantash of Mecalog and Lionel Thollon for their support.

REFERENCES

1. Thollon, L., Arnoux, P.J., Kayvantash, K., Cavallero, C. and Brunet, C., Human injury evaluation using HUMOS RADIOSS finite element model, in *Proceedings of the International Ircobi Conference on the Biomechanics of Impact*, Munich, Germany, September 18—20, 2002, pp. 369–371.
2. Robin, S., HUMOS: Human model for safety – A joint effort towards the development of refined human-like car occupant models, in *Proceedings of the International Technical Conference on the Enhanced Safety of Vehicle*, Amsterdam, The Netherlands, June 4–7, 2001, Paper No. 297.
3. Compigne, S., Caire, Y., Quesnel, T. and Verriest, J.P., Lateral and oblique impact loading of the human shoulder – 3D acceleration and force-deflection data, in *Proceedings of the International Ircobi Conference on the Biomechanics of Impact*, Bron, France, September 12–14, 2003, pp. 311–319.
4. Compigne, S., Caire, Y., Quesnel, T. and Verriest, J.P., Non-injurious and injurious impact response of the Human Shoulder – Three dimensional analysis of kinematics and determination of injury threshold, *Stapp Car Crash Journal*, vol. 48, 2004, pp. 89–125.
5. Frampton, R.J., Morris, A.P., Thomas, P. and Bodiwala, G.G., An overview of upper extremity injuries to car occupants in UK vehicles crashes, in *Proceedings of the International Ircobi Conference on the Biomechanics of Impact*, Hannover, Germany, September 24–26, 1997, pp. 37–52.
6. Lizee, E., Robin, S., Song, E., Bertholon, N., Le Coz, J.Y., Besnault, B. and Lavaste, F., Development of a 3D finite element model of the human body, *Stapp Car Crash Journal*, vol. 42, 1998, SAE Paper No. 983152, 337 pp.
7. Iwamoto, M., Miki, K., Mohammad, M., Nayef, A., Yang, K.H., Begeman, P.C. and King, A.I., Development of a finite element model of the human shoulder, *Stapp Car Crash Journal*, vol. 44, 2000, SAE Paper No. 2000-01-SC19.
8. Thollon, L., *Modélisation du membre thoracique dans le cadre d'un choc latéral. Approche expérimentale et numérique*, Thèse de mécanique, Université de la Méditerrannée, France, June 2001.
9. Thollon, L., Behr, M., Cavallero, C. and Brunet, C., Finite element modelling and simulation of upper limb with radioss, *International Journal of crashworthiness*, vol. 7 no. 3, 2002.
10. Arnoux, P.J., *Modélisation des ligaments des membres porteurs : A propos des ligaments du genou*, Thèse de mécanique, Université de la Méditerrannée, France, 2000.

FINITE ELEMENT MODEL OF HUMAN SKULL USED FOR HEAD INJURY CRITERIA ASSESSMENT

Ondřej Jiroušek[1], Josef Jíra[2], Jitka Jírová[1] and Michal Micka[1]
[1] *Institute of Theoretical and Applied Mechanics, Academy of Sciences of the Czech Republic, Prosecká 76, 190 00 Praha 9, Czech Republic; {jirousek,jirova,micka}@itam.cas.cz;* [2] *Faculty of Transportation Sciences, Czech Technical University in Prague, Konviktská 20, 110 00 Praha 1, Czech Republic; jira@fd.cvut.cz*

Abstract. Development and validation of detailed 3-D finite element model of human skull used for explicit dynamic simulation of impact conditions is presented in the paper. The FE model is based on the series of computer tomography scans of resolution 512x512 pixels taken in 1mm slices. Fully automated direct generation of the volumetric tetrahedral mesh based on the Marching Cubes Algorithm, Laplacian smoothing and Delaunay tetrahedralisation is used to develop the geometry of both the human skull and the brain.

Results obtained using this detailed FE model are compared to experimental results from a standard drop test as well as a simplified version of the skull model based on geometry of a head form used in our experiments. These experiments use a standard metal head form of variable size. Results from both the experimental and numerical modelling will be used to describe possible injury mechanisms and quantify design parameters of protective helmet related to a specific impact event.

Key words: drop test, Finite Element Method (FEM), medical imaging, mesh generation, explicit dynamics, head injury criteria.

1. INTRODUCTION

With the great advancement of the vehicle transportation in the last decades the number of traffic accidents as well as injuries has immensely grown. The annual number of traffic accidents is growing despite of the growing vehicle passive and active safety, development of the safety instruments and traffic regulations. The number of

Michael D. Gilchrist (ed.) IUTAM Proceedings on Impact Biomechanics:From Fundamental Insights to Applications, 459–467.

traffic accidents in the Czech Republic has doubled in the last few years. The numbers are alarming: every year 1200 people die owing to traffic accidents; there are almost four-times more major injuries and perhaps twenty-times more minor injuries. Serious consequences of the traffic accidents become evident also in the social domain. Young people, invalid in their years of productive lives, are no longer a rarity.

To study the impact conditions during a human head impact it is convenient to use FE modelling to save costs and time. The FE models of various parts of human musculoskeletal system can be easily developed using the information from medical imaging. It is possible to use a series of CT-scans for the dense bone material, whereas soft tissues are better visible on MRI scans. It is also possible to prescribe the material properties on relative density of the tissue. This information is directly available in the CT or MRI scans, therefore, where a relationship between the signal strength and material properties exists (e.g. cancellous bone, [1, 2]) it is possible to use inhomogeneous properties in our models.

2. METAL HEAD FORM MODEL

For the assessment of the stress conditions in human head during impact two different FE models are used. The first model is a FE model of head form given by the Czech National Standard CSN EN 960 change A1 832140. It is a parametric model allowing for different sizes of the real head form given by tables containing the dimensions in several relative planes (see Figure 1).

Material properties are prescribed based on the real head form used in respective drop test. These are made from an alloy with inertial properties corresponding to those of appropriate human head. The geometry of the computational model is given in Figure 2 and the resulting finite element model with one type of bicycle helmet is given in the same figure.

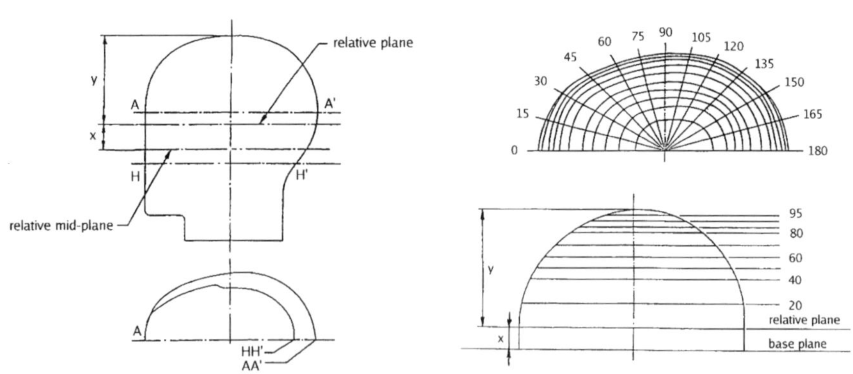

Figure 1. Input data for the parametric model given by CSN EN 960.

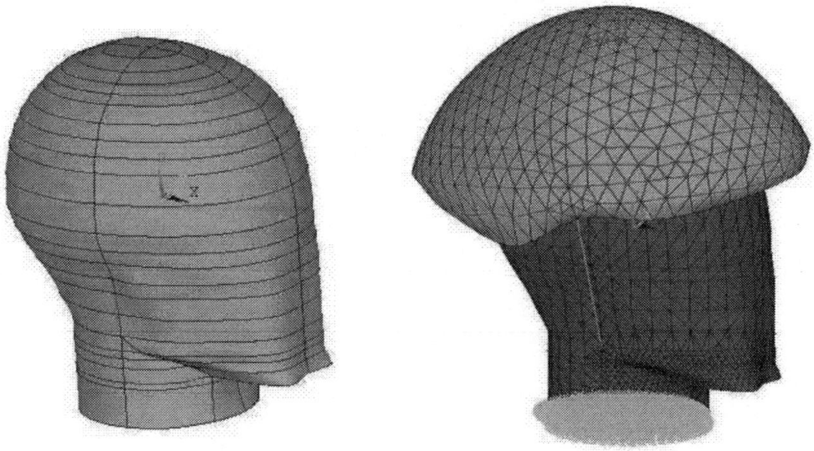

Figure 2. Solid model of the head form and example loading of FE mesh with a helmet.

Protective capacity of helmets is usually investigated using drop tests. The test consists of a controlled impact where a helmet is positioned on a metal head form and then dropped in a guided fall (see Figure 3). As the target surface, various steel test anvils (flat, edge, hemisphere or a horseshoe type) can be used to simulate different impact surfaces. The head form is instrumented with triaxial accelerometer to measure peak acceleration.

From the resultant acceleration–time curve peak acceleration and the Head Injury Criterion (HIC) are determined. HIC is calculated using following formula:

$$HIC = (t_2 - t_1)\left[\frac{1}{t_2 - t_1}\int_{t_1}^{t_2} a(t)dt\right]^{2.5}$$

where a is a resultant head acceleration and $(t_2 - t_1)$ is chosen to maximise HIC. Obtained values of HIC are then compared to values calculated using the finite element model developed using procedures described in the next section.

3. SKULL AND BRAIN MODEL FROM MEDICAL IMAGES

3.1 Overview

As a convenient alternative to experimental investigation of impact conditions during a drop test, finite element modelling can be considered. It is also easy to use FEM for evaluation of the contact stress distribution, estimation of the total deformation of each tissue, etc. The first attempts to model the behaviour of human head response by means of FE modelling were made by Hardy and Marcall [3]. These first three-dimensional models reflected only the skull, not the brain. With the advancement of better meshing techniques and more powerfull contact definitions, finite element models containing the human brain appeared. The first attempts to model

Figure 3. Drop test configuration.

the behaviour of human brain were considering its material to be linear elastic, later modelled as an inviscid fluid [4]. Viscoelastic properties of human brain were considered few years later in a number of articles, e.g. in [5] or a paper by Tawfik, Khalil and Hubbard [6].

In recent years, there were several three-dimensional, more realistic FE models of human skull and brain built using data obtained from computed tomography [7, 8]. These models usually reflect only human skull and brain, both modelled as linear elastic materials. Other models consider not only the bone and brain, but also the scalp, cerebrum, cerebellum, spinal cord and other structures, usually geometrically very simplified [9]. Only few of them [10] were validated against experimental data. In the recent years, complex models containing not only the human skull and brain, but also other very important tissues, like dura, the subarachnoid space filled with the cerebrospinal fluid [11] and even containing the cervical spine and spinal cord [12] were built. Most of the models are based on a simplified geometry of the skull and respective organs.

In this paper, we present a geometrically detailed model of human head and brain, built on the basis of series of CT and MRI scans. The geometry of an organ can be reconstructed from the medical volumetric data using procedures known from computer graphics. Here, Marching Cubes Algorithm (MCA, [13]) is used for construction of triangular representation of the organ's surface. Methods used for the development of the finite element mesh from these triangles include Schroeder's decimation algorithm [14] and Delaunay triangulation (in 3-D tetrahedralization, [15]).

For the purpose of implicit dynamic modelling material properties of the cortical bone are assumed linear elastic and homogeneous with Young's modulus of elasticity 14 GPa and Poisson's ratio 0.23. The volume of the cranial region is filled with volumetric tetrahedral elements representing the brain. The brain is modelled as viscoelastic material with density 1000 kg/m^3, shear modulus $G = 2.2$ GPa ($G_0 = 1.036$ kPa and $G_\infty = 185$ Pa, $\beta = 0.0165$ m/s).

3.2 Tissue Segmentation

For the FE models of human organs it is convenient to use medial imaging data. The first step in the development of FE model of human head was the segmentation of the cortical tissue in the set of CT images. As the input data we used 121 CT images taken at 0.33 mm intervals of 512×512 pixel resolution.

Intensity based segmentation was found suitable for the input images, since they were of high quality (every tissue of interest was well separated from each other). These algorithms rely on the fact, that pixels representing the same tissue are clustered around a mean characteristic value. For the MRI scans of human brain this was not applicable because of the inhomogeneities giving smoothly varying, nonlinear fields in the images. Therefore, more complicated segmentation, based on edge detection technique was employed to achieve this task. Edge detection techniques make use of standard mathematical operators with a small spatial extent to detect local edges in the image. The approximate mathematical gradient operators, such as Gradient, Laplacian, Sobel, Roberts, Prewitt or Kirsch, are used to highlight the appropriate portions of the image.

4. SURFACE RECONSTRUCTION FROM SEGMENTED TISSUE

To find a surface of an organ in the whole sequence of CT slices, a surface reconstruction method utilising the Marching Cubes Algorithm [13] was implemented in the surface reconstruction tool. From the nature of the MCA it is clear, that the triangular representation of the surface obtained by this algorithm is too dense to be used as a surface mesh in the finite element method. It is therefore necessary to reduce the number of triangles on the surface to a reasonable number.

To achieve this resolution reduction, out-of-plane distance decimation and border-distance decimation introduced by Schroeder et al. [14] was used for repetitive removal of vertices from a triangular mesh in order to reduce the density of the mesh in places where they are not necessary for the surface description. Surface mesh optimisation using energy minimisation approach was used to optimise the final mesh. This approach helps to balance two contradictory goals - resulting mesh with fewer vertices than the original one while keeping the geometry close to the original model.

4.1 Mesh Generation and Mesh Optimisation

To describe the human skull and the brain as a set of finite elements, it is necessary to fill the surfaces obtained in the previously described section with tetrahedral

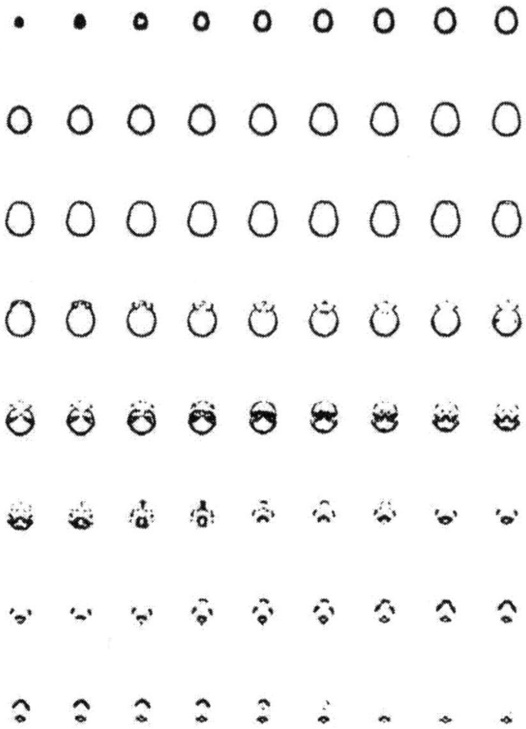

Figure 4. Segmentation of the cortical bone from a series of CT scans.

elements. For the discretisation of the volume Delaunay triangulation [15] in 3-D is used. Delaunay triangulation does not guarantee elements with good aspect ratios. The resulting mesh can contain elements of bad shape that can slow down the solution or produce undesirable errors. Mesh optimisation therefore follows to avoid *sliver* elements (elements with very small spatial angle between two sides) as well as other poorly-shaped elements. Three methods used to improve the quality of existing tetrahedral mesh are: (i) Laplacian-based mesh smoothing, (ii) nodal points insertion and deletion and (iii) local re-meshing.

The FE mesh of human brain was obtained using procedures as for the reconstruction of human skull. Resulting FE mesh of the brain is given in detail in Figure 6. For the purpose of explicit dynamics modelling it is suitable to have all the elements in the mesh of equal size. In the case of such complex geometries of our case, this was not possible to guarantee and there is a great difference between the smallest and the largest element in the final mesh. This has the consequence in the larger critical time step and consequently in larger computational time needed for the whole analysis.

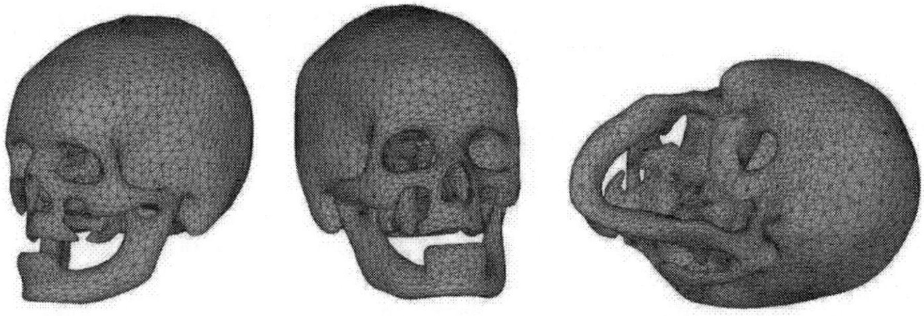

Figure 5. Resulting FE model of human skull.

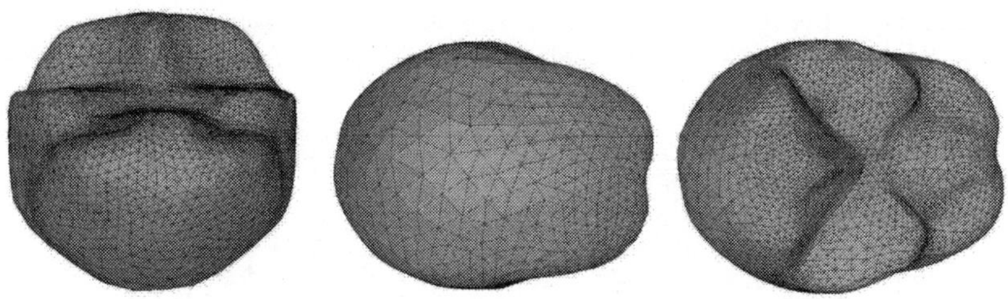

Figure 6. Resulting FE model of human brain.

5. RESULTS AND CONCLUSIONS

Procedures described in the paper are used to generate FE models of the human skull and brain from medical imaging data. Models with a different total number of degrees of freedom were developed. The presented FE models of human skull allow for comparison of Head Injury Criterion (HIC) values obtained from experiments and FE simulations. For the FE modelling of the impact conditions, explicit dynamics solver LS-DYNA is used.

Various impact conditions (according to the drop tests used in the HIC evaluation) were studied. As an example analysis, simple drop test (human skull impacting a rigid flat obstruction) is considered to demonstrate the capabilities of the explicit solver. The linear viscoelastic material model for the brain is defined by:

$$G(t) = G_{\infty} + (G_0 - G_{\infty})\, e^{-\beta t} \tag{1}$$

The solution of an explicit problem is stable only if the time increment Δt is smaller then the critical one. The critical time increment can be calculated based on the Courant-Friedrichs-Levy criterion and for stability reasons is lowered by a scale factor of 0.9 to $\Delta t = 0.9(l/c)$, where l is the characteristic length of an element and c is the propagation velocity. Reduced (one-point) integration plus viscous hourglass control was used for faster element formulation. The use of one-point integration is advantageous due to savings on computer time and in case of large deformations of the brain (high-speed impacts). Contact between the surface of the brain and the inner surface of the dura was accomplished using the surface-to-surface contact elements with friction. The coefficient of friction is assumed to be dependent on the relative velocity, v_{rel} of the contacting surfaces and on the static and dynamic coefficients of frictions:

$$\mu_c = \mu_d + (\mu_s - \mu_d)e^{-v_{rel}} \tag{2}$$

Central difference time integration is used to solve for accelerations instead of displacements.

For comparative purposes, an explicit analysis of a guided fall with boundary conditions similar to those used in the mechanical testing, was considered. Contact conditions were prescribed between the brain and inner surface of the skull as well as between the outer surface of the skull and the rigid plate. As an illustrative example of the results the fields of displacements and accelerations are given in Figure 7, while compressive stresses in the brain matter are given in Figure 8.

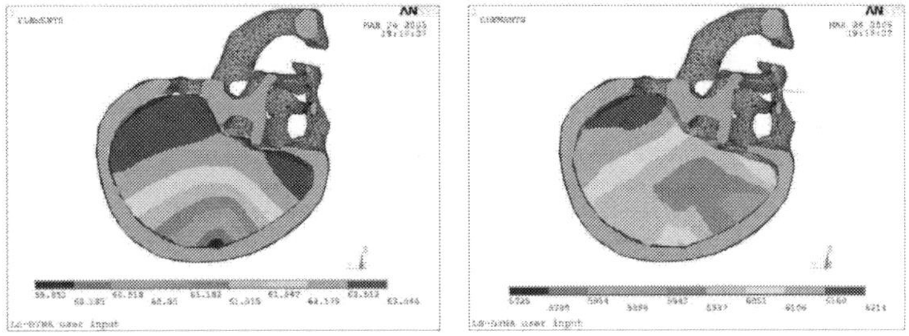

Figure 7. Fields of displacements and accelerations (2 m guided fall).

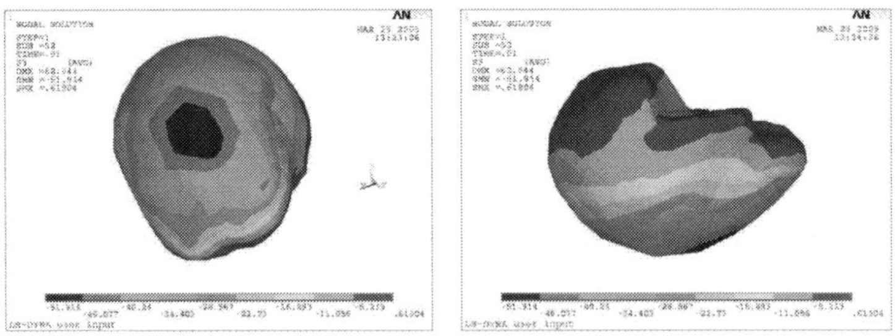

Figure 8. Third principal stresses (σ_3) from the FEM simulation of the drop test.

This study was aimed at the development and validation of high resolution, three-dimensional FE model of human skull and brain. Using the explicit dynamics code it is possible to simulate the contact conditions between these materials during a short-time impact. After profound validation of the model, it will be used in injury criteria assessment and for evaluation of design parameters important for the protective helmet design.

ACKNOWLEDGEMENT

This research was supported by the Grant Agency of the Czech Republic (grant No. 103/05/1020).

REFERENCES

1. Cowin, S.C. and Turner, C.H., On the relationship between the orthotropic Young's moduli and fabric, *Journal of Biomechanics*, vol. 5, June 1992, 1493–1494.
2. Rice, J.C., Cowin, S.C., and Bowman, J.A., On the dependence of the elasticity and strength of cancellous bone on apparent density, *Journal of Biomechanics*, vol. 21, March 1988, 155–168.
3. Hardy, C.H. and Marcal, P.V., Elastic analysis of a skull, *ASME Transcations*, 1983, 838–842.
4. Shugar, T.A. and Katona, M.G., Development of finite element head injury model, *J. ASCE*, vol. 101, 1975, 223–239.
5. Wahi, K.K. and Merchang, H.C., Mechanical response of a head injury model with viscoelastic brain tissue, *Annals of Biomedical Engineering*, vol. 5, 1977, 303–321.
6. Khalil, T.B. and Hubbard, R.P., Parametric study of head response by finite element modeling, *Journal of Biomechanics*, vol. 10, no. 2, 1977, 119–132.
7. Johnson, E.A.C. and Young, P.G., On the use of a patient-specific rapid-prototyped model to simulate the response of the human head to impact and comparison with analytical and finite element models, *Journal of Biomechanics*, vol. 38, January 2005, 39–45.
8. Jirousek, O., Jirova J. and Jira J., Multiresolutional finite element models of human skull based on data from computer tomography, *Acta of Bioengineering and Biomechanics*, vol. 5, September 2003, 214–217.
9. Kleiven, S. and von Holst, H., Consequences of head size following trauma to the human head, *Journal of Biomechanics*, vol. 35, February 2002, 153–160.
10. Ruan, J.S., Khalil, T. and King, A.I., Human head dynamic response to side impact by finite element modeling, *Journal of Biomechanical Engineering*, vol. 3, no. 113, 1991, 276–283.
11. Willinger, R., Kang, H.-S. and Diaw, B., Three-dimensional human head finite-element model validation against two experimental impacts, *Annals of Biomedical Engineering*, vol. 27, no. 3, 1999, 403–410.
12. Zong, Z., Lee, H.P. and Lu, C., A three-dimensional human head finite element model and power flow in a human head subject to impact loading, *Journal of Biomechanics*, to appear, 2005.
13. Lorensen, W.E. and H.E. Cline, H.E., Marching cubes: A high resolution 3D surface construction algorithm, *Computer Graphics*, vol. 21, July 1987, 155–168.
14. Schroeder, W.J. and Zarge, J.A., Decimation of triangle meshes, *Computer Graphics*, vol. 26, August 1992, 65–70.
15. Fang, T.P. and Piegl, L.A., Delaunay triangulation in three dimensions, *IEEE Computer Graphics and Applications*, vol. 15, September 1995.

A COMPUTATIONAL STUDY OF THE OPTIC NERVE EVULSION

S. Cirovic, R.M. Bhola, D.R. Hose, I.C. Howard, P.W. Lawford and M.A. Parsons
The University of Sheffield, Mappin Street, Sheffield, S1 3JD, Sheffield,UK;
tel.: ++(0)144 222 5878, fax: ++(0)144 222 7890, e-mail: S.Cirovic@sheffield.ac.uk

Abstract. Optic nerve evulsion is a traumatic lesion of the optic nerve which is usually caused by high-speed traffic accidents, or a blunt object injury. The direct causes leading to nerve damage are not clear. We used a finite element model of the eye and the orbit to investigate the possible mechanisms of injury. The model, which involved the globe, optic nerve, recti muscles, and the orbital fat, was subjected to triangular acceleration pulses of peak intensity between 100 and 450g. Collision with a foreign object of cylindrical shape 6mm in diameter was used to simulate blunt object injury. The results point to rapid rotation of the globe, and a steep pressure gradient between the orbital fat and the soft tissue within the optic nerve as possible factors contributing to the optic nerve damage.

Key words: optic nerve, evulsion, Finite Element Method (FEM), impact.

1. INTRODUCTION

The term "optic nerve evulsion" or "optic nerve avulsion" describes a traumatic injury in which the optic nerve is forcibly disinserted from the retina choroid and vitreous, and the lamina cribrosa is retracted from the scleral rim. Complete evulsion results in a blind eye, while partial evulsion (involving only a segment of the optic disc) causes variable degrees of visual loss [1]. The most common cases of evulsion are high speed traffic accidents, and sports injuries, in particular those which involve a blunt object (e.g. finger poking in basketball). Evulsion of the nerve may be accompanied

Michael D. Gilchrist (ed.), IUTAM Proceedings on Impact Biomechanics: From Fundamental Insights to Applications, 469–476.

by evulsion of the globe. However, most cases of nerve evulsion occur with the globe in place and extraocular muscles intact [2]. The most commonly proposed mechanisms of injury are: extreme and rapid rotation of the globe caused by a blunt object intruding between the eye and the orbital wall, sudden rise in the intraocular pressure, and sudden anterior displacement of the globe [1]. However, there is no solid evidence for any of the above listed possible mechanisms.

The results of our recent computational study revealed that in the situation where the attachment of the globe to the orbital fat is lost due to rapid deceleration and/or rupture of the conjuctiva, the anterior displacement of the globe from the orbit is mainly prevented by the optic nerve, rather than the extraocular muscles. In this event, which would result in the evulsion of the globe, the optic nerve is likely to be damaged by the inertia of the eyeball [3]. In this study we wish to examine the case where the evulsion of the nerve occurs without the permanent dislocation of the globe.

2. THE FINITE ELEMENT MODEL

The finite element model of the eye and the orbit is shown in Figure 1. The geometry of the orbit is a simplified representation of a realistic geometry obtained from the MRI scans of an adult volunteer. The opening of the orbital space is 40mm wide both in the superio-inferior, and anterio-lateral directions, and it tapers down to 5mm width at the apex of the orbit. The axis of the orbit, which connects the apex and the centre of the globe, forms an angle of 22.5° with the plane of symmetry of the head. The depth of the orbit measured along its axis is 42mm. The globe is a sphere of 12mm in radius. A 1mm thick outer shell with a Young's modulus of 50MPa and Poisson's ratio of 0.45 [4] represents the sclera. The inner space is occupied by a soft, nearly incompressible, solid representing the vitreous and aqueous humour. The Young's modulus of 42kPa and Poisson's ratio of 0.49 for the vitreous were taken from Power et al. [5]. The orbital space is filled with the fat which was represented as a solid with a bulk modulus of 2000MPa and a shear modulus of 5kPa [6]. The globe is submerged into the fat up to 4mm (one third of the radius) over the equator.

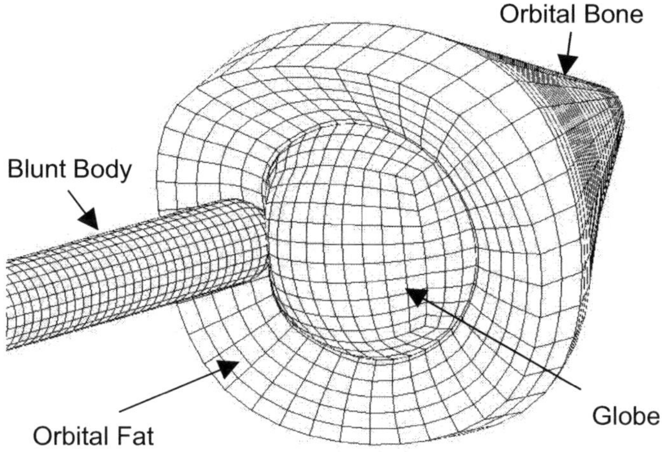

Figure 1. Finite element model of the eye and the orbit.

The geometry of the extraocular muscles was taken from Robinson [7]. Each of the six extraocular muscles was represented by five fibres in parallel. The fibres were composed of one–dimensional Hill-type muscle elements with passive components in parallel with an active component that was capable of generating an isometric force. Following Winters [8] we used an exponential law to express the force-elongation characteristics for the passive component of each muscle, which can be expressed as:

$$F = k \exp [a(L-1)] \qquad (1)$$

where F is the force in Newtons, L is the engineering strain, and $k = 3.33 \times 10^{-2}$ and $a = 13.8$ are constants determined to fit the experimental data reported by Scott [9]. The force in a single fibre equals the total passive muscle force divided by the number of fibres in parallel. The optic nerve was represented by a tubular membrane shell of 1.5mm cross-sectional radius of the same thickness and material properties as the sclera. The nerve stretches in a straight line from the apex of the orbit to the posterior pole of the globe and it is filled with a soft tissue with the same material properties as the orbital fat. The fat, the optic nerve, and the interior of the nerve were meshed together. To account for the fact that in reality the nerve is longer than the shortest distance between the orbital apex and the globe, the posterior end of the optic nerve is connected to the orbit with a series of cables, which have zero stiffness when the displacement is less than 8mm, and the stiffness equal to the axial stiffness of the nerve for larger displacements. A rigid cylinder of 6mm in radius, with a spherical tip was used to represent a blunt body (e.g. finger). The axis of the cylinder is parallel with the anterio-posterior axis, and located 2mm medially from the centre of the globe, so that it impacts the surface between the globe and the edge of the orbit. The fat, vitreous and the interior of the nerve were meshed with brick elements, and the sclera, optic nerve and the orbital bone with quadrilateral shell elements. In total, 13100 solid elements, 2780 shell elements and 16629 nodes were used to construct the model. Modelling and meshing were performed in ANSYS. We used a contact algorithm to tie the optic nerve and the orbital fat to the globe. All the materials were modelled as linear elastic with density of 1000 kg/m^3 except for the orbital bone which was modeled as a rigid body.

2.1 Simulations

Blunt trauma, involves various scenarios, and a variety of foreign objects (fingers, snooker cues, golf clubs, paint balls, etc.). Furthermore, clinical reports of trauma usually do not provide sufficient information to enable accurate biomechanical reconstruction of a specific event. Taking this into account, we decided to focus on hypothetical impact scenarios, with the goal of gaining a general understanding of the potential injury mechanisms. This was done by applying simple deceleration pulses to the orbit or the blunt body. Only a small number of critical parameters (peak acceleration, depth of blunt body penetration) were varied within a plausible range. Since the motion was prescribed, the impact conditions did not depend on the velocity, mass and properties of the colliding bodies. We simulated two types of traumatic events; in the first type (head impact), the orbit was initially moving while the foreign object was stationary, while in the second (blunt object impact), the opposite was the case. In both cases the triangular acceleration pulse was prescribed to either the orbital bone, or the foreign object to bring it to rest. The maximal acceleration was varied from 100g (g=9.81m/s^2) to 450g. The shape and the intensity of the acceleration pulses correspond roughly to the pulses produced in head impact upon fall from a small height [10]. The upper limit of 450g was chosen because our earlier study [3] suggested that higher values of acceleration may lead to the evulsion of the whole globe, even without major orbital trauma or the involvement of a foreign body. For the scenario which involved the peak acceleration of 450g and the orbit initially moving, we also simulated impact without blunt body penetration in order to isolate the effect of inertia of the tissue within the orbit. The initial velocity of the moving object (2.42-7.28m/s), and the duration of the acceleration pulse (2.3-4.9ms) were chosen so that the orbit is brought to rest when the blunt body penetrates 4-8mm beyond the orbital rim. Higher levels of penetration were considered likely to cause tearing of the orbital tissue, and perhaps a direct contact between the foreign object and the nerve. From the purely practical point of view, higher penetration would introduce an excessive distortion of the mesh. The point of impact was always located medially, between the globe and the orbital rim, and in the horizontal plane of symmetry of the globe. The velocity of the moving object was always aligned with the anterio-posterior axis of the head. The simulations were performed using the LS-DYNA (Livermore Software Technology Corporation) commercial finite element package.

2. RESULTS

Figure 2 shows the deformed mesh in the horizontal plane, for the case where the peak acceleration was 100g. It is clear from the figure that there is a large displacement of the globe in the lateral direction. The maximal lateral velocity of the globe was 1.5m/s, whereas the maximal angular velocity in the horizontal plane was 30 rad/s. The orbital fat bulges at the lateral side to compensate for the penetration of the blunt body (not shown in the figure) on the medial side. This is due to the near incompressibility of the orbital fat, which has to maintain approximately constant volume. The wave on the surface of the orbital fat, which propagates from the penetration point, creates the movement of the globe. For the simulations which involved collision with a foreign body, the build-up of the stresses in the walls of the optic nerve was on the lateral and superior side, with the peak value of the maximal principal of the order of 1MPa. The location of peak stresses just beneath the posterior pole of the globe is consistent with the fact that the injuries most commonly involve the optical disk [1]. It was found that the stress at the insertion of the optic nerve into the sclera is higher when the orbit is the moving body and the blunt object is stationary, than in the opposite case. The stresses in the orbital tissue are then influenced both by the deceleration of the orbit, and the intrusion of the foreign body.

The effect of inertial forces alone (deceleration without the collision with a foreign body) is illustrated in Figure 3. The figure gives the pressure in the fat and in the soft tissue within the optic nerve at the posterior pole of the globe, for the maximal acceleration of 450g. The pressure in the fat is nearly hydrostatic, and closely follows the shape of the deceleration pulse. There is a large difference between the pressure distribution in the fat and within the optic nerve. This pressure gradient is supported by the wall of the nerve.

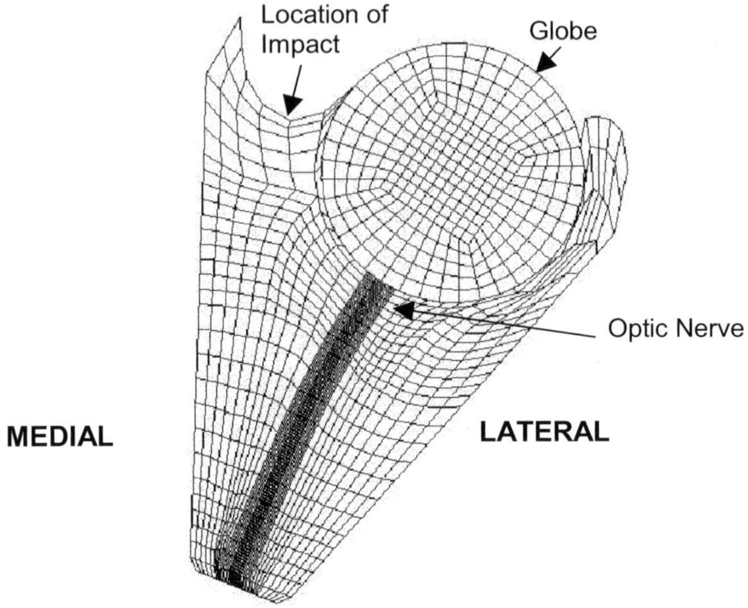

Figure 2. Horizontal cross-section of the deformed mesh for the collision with the blunt body (not shown) and the maximal impact acceleration of 100g. The globe is displaced laterally, and the orbital fat bulges at the lateral side, opposite to the site of impact.

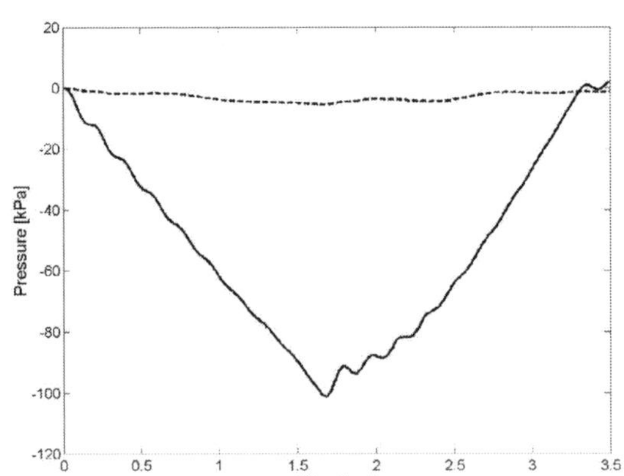

Figure 3. Pressure in the fat (solid line) and in the soft tissue within the optic nerve (dashed line) at the posterior pole of the globe due to the inertial effects alone. A triangular deceleration pulse (450g peek, 2.3ms duration) was prescribed to the orbital bone to bring it to rest from the initial velocity of 5.15m/s. There was no collision with a foreign body.

4. DISCUSSION

The study points to two possible mechanisms for the evulsion of the optic nerve which are both related to the incompressibility of the intra-orbital content. The first is the rapid movement of the globe which involves both lateral motion and rotation. This globe movement is generated by the movement of the fat which has to maintain a nearly constant volume and at the same time accommodate the penetration of a foreign object, a wave on the surface of the orbital fat causes the motion of the globe, even if there is no direct contact between the globe and the blunt body. The effect of the globe rotation has already been recognized as factor in the evulsion of the optic nerve. The second potential mechanism is the large difference between the pressure in the orbital fat and the pressure in the soft tissue within the nerve. We attribute this effect to the fact that the fat has an open surface at the anterior side of the orbit, whereas the soft tissue inside the nerve is confined within a stiff shell on all sides. Thus, the same inertial load results in different stress distribution due to different boundary conditions. This result is somewhat similar to the hypothesis that the build-up in the intraocular pressure causes the damage of the optic nerve. This should explain why the stress concentration near the optic nerve head is higher in the case where the acceleration is applied to the orbit, than when the orbit is stationary. From the current results, this mechanism seems to be secondary to the globe rotation. It should be said however, that we did not model the connection of the optic nerve with the cranium. The transfer of pressure from the cerebrospinal fluid may be a significant factor in the events where the head is rapidly accelerated, and may further accentuate the effect of this mechanism. A model which involves both the cranium and the eye-orbit system is required to fully investigate this possibility.

ACKNOWLEDGEMENT

This work was funded by the Sheffield Hospitals Charitable Trust grant.

REFERENCES

1. Bradley, S.S., Glenville, M.A., Lucarelli, M.J., Samiy, N. and Lessel, S., Optic nerve avulsion, *Ophthalmic Plastic and Reconstructive Surgery*, **18**, 2002, 261—267.
2. Morris, W.R., Osborn, F.D. and Fleming, J.C., Traumatic evulsion of the globe, *Archives of Ophthalmology*, **115**, 1997, 623—630.
3. Cirovic, S., Bhola, R.M., Hose, D.R., Howard, I.C., Lawford, P.V. and Parsons, M.A., A finite element model for the simulation of non-accidental injury in children, in

Proceedings of the 14th European Society of Biomechanics Conference, Hertogenbosh, July 4-7, 2004.

4. Iomdina, E.N., A comparative study of biomechanical properties of the cornea and sclera, in *Proceedings of the 14th European Society of Biomechanics Conference*, Hertogenbosh, July 4-7, 2004.
5. Power, E.D., Dume, S.M., Stitzel, J.D., Herring, I.P., West, R.L., Bass, C.R., Crowly, J.S. and Brozoski, F.T., Computer modeling of the airbag-induced ocular injury in pilots wearing night vision goggles, *Aviation Space and Environmental Medicine*, **73**, 2002, 1000-1006.
6. Schoemaker, I., Hoefnagel, P.P.W, Mastenbroek, T.J., Kolff, F.C., Picken, S.J., van der Helm, F.C.T. and Simonsz, H.J., 2004. Elasticity, viscosity, and deformation of rertobulbular fat in eye rotation, *in Proceedings of the Association for Research in Vision and Ophthalmology 2004 Annual Meeting,* Fort Lauderdale, Florida, USA.
7. Robinson, D.A., A quantitative analysis of the extraocular muscle cooperation and squint, *Investigative Ophthalmology*, **14**, 1975, 801-825.
8. Winters, J.M., Hill-based muscle models: a system engineering perspective, in *Multiple Muscle Systems: Biomechanics and Movement Organization*, Winters, J.M. and Woo S.L.-Y. (eds), Springer-Verlag, 1990.
9. Scott A.B., Extraocular muscle forces in strabismus, in *The Control of Eye Movements,* Bach-Y-Rita, P. and Collins, C.C. (eds), Academic Press, New York and London, 1971, pp. 327-342.
10. Goldsmith W. and Plunkett J., A biomechanical analysis of the causes of traumatic brain injury in infants and children, American Journal of Forensic Medicine and Pathology, 25, 2004, 89-100.

AN ANISOTROPIC VISCOUS HYPERELASTIC CONSTITUTIVE MODEL OF THE POSTERIOR CRUCIATE LIGAMENT SUITABLE FOR HIGH LOADING RATE SITUATIONS

Theoretical Framework and Material Identification

Georges Limbert and John Middleton
FIRST Numerics Ltd, Cardiff Medicentre, Heath Park, Cardiff CF14 4UJ, United Kingdom, Tel: +44 (0)2920 682162, Fax: +44 (0)2920 682161, www.firstnumerics.com, g.limbert@firstnumerics.com, middletonj2@cardiff.ac.uk
Biomechanics Research Unit, Wales College of Medicine, Cardiff University, Cardiff Medicentre, Cardiff CF14 4UJ, United Kingdom

Abstract. A structurally motivated phenomenological constitutive law to describe the anisotropic viscous hyperelastic behaviour of the human posterior cruciate ligament (PCL) at high strain rate is proposed. The mechanical formulation is based on the definition of a Helmholtz free energy function containing a hyperelastic and a viscous potentials which depend explicitly, through tensorial invariants, on the right Cauchy–Green deformation tensor, its rate and a structural tensor characterising the directional properties. The constitutive parameters are identified with experimental data and the ability of the model to capture accurately the key mechanical features of the PCL at high strain rates is demonstrated.

Key words: constitutive modeling, hyperelasticity, viscoelasticity, anisotropy, strain rate effect, tensorial invariants, posterior cruciate ligament, soft tissue.

1. INTRODUCTION

Knowledge of the mechanical behaviour of the PCL in highly dynamic situations is relevant for understanding the traumatic consequences of car crash or sport injuries on the whole knee joint. Finite element method analyses are useful for gaining

Michael D. Gilchrist (ed.), IUTAM Proceedings on Impact Biomechanics: From Fundamental Insights to Applications, 477–483.

insight into these biomechanical phenomena. It is therefore important to develop accurate constitutive models of biological tissues that can be implemented into numerical codes. The PCL provides 93% of the total restraining force to straight posterior translation of the tibia relative to the femur in the extended knee. At 75 to 80 degrees of flexion, the displacement of the tibia, accompanied with a posterior translational force, is maximal. However sectioning of the PCL increases posterior translation at all angles of flexion [1].

The objectives of this study are to propose a structurally-based phenomenological constitutive law capable of representing the key mechanical features of the PCL (finite strain, anisotropy, nearly incompressibility, negligible stiffness in compression along the fibre orientation) whilst capturing accurately the strain rate effect unlike the Quasi-Linear Viscoelasticity Theory [2]. This viscoelastic effect is to relate to the short-term memory effect [3]. The shape of the stress-strain curves at which traction tests are performed is affected by the loading rate [4–6] and it appears that stresses increase with strain rate [7]. Constitutive parameters of the new constitutive law are identified with experimental data for the human PCL.

2. CONSTITUTIVE FRAMEWORK

The general constitutive framework for transversely isotropic viscohyperelasticity has been described by Limbert and Middleton [8]. We denote by $\mathbf{F}(\mathbf{X}, t)$ the gradient of the deformation of a continuum body and $\mathbf{C}(\mathbf{X}, t)$ the right Cauchy–Green deformation tensor defined as follows:

$$\mathbf{C}(\mathbf{X}, t) := \mathbf{F}^T \cdot \mathbf{F} \tag{1}$$

where the superscript "T" denotes the transpose operator, $\mathbf{X}$ the material position of the material point and t is the time parameter. One also defines $\mathbf{1}$ as the second-order identity tensor and *det*, the determinant of a second-order tensor. The material rate of deformation is:

$$\dot{\mathbf{C}}(\mathbf{X}, t) = \frac{\partial \mathbf{C}}{\partial t} = \dot{\mathbf{F}}^T \cdot \mathbf{F} + \mathbf{F}^T \cdot \dot{\mathbf{F}}. \tag{2}$$

Because ligaments can be considered as biological composite structures made of one family of collagen fibres continuously distributed in a highly compliant isotropic solid matrix also called ground substance [2], it is reasonable to use transversely isotropic material symmetry to describe their mechanical behaviour.

Following the approach described by Spencer [9] and to reflect the local structural arrangement of the fibres it is convenient to introduce the symmetric structural tensor $\mathbf{N_0} = \mathbf{n_0} \otimes \mathbf{n_0}$. To define the general viscohyperelastic behaviour the existence of a free Helmholtz energy function $\psi(\mathbf{X}, \mathbf{C}, \dot{\mathbf{C}}, \mathbf{N_0})$ is postulated [8]. Furthermore, it is assumed that the elastic and viscous features of the material derive respectively from a strain energy function $\psi^e[\{I_\alpha(\mathbf{X}, \mathbf{C}, \mathbf{N_0})\}_{\alpha=1\ldots5}]$ and a viscous potential $\psi^v[\{J_\alpha(\mathbf{X}, \mathbf{C}, \dot{\mathbf{C}}, \mathbf{N_0})\}_{\alpha=1\ldots12}]$. I_α and J_α are tensorial invariants associated respectively with the elastic (tensor $\mathbf{C}$) and viscous response (tensor $\dot{\mathbf{C}}$) [8]. These invariants form the irreducible integrity bases of the tensors $\mathbf{C}$, $\dot{\mathbf{C}}$ and $\mathbf{N_0}$ [10].

$$\psi = \psi^e[\{I_\alpha(\mathbf{X}, \mathbf{C}, \mathbf{N_0})\}_{\alpha=1\ldots5}] + \psi^\nu[\{J_\alpha(\mathbf{X}, \underline{\mathbf{C}}, \dot{\mathbf{C}}, \mathbf{N_0})\}_{\alpha=1\ldots12}]. \tag{3}$$

The *principle of frame indifference* is automatically satisfied because the free Helmholtz energy function is defined in terms of quantities associated with the reference state (Lagrangean configuration). The complete set of invariants defining $\psi(\mathbf{X}, \mathbf{C}, \dot{\mathbf{C}}, \mathbf{N_o})$ is given as follows [8]:

$$I_1 = \mathbf{1} : \mathbf{C}, \quad I_2 = \frac{1}{2}[I_1^2 - (\mathbf{1} : \mathbf{C}^2)], \quad I_3 = det(\mathbf{C})$$

$$I_4 = \mathbf{N_0} : \mathbf{C}, \quad I_5 = \mathbf{N_0} : \mathbf{C}^2 \tag{4}$$

$$J_1 = \mathbf{1} : \dot{\mathbf{C}}, \quad J_2 = \frac{1}{2}((\mathbf{1}) : \dot{\mathbf{C}}^2), \quad J_3 = det(\dot{\mathbf{C}}),$$

$$J_4 = \mathbf{N_0} : \dot{\mathbf{C}}, \quad J_5 = \mathbf{N_0} : \dot{\mathbf{C}}^2,$$

$$J_6 = \mathbf{1} : (\mathbf{C} \cdot \dot{\mathbf{C}}), \quad J_7 = \mathbf{1} : (\mathbf{C} \cdot \dot{\mathbf{C}}^2), \quad J_8 = \mathbf{1} : (\mathbf{C}^2 \cdot \dot{\mathbf{C}}), \quad J_9 = \mathbf{1} : (\mathbf{C}^2 \cdot \dot{\mathbf{C}}^2),$$

$$J_{10} = \mathbf{1} : (\mathbf{N}_0 \cdot \mathbf{C} \cdot \dot{\mathbf{C}}), \quad J_{11} = \mathbf{1} : (\mathbf{N}_0 \cdot \mathbf{C}\dot{\mathbf{C}}^2), \quad J_{12} = \mathbf{1} : (\mathbf{N}_0 \cdot \mathbf{C}^2 \cdot \dot{\mathbf{C}}) \tag{5}$$

"." and ":" denote respectively the scalar product and double-contracted product of two tensors.

Based on thermodynamics considerations [8], the expression of the second Piola–Kirchhoff stress tensor is obtained by differentiation of the Helmholtz free energy function with respect to its arguments. The stress response can be separated into an elastic and a viscous contribution $\mathbf{S}^e$ and $\mathbf{S}^\nu$:

$$\mathbf{S} = \underbrace{2\frac{\partial\psi^e}{\partial I_\alpha}\frac{\partial I_\alpha}{\partial \mathbf{C}}}_{\mathbf{S}^e} + \underbrace{2\left(\frac{\partial\psi^\nu}{\partial J_\beta}\frac{\partial J_\beta}{\partial \dot{\mathbf{C}}}\right)}_{\mathbf{S}^\nu} \quad \alpha = \{1\ldots5\}, \quad \beta = \{1\ldots12\} \tag{6}$$

3. MODELLING OF THE POSTERIOR CRUCIATE LIGAMENT

Superposition of the elastic and viscous stress responses is assumed as it has been shown to be a valid assumption [6]. The additive decomposition of stress corresponds to a Kelvin–Voigt rheological model in which a spring and a dashpot are arranged in parallel. The function ψ is defined as:

$$\psi = \psi^e + \psi^\nu. \tag{7}$$

Justification for the choice of viscosity invariants and functional form of the elastic and viscous potentials can be found elsewhere [8] and are not reported here for sake of brevity.

3.1 Elastic Helmholtz Free Energy Function ψ^e

We propose the following form for ψ^e:

$$\psi^e = (I_1, I_4) = \psi^e_m(I_1) + \psi^e_f(I_4)$$
$$\psi^e_m = c_1(I_1 - 3)$$
$$\psi^e_f = = \frac{c_2}{2c_3}[e^{c_3\langle I_4-1\rangle^2} - 1] \tag{8}$$

ψ^e_m is represented by an incompressible neo-Hookean strain energy density [11]. The typical non-linear stiffening of biological soft connective tissues under tension is provided by the progressive recruitment of the collagen fibres which unfold under tensile loads applied at the tissue level [2]. An exponential form is suitable to represent this behaviour (Equation (8)-3). Collagen fibres do not support significant compressive loads and buckle under negligible compressive forces. We can therefore assume that collagen fibres can only carry tensile loads ($\langle I_4 - 1\rangle = 0$ if $I_4 < 1$).

3.2 Viscous Helmholtz Free Energy Function ψ^ν

One can consider that the viscous response of the material is provided simultaneously by the ground substance and from the collagen fibres interacting with it. For the purpose of a *phenomenological* model, and in the absence of relevant experimental data, and based on superposition principles, it is reasonable to assume that the viscous Helmholtz free energy function ψ^ν can be decomposed into the sum of two viscous potentials, ψ^e_m and ψ^ν_f, characterising respectively the isotropic and anisotropic viscous response of the tissue. It follows:

$$\psi^\nu = \underbrace{\eta_1\, J_2(I_1 - 3)}_{\psi^\nu_m} + \underbrace{\frac{1}{2}\eta_2\, J_5\langle I_4 - 1\rangle^2}_{\psi^\nu_f}. \tag{9}$$

η_1 and η_2 are material parameters which warrant convexity of the viscous potentials ψ^ν, provided that they are positive. It is relevant to note that coupling between elastic and viscous effects are taken into account by incorporating elastic invariants (I_1 and I_4) in the viscous potential ψ^ν.

3.3 Second Piola–Kirchhoff Elastic and Viscous Stress Tensors

As a reaction to the kinematic constraint of incompressibility ($I_3 = 1$), an arbitrary pressure p enters the stress under the form of a Lagrangean multiplier. The second Piola–Kirchhoff elastic and viscous stress tensors are obtained by differentiation respectively with respect to $\mathbf{C}$ and $\dot{\mathbf{C}}$:

$$\mathbf{S}^e = 2[c_1\mathbf{1} + c_2\, e^{c_3(I_4-1)^2}\langle I_4 - 1\rangle\mathbf{N_0}] + p\mathbf{C}^{-1}$$
$$\mathbf{S}^\nu = 2[\eta_1(I_1 - 3)\dot{\mathbf{C}} + \frac{1}{2}\eta_2\langle -1\rangle^2(\mathbf{n_0} \otimes \dot{\mathbf{C}} \cdot \mathbf{n_0} + \mathbf{n_0} \cdot \dot{\mathbf{C}} \cdot \otimes\mathbf{n_0})] \tag{10}$$

4. IDENTIFICATION OF MATERIAL PARAMETERS

Material coefficients of the transversely isotropic hyperviscoelastic constitutive law were identified with experimental data related to tensile tests (in the natural fibre orientation) of the human PCL [6] using a least-square non-linear constrained optimisation procedure. In order to investigate the relative viscous contribution of the matrix and that of the fibres two material identifications were performed. The first assumed that the viscous effects were only due to the interactions of the fibres with the ground substance when they extend and start to carry load. This was done by constraining the parameter η_1 to be null ($c_1 = 1 \pm 0.1930$ (MPa); $c_2 = 0.9749 \pm 0.2502$ (MPa); $c_3 = 33.0106 \pm 6.3714$; $\eta_1 = 0$ (MPa.s); $\eta_2 = 22.2748 \pm 0.1684$ (MPa.s). The second identification did not constrain η_1 nor ($c_1 = 1 \pm 0.1930$ (MPa); $c_2 = 0.9749 \pm 0.2502$ (MPa); $c_3 = 33.0106 \pm 6.3714$; $\eta_1 = 0.9719 \pm 5.6664$ (MPa.s); $\eta_2 = 21.3852 \pm 5.1893$ (MPa.s). The graphic results of the identification are represented on Figure 1 (for material parameter set 2 only as they are virtually identical for material set 1).

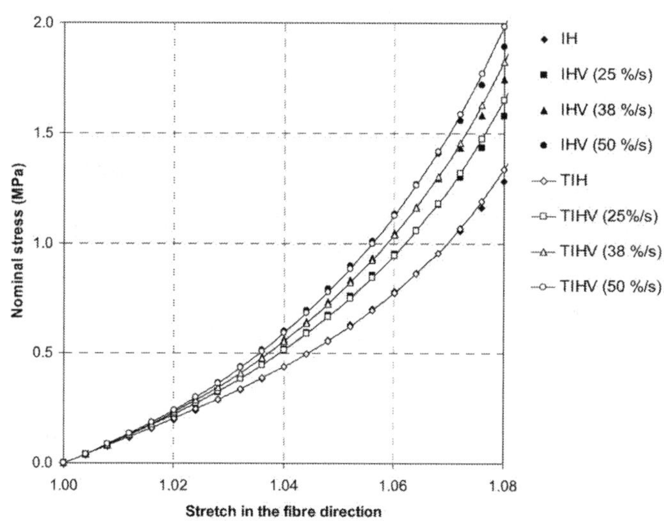

Figure 1. Analytical stress-stretch curves obtained at four different elongation rates for the Transversely Isotropic Hyperviscoelastic law (TIHV) by identification with the analytical-experimental curves from Pioletti's model (IHV).

For the two set of material parameters, very good fit between analytical and experimental values was found although the analytical curves tend to slightly overestimate the experimental values beyond stretch of 1.07. Refining the choice of the energy functions could possibly correct this limitation. The strain rate effect is clearly accounted for (Figure 1).

In the second identification procedure, a small value of the isotropic viscosity coefficient η_1 was found and this has led to a dominant influence of the anisotropic

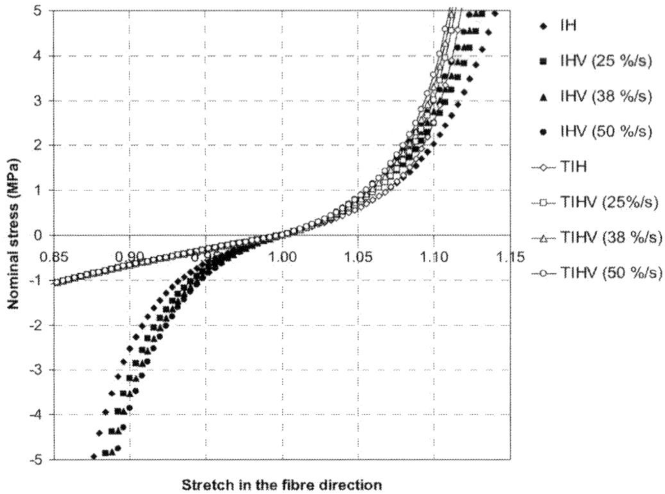

Figure 2. Comparison between the present (TIHV) model and Pioletti's (IHV) model for uniaxial compression-extension along the fibre direction (material parameter set 1).

viscosity coefficient η_2 on the viscous response of the transversely isotropic hyperelastic material in uniaxial extension along the fibre direction. The viscous contribution is also small in the case of uniaxial compression along the fibres. However, the standard deviation is high for η_1 indicating a weak correlation with respect to the isotropic response. This suggests that additional experimental tests are necessary to ascertain the respective contribution of the ground substance and that of the fibres on the viscous response of the ligamentous tissue.

5. CONCLUDING REMARKS

In this study, the constitutive framework developed by Limbert and Middleton [8] has been applied to the modelling of the PCL at the finite strain regime and for high loading rates. The advantage of the *transversely isotropic* hyperviscoelasticity formulation over the *isotropic* hyperviscoelasticity formulation from Pioletti et al. [6] had already been demonstrated [8] and is highlighted again in the development of a constitutive model of the PCL.

The new constitutive law has been developed to take into account simultaneously the fibrous and directional viscous characteristics of the PCL while remaining thermodynamically admissible.

Although phenomenological viscohyperelastic constitutive laws featuring transverse isotropy were already available [12], none of them was describing faithfully the very short-term stiffening effect associated with strain rate, as observed in ligament mechanics [13].

The present phenomenological model of the PCL has been implemented into an explicit finite element code and provides a numerical framework for the study of the mechanics of the knee joint in high loading rate situations.

REFERENCES

1. Fukubayashi, T., Torzilli, P.A., Sherman, M.F. and Warren, R.F., An in vitro biomechanical evaluation of anterior-posterior motion of the knee. Tibial displacement, rotation, and torque. *Journal of Bone and Joint Surgery*, 64A(2), 1982, 258–64.
2. Fung, Y.C., *Biomechanics: Mechanical Properties of Living Tissues*, Springer, New York, 1993.
3. Pioletti, D.P., Viscoelastic properties of soft tissues: Application to knee ligaments and tendons, PhD Thesis, Ecole Polytechnique Fédérale de Lausanne, 1997.
4. Ticker, J.B., Bigliani, L.U., Soslowsky, L.J., Pawluk, R.J., Flatow, E.L. and Mow, V.C., Inferior glenohumeral ligament: geometric and strain-rate dependent properties, *Journal of Shoulder and Elbow Surgery*, 5, 1996, 269–279.
5. Haut, R.C., Age-dependent influence of strain rate on the tensile failure of rat-tail tendon, *Journal of Biomechanical Engineering*, 105, 1983, 296–299.
6. Pioletti, D.P., Rakotomanana, L.R., Benvenuti, J.F. and Leyvraz, P.F., Viscoelastic constitutive law in large deformations: application to human knee ligaments and tendons, *Journal of Biomechanics*, 31(8), 1998, 753–757.
7. Haut, R.C. and Little, R.W.A., A constitutive equation for collagen fibers, *Journal of Biomechanics*, 5, 1972, 423–430.
8. Limbert, G. and Middleton, J., A transversely isotropic viscohyperelastic material. Application to the modelling of biological soft connective tissues, *International Journal of Solids and Structures*, 41(15), 2004, 4237–4260.
9. Spencer, A.J.M., *Continuum Theory of the Mechanics of Fibre-Reinforced Composites*, Springer-Verlag, New York, 1992.
10. Boehler, J.P., *Applications of Tensor Functions in Solid Mechanics*, Springer, Wien, 1987.
11. Ogden, R.W., *Non-Linear Elastic Deformations*, Ellis Horwood Ltd., West Sussex, England, 1984.
12. Puso, M.A. and Weiss, J.A., Finite element implementation of anisotropic quasi-linear viscoelasticity using a discrete spectrum approximation, *Journal of Biomechanical Engineering*, 120(1), 1998, 62–70.
13. Woo, S.L.Y., Gomez, M.A. and Akeson, W.H., The time and history-dependent viscoelastic properties of the canine medial collateral ligament, *Journal of Biomechanical Engineering*, 103, 1981, 293–298.

AUTHOR INDEX

SUBJECT INDEX

Kolibri 360 GmbH
Ferdinand-Jühlke-Straße 7
99095 Erfurt, Deutschland
produktsicherheit@kolibri360.de